Cecil Hobart Peabody, Edward Furber Miller

Steam-boilers

Cecil Hobart Peabody, Edward Furber Miller

Steam-boilers

ISBN/EAN: 9783743467248

Manufactured in Europe, USA, Canada, Australia, Japa

Cover: Foto ©berggeist007 / pixelio.de

Manufactured and distributed by brebook publishing software (www.brebook.com)

Cecil Hobart Peabody, Edward Furber Miller

Steam-boilers

STEAM-BOILERS.

BY

CECIL H. PEABODY AND EDWARD F. MILLER

*Professor of Marine Engineering
and Naval Architecture,*
*Assistant Professor of Steam
Engineering,*
Massachusetts Institute of Technology.

FIRST EDITION.

FIRST THOUSAND.

NEW YORK:
JOHN WILEY & SONS.
LONDON: CHAPMAN & HALL, LIMITED.
1897.

ROBERT DRUMMOND, ELECTROTYPER AND PRINTER, NEW YORK

PREFACE.

In this book we have attempted to give a clear and concise statement of facts concerning boilers, and of methods of designing, making, managing, and caring for boilers. Though the book is intended primarily for the use of students in technical schools and colleges, it is hoped that it may be found useful to engineers in general.

There is given a description of various types of boilers in common use. Following this is a discussion of combustion, corrosion, and incrustation, with a statement of the most recent investigations and conclusions on these important subjects. We are fortunately able to give a satisfactory table of the compositions of American fuels—the first, so far as we are aware, that has been published.

A statement is given of the proper and of the customary sizes and form of furnaces, and of the methods of firing. In the present unsatisfactory condition of the chimney problem we have contented ourselves with giving the ordinary theory and pointing out its defects, together with the common ways of proportioning chimneys.

Tables of grate-areas and heating-surfaces, and of other proportions of furnaces and boilers, have been made up from the best current practice for stationary, locomotive, and marine boilers.

In the chapter on strength of boilers we have given briefly the methods and conditions for testing materials and for making boilers, and the properties which such materials should have. Especial attention is given to the properties and proportions of riveted joints, deduced by Professors Lanza and Schwamb from tests at the Watertown Arsenal. Simpler calculations of stresses in the members of boilers are explained, and more complex ones, depending on the theory of elasticity and theories of beams and continuous girders, are illustrated by examples.

A description is given of staying and other details affecting the design and construction of boilers, and of such accessories as safety-valves, gauges, and steam-traps. In order to give a conception of the methods and conditions of boiler-making, we have given a description of a modern boiler-shop and the machinery and processes used in it.

In the chapter on boiler-testing we have given the methods used in the laboratories of the Massachusetts Institute of Technology, including gas analysis, measurement of air used, and temperature, determinations in the furnace and chimney.

Finally, the principles and methods set forth in the earlier chapters are brought together and illustrated by applying them to the design of a boiler of a common type. For our own students this chapter serves as an introduction to a course in machine design given by Professor Schwamb, who has kindly furnished us with methods and materials which he has collected and developed in connection with the designing of boilers.

In the appendix are given various useful tables, such as logarithms, natural trigonometric functions, areas and circumferences of circles, proportions of rods and screws, and properties of saturated steam.   C. H. P. and E. F. M.

BOSTON, February 1, 1897.

CONTENTS.

CHAPTER I.
Types of Boilers.. 1

CHAPTER II.
Fuels and Combustion... 37

CHAPTER III.
Corrosion and Incrustation....................................... 65

CHAPTER IV.
Settings, Furnaces, and Chimneys............................... 91

CHAPTER V.
Power of Boilers... 130

CHAPTER VI.
Staying and Other Details.. 148

CHAPTER VII.
Strength of Boilers.. 170

CHAPTER VIII
Boiler Accessories... 235

CHAPTER IX.

SHOP-PRACTICE 272

CHAPTER X.

TESTING BOILERS 300

CHAPTER XI.

BOILER DESIGN 323
APPENDIX 357
INDEX 369

STEAM-BOILERS.

CHAPTER I.

TYPES OF BOILERS.

STEAM-BOILERS may be classified according to their form and construction or according to their use. Thus we have horizontal and vertical boilers, internally and externally fired boilers, shell-boilers and sectional boilers, fire-tube and water-tube boilers: the several features mentioned may be combined in various ways so as to give rise to a large number of kinds and forms of boilers. Again, we have stationary, locomotive, and marine boilers, together with a variety of portable and semi-portable boilers. Locomotive boilers are always shell-boilers, internally fired, and with fire-tubes; and the restrictions of the service have developed a form that has changed little from the beginning, except in the direction of increased size and power. Marine boilers present a much larger variety of form and construction, depending on the steam-pressure used and the size and service of the vessel to which they are supplied. The Scotch or drum boiler is more widely used than any other form at present, but the tendency to use high-pressure steam has led to the introduction of various forms of water-tube boilers for marine work. The variety of forms and methods of construction of stationary boilers is very wide: each country and section of a country is likely to have its own favorite type. Thus in New England, where the water is good, cylindrical tubular boilers are largely used; in some of the Western States, where water contains mineral impurities, flue-boilers are preferred; and in England, the Lancashire and Galloway boilers are favored; and again, various forms of sectional and water-tube boilers are now widely used.

Cylindrical Tubular Boiler.—This type of boiler is shown by Fig. 1 and by Plate I. It consists essentially of a cylindrical shell closed at the ends by two flat *tube-plates*, and of numerous *fire-tubes*, commonly having a diameter of three or four inches. About two thirds of the volume of the boiler is filled with water, the other third being reserved for steam. The water-line is six or eight inches above the top row of tubes. The tube-plates below the water-line are sufficiently stayed by the tubes; above the water-line the flat plates are stayed by *through rods* or *stays* as in Plate I, by diagonal stays like those shown by Fig. 52, page 154, or otherwise. A pair of cylindrical boilers in brick setting are shown by Figs. 36 and 37, on pages 92 and 93, with the furnaces under the front (right-hand) end. The products of combustion pass back over a *bridge-wall*, limiting the furnace, to the *back end*, then forward through the *tubes* and up the *uptake* to the flue which leads to the chimney.

The shell commonly extends beyond the front tube-plate, as shown at the right in Fig. 1, and is cut away to facilitate the arrangement of the uptake. The boiler is usually supported by cast-iron brackets riveted to the shell; the front brackets may rest on or be fixed to the supporting side walls, but the rear brackets should be given some freedom to avoid unduly straining the boiler by expansion. Thus the rear brackets may rest on rollers, which in turn bear on a horizontal iron plate. The expansion takes place toward the back end of the boiler, and to allow for this expansion a space is left between the back tube-sheet, and the arch of fire-brick back of the boiler.

TYPES OF BOILERS.

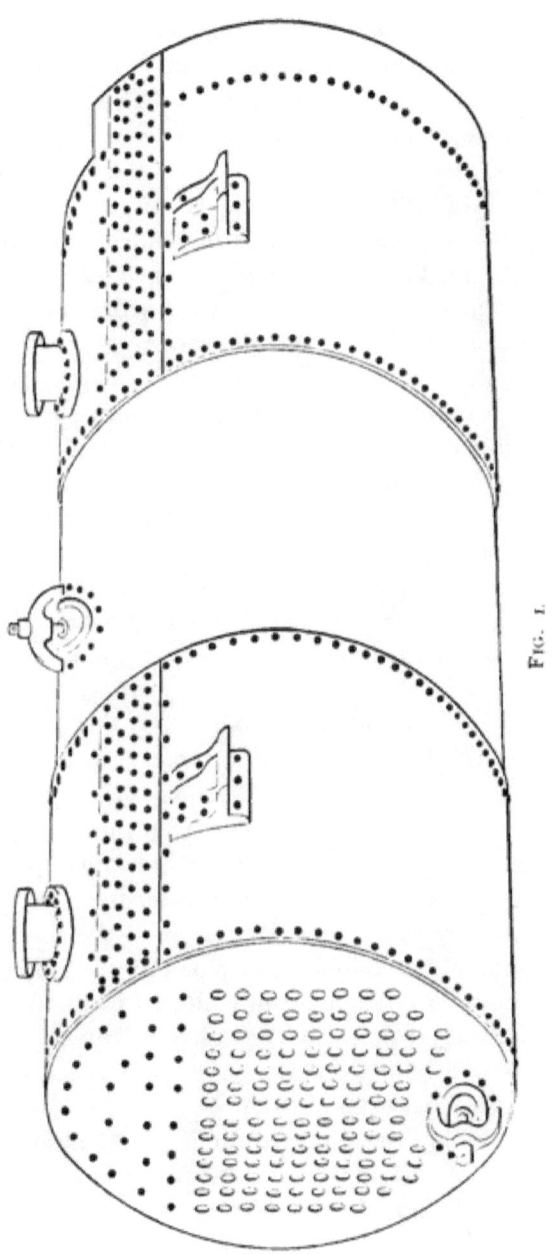

Fig. 1.

The boilers shown by Fig. 1 and by Plate I each have two steam-nozzles, one near each end. The safety-valve is usually attached to the front nozzle, which is above the furnace. The steam-pipe leading steam from the boiler is attached to the rear nozzle, which is over the back end of the boiler, where ebullition is less violent, and consequently there is less danger that water will be thrown into the steam-pipe.

Boilers of this type commonly have a *manhole* on top near the middle, and a *hand-hole* near the bottom of each tube-sheet, as shown on Plate I, to give access to the interior of the boiler and to facilitate washing out. Many boilers are now made with a manhole near the bottom of the front tube-sheet, in addition to the one on top. All parts of the boiler can then be cleaned and inspected whenever desirable. Some of the lower tubes must be left out when there is a manhole in the tube-sheet, but this is of small consequence, as the lower tubes are not efficient, and enough heating-surface can be provided elsewhere. The omission of the lower tubes requires also special stays for the portion of the tube-sheet left unsupported.

The *feed-pipe* for the boiler shown by Plate I enters the front head at the left, below the water-line, and runs toward the back end of the boiler, where it may end in a perforated pipe leading across the boiler. The feed-pipe may enter the top of the boiler, near the back end, and terminate in a similar perforated transverse pipe below the water-line.

A *blow-off pipe* leads from the bottom of the shell near the back tube-sheet. On the blow-off pipe there is a plug or valve which may be opened when steam is up, to blow out mud and soft scale that may collect in the boiler. The boiler is commonly set with a slight inclination toward the rear so that mud may collect near the blow-off pipe. The boiler may be emptied by allowing the water to run out at the blow-off pipe.

About half of the shell, two thirds of the back tube-sheet, and all the inside surface of the tubes come in contact with the products of combustion and form the *heating-surface;* all the heating-surface is below the water-line.

The boiler-setting, shown by Figs. 36 and 37 on pages 92 and 93, is made of brick laid in cement or mortar; all parts that are directly exposed to the fire are lined with fire-brick. The walls have confined air-spaces to reduce transmission of heat. The *boiler front* is commonly made of cast iron, and has *fire-doors* leading to the furnace, and *ash-pit doors* opening from the *ash-pit,* or space below the grate; there are also large doors giving access to the tubes through the *smoke-box* at the front end of the boiler. The furnace is formed by the side walls, the bridge, and the lower part of the boiler front, which latter is lined with fire-brick above the grate. Doors through the rear wall give access to the space back of the bridge. The top of the boiler is covered by a brick arch or by non-conducting material.

Two-flue Boiler.—The cylindrical flue-boiler differs from the tubular boiler mainly in replacing the fire-tubes by one or more large flues. Fig. 2 shows such a boiler with two

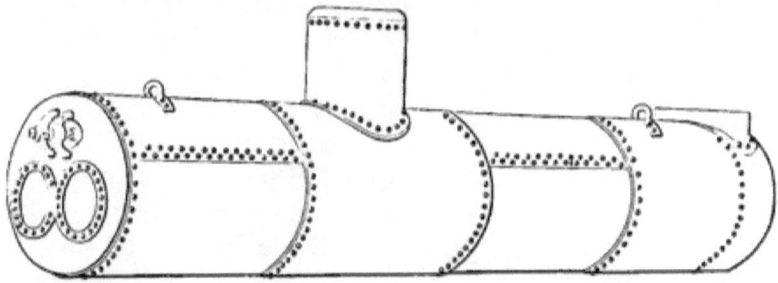

FIG. 2.

flues. This type of boiler is usually longer than a tubular boiler, but even so it has less heating-surface and is less efficient in the use of coal. Nevertheless the greater simplicity and accessibility for cleaning recommend it where feed-water is bad.

The setting of a flue-boiler resembles that for the cylindrical tubular-boiler. The figure shows two loops at the top of the shell for hanging the boiler; a crude method of supporting, suitable only for small and short boilers.

Plain Cylindrical Boiler.—In places where fuel is very cheap, especially where it is a waste product, as at sawmills, the plain cylindrical boiler is frequently used. Its external appearance is similar to that of the two-flue boiler (Fig. 2), except that there are no flues and the ends are commonly hemispherical or else curved to a radius equal to the diameter of the shell. Such plain cylindrical boilers are also employed to utilize the waste gases from blast-furnaces. They are commonly 30 to 42 inches in diameter and from 20 to 40 feet long. They have been made 70 feet long. With such extreme lengths special care must be taken to insure equal distribution of the weight to the supports and to provide for expansion.

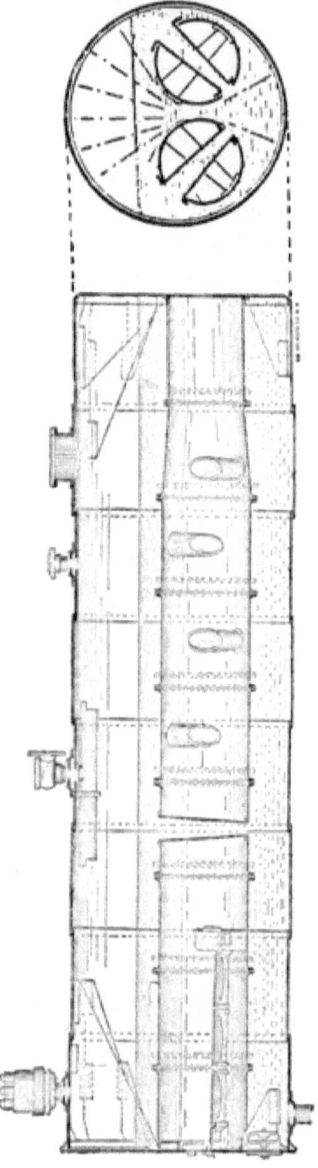

Fig. 3.

Lancashire Boiler.—This boiler, shown by Fig. 3, is a two-flue shell-boiler with furnaces in the tubes; it is therefore an internally-fired boiler, in which it differs from the two preceding types, which are externally-fired. The chief difficulty in the design of these boilers is to provide sufficiently large furnaces without making the external shell too large. As compared with the cylindrical tubular boiler, this boiler will be sure to have long, narrow grates, with a shallow ash-pit and a low furnace-crown: the boiler also appears to be deficient in heating-surface. In compensation, radiation and loss of heat from the furnace are almost entirely done away with, and the thick outside shell, with its riveted joints, is not exposed to the fire, as with the tubular boiler. The flues are made in short sections riveted together at the ends, thus forming a series of stiffening rings that add very much to the strength of the flues against collapsing. Conical through-tubes, vertical or inclined, give increased heating-surface, break up the currents of the hot gases, improve the circulation of the water, and strengthen the flues. These tubes are small enough at the lower end to pass through the hole cut in the flue for the upper end, and thus are readily put in or taken out for repairs.

The flat plates at the ends of the shell are stayed by *gusset-stays* or triangular flat plates to the shell of the boiler. The boiler is provided with a manhole near the back end and a safety-valve near the front end. Steam is taken through a horizontal dry-pipe, perforated on the top.

Galloway Boiler.—This boiler has two furnace-flues at the front end, like the Lancashire boiler. Beyond the furnace the two flues merge into one broad flue, having the upper and lower surfaces stayed by numerous conical through-tubes, like those shown in Fig. 3 for the Lancashire boiler.

Cornish Boiler.—This boiler was developed in conjunction with the Cornish engine, and both boiler and engine long had a reputation for high efficiency. It differed from the Lancashire boiler in that it had but one flue; it formerly did not have cross-tubes. The one furnace of the Cornish boiler, with a given diameter of shell, can have better proportions than the two furnaces of the Lancashire boiler, but there is even greater difficulty to get sufficient grate-area and heating-surface. The high economy shown by these boilers when used with the Cornish pumping-engine was due to a slow rate of combustion, and to the skill and care of the attendant, who was usually both engineer and fireman, and who was stimulated by a system of competition and awards, maintained by the mine-owners in that district.

The Lancashire and the Cornish boilers are set in brickwork which forms flues leading around the outside shell, thus making the shell act as heating-surface. Fig. 4 gives a cross-sec-

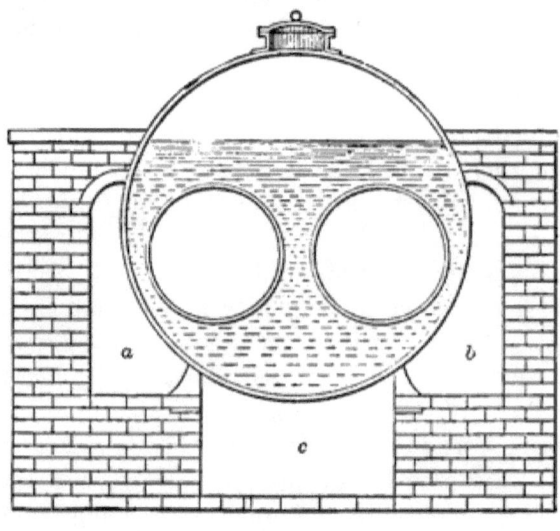

FIG. 4.

tion of the Lancashire boiler and its setting. After the gases from the fires leave the internal flues they are directed into the flue a and come forward; then they are transferred to the flue b and pass backward; finally they come forward in the flue c, and are then allowed to pass to the chimney. This forms what is known as a *wheel-draught*. In some cases the gases divide at the rear and come forward through both side flues *a* and *b*, and uniting pass back through *c* and thence to the chimney, forming a *split-draught*.

Vertical Boilers.—Boilers of this type have a cylindrical shell with a fire-box in the lower end, and with fire-tubes running from the furnace to the top of the boiler. Large vertical boilers have a masonry foundation and a brick ash-pit; small vertical boilers have a cast-iron ash-pit that serves as foundation. Vertical boilers require little floor-space; if properly designed they give good economy, or they may be made light and powerful for their size, when economy is not important.

Fig. 5 shows a large vertical boiler designed by Mr. Manning. It is made 20 to 30 feet high, so that there is a large heating-surface in the tubes. The shell is enlarged at the fire-box to provide a larger furnace and more area on the grate. The internal shell which forms the fire-box is joined to the external shell by a welded iron ring called the foundation-ring. This internal shell should be made of moderate thickness to avoid burning or wasting away under the action of the fire. Being under external pressure, the shell of the fire-box must be stayed to avoid collapsing. For this purpose it is tied to the outside shell at intervals of four or five inches each way, by bolts that are screwed through both shells and riveted over cold, on both ends. The stays near the bottom have each a hole drilled from the outside nearly through to the inside end. Should any stay break or become cracked, steam will escape and give warning to the fireman.

The tubes are arranged in concentric circles, leaving a space about ten inches in diameter at the middle of the crown-sheet; the corresponding space in the upper tube-sheet provides for the attachment of the nozzle for the steam outlet.

There are numerous hand-holes in the shell outside of the fire-box, some near the crown-sheet, and some near the foundation-ring, and these are the only provision for cleaning the

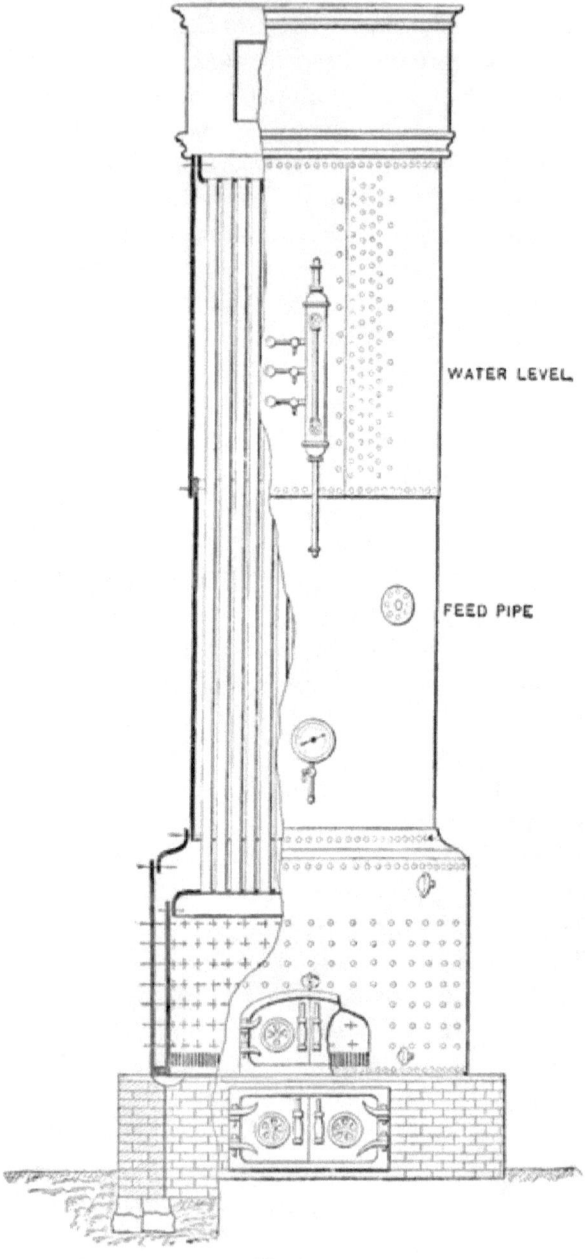

Fig. 5.

boiler, which consequently is adapted for the use of good feed-water only. The feed-pipe enters the shell at one side and extends across the boiler; it is perforated to distribute the feed-water.

The sides of the fire-box, the remaining surface of the tube-sheet allowing for the holes for the tubes, and the inside

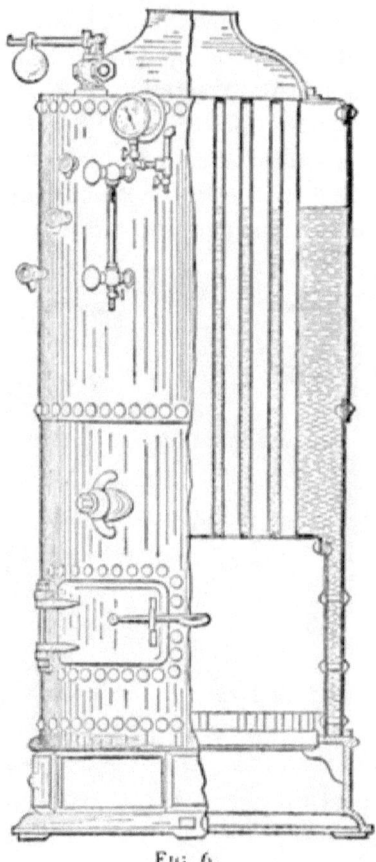

FIG. 6.

of the tubes up to the water-line form the heating-surface: the inside of the tubes above the water-line form the *super-*

heating-surface, since it transmits heat from the gases to the steam and superheats it.

This type of boiler has found favor at factories where floor-space is valuable, since a powerful battery of boilers may be placed in a small fire-room.

A small vertical boiler adapted for hoisting, pile-driving, and other light work is shown by Fig. 6. It commonly has a short smoke-pipe, into which the exhaust steam from the engine is turned to form a forced draught and give rapid combustion. Under this treatment the upper ends of the tubes frequently give trouble by leaking. To avoid this difficulty the tubes are sometimes ended in a sunken or submerged tube-sheet which is kept below the water-line, as shown by Fig. 7. The space between the edge of the tube-sheet

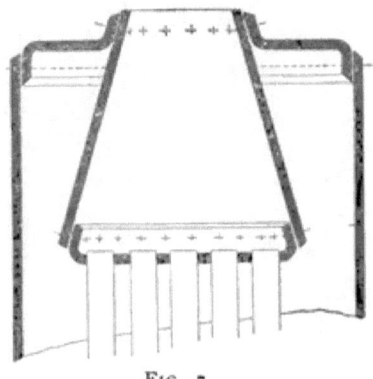

FIG. 7.

and the outside shell is likely to be contracted, and not to give proper exit for the steam formed on the tubes and crown-sheet. Furthermore, the cone forming the smoke-chamber above the tube-sheet is subjected to external pressure and is likely to be weak.

A form of vertical boiler having a sunken tube-plate is shown by Fig. 8. It was at one time much used for steam fire-engines, but to save weight it was so crowded with tubes and the water-spaces were so contracted that it gave much trouble when forced, as at a fire.

Fire-engine Boiler.—A boiler for a steam fire-engine should be light and compact, able to make steam quickly and

Fig. 8.

to steam freely when urged. They have small water-space and large heating-surface for their size, but are not economical in the use of fuel. It is customary to use cannel-coal for fire-engines, as it burns freely without clogging. A forced draught is obtained by exhausting steam up the smoke-pipe. When standing in the engine-house ready for duty the boilers are kept hot by connecting them to a heating-boiler in the basement. The connection is so made with snap-valves that it is broken by pulling the fire-engine out of position.

Figs. 9 and 10 show a vertical section and two half-horizontal sections of the Clapp fire-engine boiler. The boiler has a cylindrical shell and a deep internal fire-box. From the crown-sheet a number of fire-tubes lead through the water and steam space to the upper tube-sheet. In the upper part of the fire-box there are a number of water-tubes that start from the side of the fire-box, make several helical coils, and then open into the water-space above the crown-sheet.

There are three concentric sets of these helical coils, leaving a cylindrical space in the centre, which is occupied by a series of castings, shown in perspective and partly in section by Fig. 11. The casting is formed of an annular torus with a cross-tube, and an inverted U tube above. Water enters at the middle of the cross-tube, passes into the torus, and then up and out at the top of the U.

The left half of Fig. 10 shows the helical tubes from above; the right half shows the arrangement of the fire-tubes and the openings of the water-tubes.

Marine Boilers.—A single-ended three-furnace Scotch marine boiler is shown in perspective by Fig. 12; Fig. 13 gives the working drawings of a similar boiler with two furnaces. The arrangement of the furnaces in the flues, is similar to that for the Lancashire boiler, shown, by Fig. 3. The furnace-flue leads into a combustion-chamber, from which the products of combustion pass through fire-tubes to the uptake, which is bolted onto the front end of the boiler.

The flues are from three and a half to four and a half feet in diameter; the size of the boiler depends on the number and size of the flues. Large boilers have as many

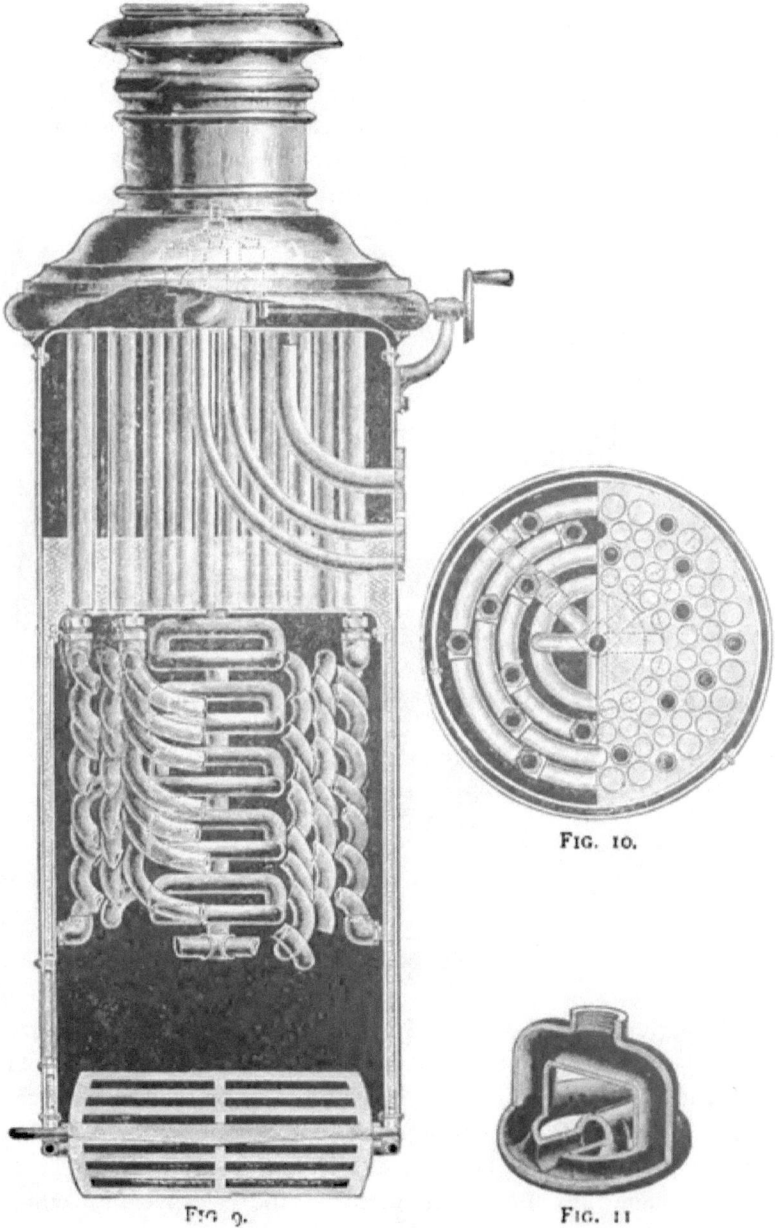

Fig. 9.

Fig. 10.

Fig. 11.

as four flues. A three-furnace boiler commonly has three combustion-chambers, while a four-furnace boiler may have two, into each one of which two furnaces lead. Double-ended boilers have furnaces at each end, and resemble two single-ended boilers placed back to back. A double-ended boiler is lighter, cheaper, and occupies less space than

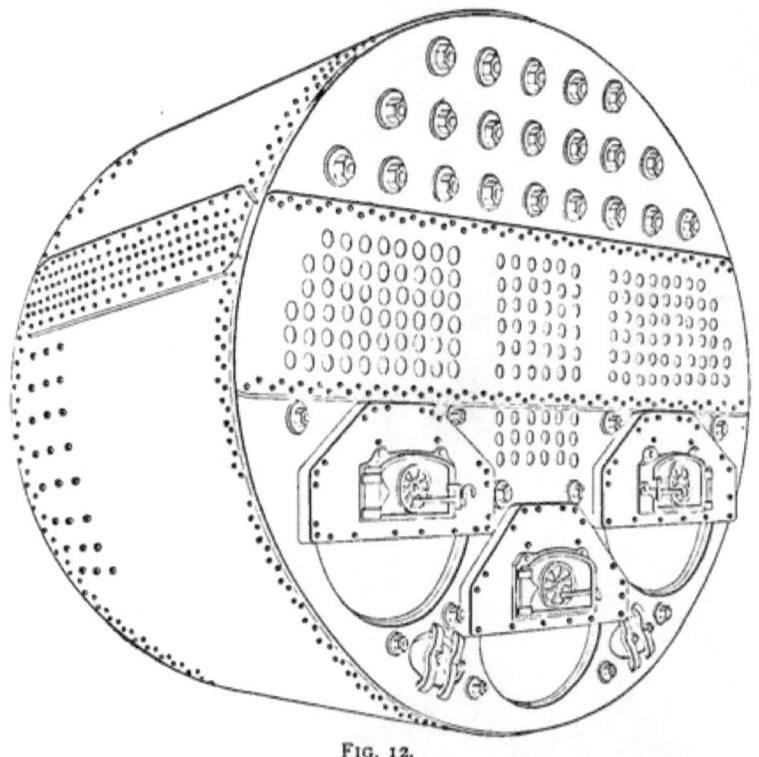

FIG. 12.

two single-ended boilers. In the best practice there are two distinct sets of combustion-chambers for the two sets of furnaces. To still further lighten double-ended boilers, common combustion-chambers for corresponding furnaces at the two ends have been used. The results from such boilers have not been satisfactory, more especially when

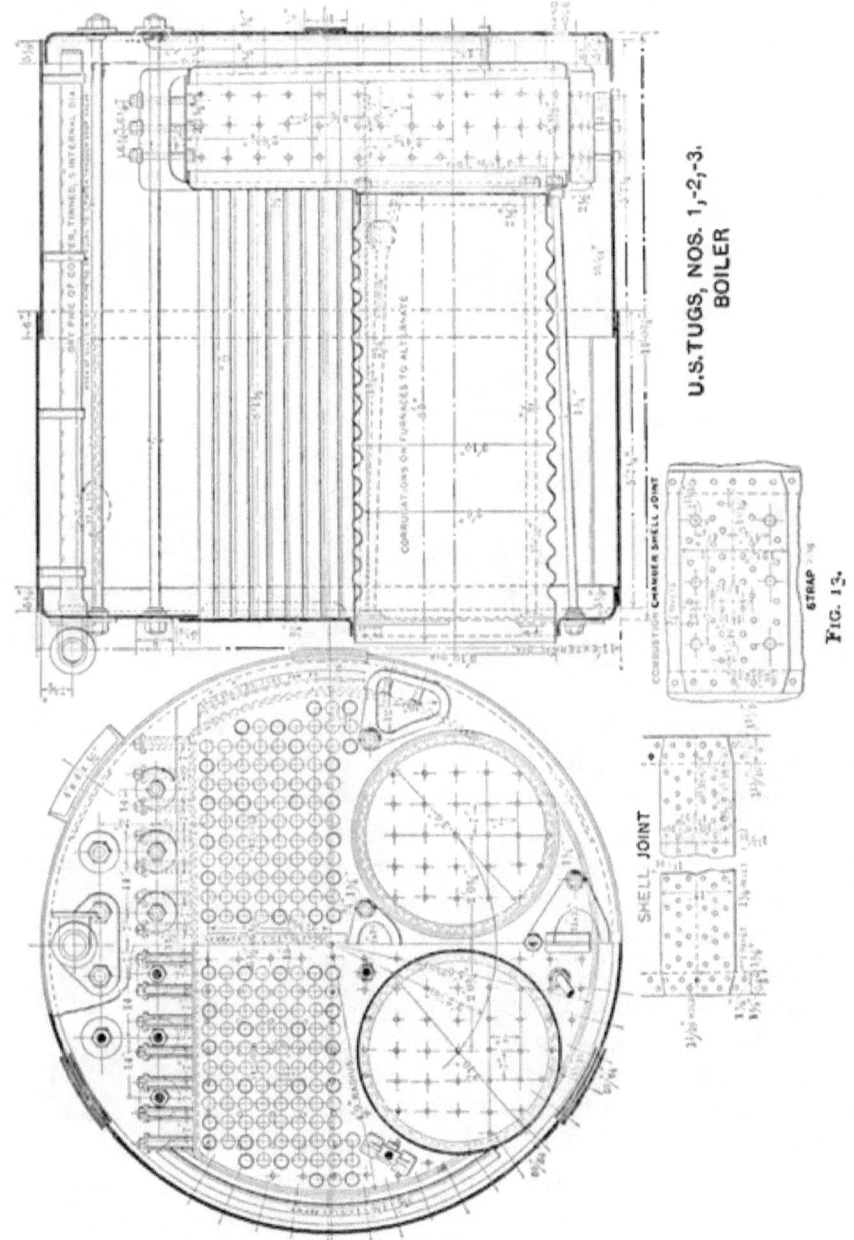

Fig. 13.

used under forced draught in the closed stoke-holes of warships; there has been so much trouble from leaky tubes under such conditions that forced draught has been abandoned in many cases, and ships have consequently failed to make the speed anticipated.

The circulation of water is defective in all Scotch boilers, and more especially in double-ended boilers. Considerable time—three or four hours—is always allowed for raising steam. Frequently some arrangement is made for drawing cold water from the bottom of the boiler and returning it near the water-line, while steam is raised. Haste and lack of care are liable to cause leakage from unequal expansion. The flue has the highest temperature of any part of the boiler and consequently expands the most, so that some allowance for expansion must be made or it will strain the tube-sheets and cause leaks. The methods of providing for expansion and at the same time stiffening the flues against collapsing under external pressure are shown on pages 210 to 216, and will be described in detail later on.

Gunboat Boilers.—Some gunboats and other small naval vessels have not room under the deck for Scotch boilers. The form shown by Fig. 14 has been used on such vessels; it has two furnace-flues, leading to a common combustion-chamber, from which fire-tubes lead to the back end of the boiler.

Locomotive-boilers.—The typical American locomotive-boiler is shown by Plate II. Fig. 15 gives a perspective view of a boiler of the locomotive type used for small factories, or where steam is required temporarily; it has no permanent foundation, but is supported on brackets at the fire-box and by a pedestal-bearing on rollers near the back end.

The locomotive-boiler consists essentially of a rectangular fire-box and a cylindrical barrel through which numerous tubes pass from the fire-box to the smoke-box, which forms a continuation of the barrel, and from which the products of combustion pass up the smoke-stack.

TYPES OF BOILERS.

The fire-box is joined to the outer shell at the bottom by a forged rectangular foundation-ring, similar (except in shape)

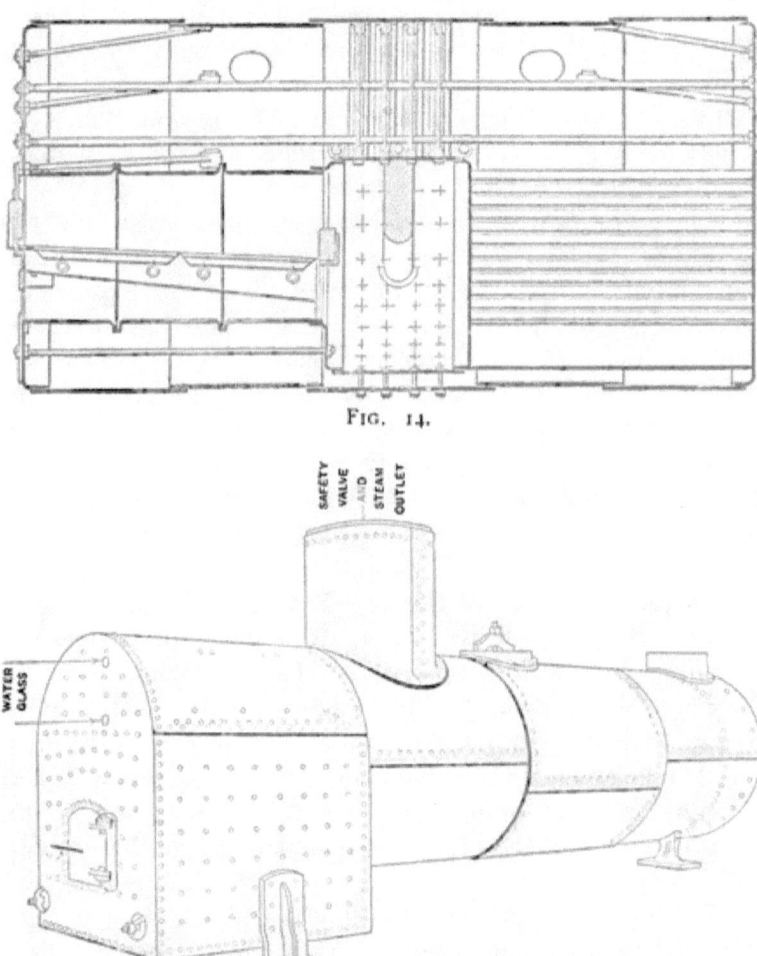

FIG. 14.

FIG. 15.

to the foundation-ring of a vertical boiler. Near this ring are several hand-holes for clearing out the space between the fire-box and the shell, commonly called the *water-leg*. The boiler also has a manhole at the top of the barrel. The water-leg is stayed by screwed stay-bolts riveted cold at the ends.

The flat crown-sheet is stayed to a system of *crown-bars* which rest on the side sheets of the fire-box and are also slung from the shell. Plate III shows a locomotive-boiler with a flattened top over the fire-box, to which the crown-sheet is stayed by through-bolts. Other methods and details of staying crown-sheets will be given later.

The tubes for a locomotive-boiler are smaller than for stationary boilers (about two inches in diameter) and are spaced much more closely. This is to obtain a large heating-surface required by the high rate of combustion, which often exceeds one hundred pounds of coal per square foot of grate-surface per hour. The boiler works under a strong forced draught, produced by throwing the exhaust up the smoke-stack.

The boiler is fastened rigidly to the frame of the locomotive at the smoke-box end; a small longitudinal motion on the frame at the fire-box end is provided by *expansion-pads*, shown by Fig. 4, Plate II.

Locomotive Type of Boiler—Reference has already been made in connection with Fig. 15 to a boiler of locomotive type used for stationary purposes. Plate IV shows a modification of the locomotive type designed by Mr. E. D. Leavitt to give high evaporative efficiency. The boiler represented has a barrel 90 inches in diameter, and it is 34 feet 4 inches long over all. It supplies steam at 185 pounds pressure to the square inch to the high-duty pumping-engine at Chestnut Hill Reservoir, Boston.

The fire-box of this boiler is spread at the bottom to give increased grate-area, and contains two separate furnaces, shown by the section AA on Plate IV. The products of combustion pass through openings, shown by section BB, into a combustion-chamber, which has the section shown at CC. From the combustion-chamber, the gases pass through tubes to the smoke-box and uptake. As far as the combustion-chamber the top of the boiler is flattened to facilitate the staying of the crown-sheets of the furnace, passages, and combustion-chamber; the barrel of the boiler beyond the combustion-chamber is cylindrical.

The boiler is somewhat complicated in construction and staying, and must be handled with care, especially in starting, to avoid straining from unequal expansion. It is adapted for the use of good feed-water only.

Boilers of the locomotive type were at one time used for torpedo-boats. The fire-box was made shallower than for locomotive-boilers, and forced draught in a closed stoke-hole was used, the rate of combustion being even higher than on locomotives. Whatever may have been the reasons, it was a fact that this type of boiler, which is very reliable on locomotives, gave much trouble in torpedo-boats.

Water-Tube Boilers.—The boilers thus far considered have an external shell containing a large body of water. Heat is communicated to the water through the shells or through the sides of internal furnaces, and also by carrying the gases through tubes or flues. The boilers and water contained, are heavy and cumbersome, and the shells under high pressure must be made very thick. If the boiler fails either through some defect or through carelessness of attendants, a disastrous explosion is likely to take place. If properly designed and made and if cared for by competent and careful attendants they are safe, reliable, and durable. The large mass of hot water tends to keep a steady pressure, though at the expense of rapidity of raising steam or of meeting a sudden demand for more steam.

A large number of water-tube boilers of all sorts of shapes and methods of construction has been devised to overcome the admitted defects of shell-boilers. They all have the larger part of their heating-surface made up of tubes of moderate size filled with water. They all have some form of separators, drum, or reservoir in which the steam is separated from the water; some of these boilers have a shell of considerable size, thus securing a store of hot water and a good free-water surface for disengagement of steam. Such shell, drum, or reservoir is either kept away from the fire or is reached only by gases that have already passed over the surface of water-tubes.

The tubes are of moderate or small diameter, and so can be abundantly strong even when made of thin metal. Even if a tube fails through defect in manufacture or through wasting during service, it will not cause a true explosion; and yet the failure of a tube in a confined boiler or fire-room has frequently caused death by scalding.

Water-tube boilers may be made light, powerful, and compact, and are well adapted for use with forced draught. Steam may be raised rapidly from cold water, but pressure falls as rapidly if the fire loses intensity, and fluctuations in pressure are likely to occur. The two greatest difficulties are to secure a proper circulation of water through the tubes and to properly separate the steam from the water. There are many joints that may give trouble by leaking, and some types have numerous hand-holes for cleaning the tubes, which may further increase the chances of petty leaks.

A few water-tube boilers will be described as illustrations; many others equally good will be passed by, since it will be impossible to describe all.

Babcock and Wilcox Boiler.—This boiler, which is shown by Figs. 16 and 17, is a water-tube boiler having a cylindrical shell to furnish steam-space, and in which is the free-water surface for the disengagement of steam. The tubes are expanded into vertical headers at each end; the front-end headers open into a cross-connection in communication with the cylindrical shell, while the back-end headers are connected with a similar cross-connection by slightly inclined pipes. The tubes in each section are staggered so that the tubes taken as a whole are in horizontal rows, but not in ver-

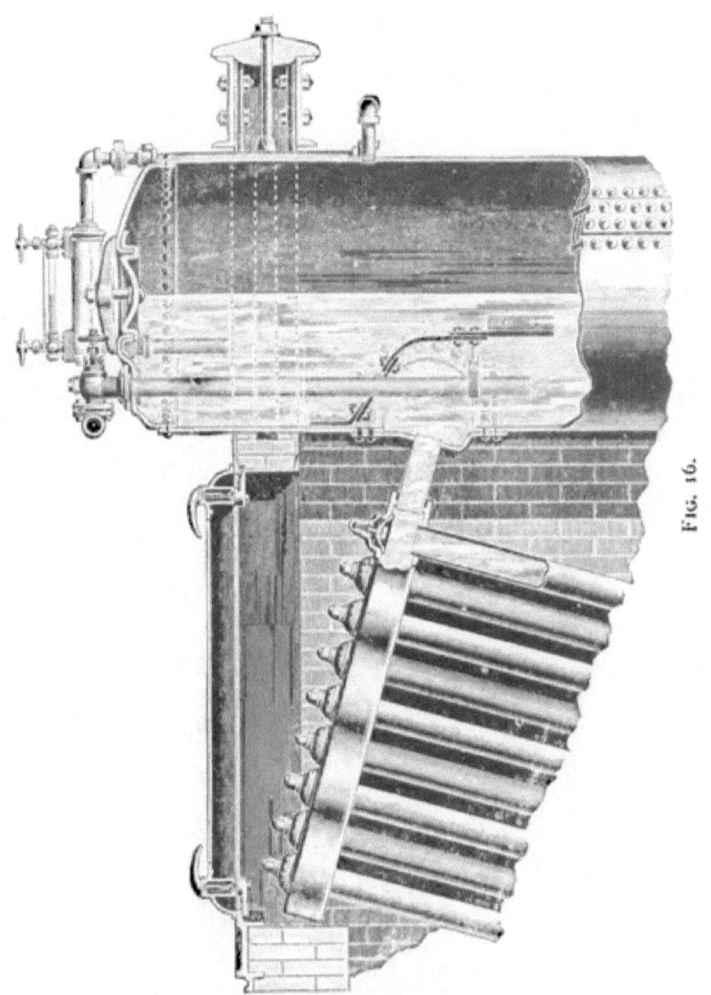

Fig. 16.

the tube when it needs cleaning or scaling. By the aid of a brick bridge-wall at the end of the furnace and a continuation of this wall formed of special tiles through the tubes, together

24   STEAM-BOILERS.

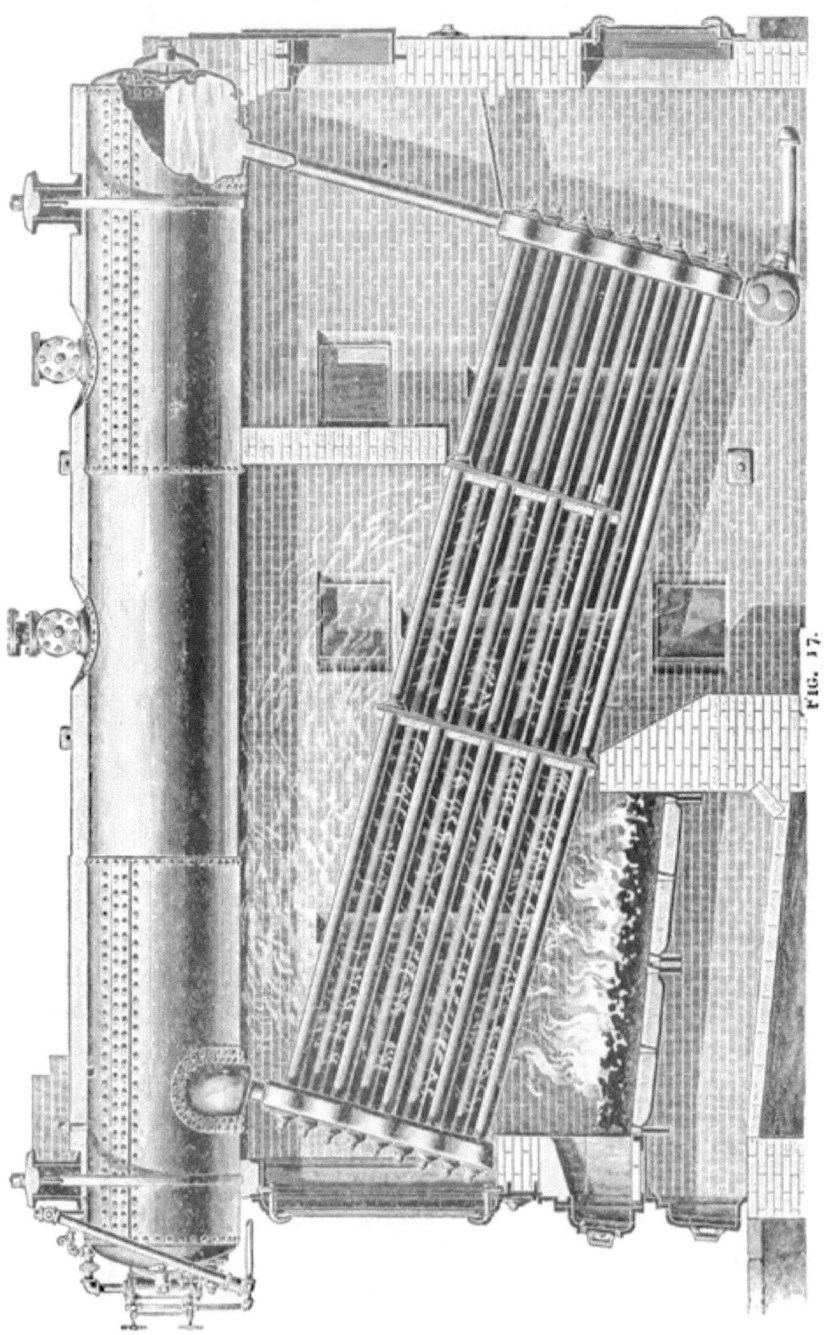

FIG. 17.

with a hanging bridge-wall similarly continued through the tubes, the products of combustion pass over the tubes three times on the way to the uptake at the back end of the boiler. The lower half of the cylindrical shell serves as heating-surface, but it is at such a height above the fire and is so shielded by the water-tubes that it is not liable to be overheated. The boiler is hung from cross-girders front and back, which in turn are supported on iron columns, and the brick setting is only a screen to retain the heat.

The circulation of the water in the boiler is down from the shell at the rear to the water-tubes, forward and upward through the tubes, in which course it is partially vaporized and consequently has a less average density, then up into the shell, at the front where the steam and water separate; the water in the shell flows continually from the front to the rear to supply the current through the tubes.

The Heine Boiler, shown by Fig. 42, page 106, resembles the Babcock and Wilcox boiler in general arrangement, but differs in that the tubes are expanded into one large header at each end, made of plate, properly stayed and provided with hand-holes. Again, the gases from the fire are constrained to pass along the tubes instead of across them, for which purpose there are floors or nearly horizontal bridges of tiles, laid on two or three layers of tubes, instead of the nearly vertical bridges of tiles used in the Babcock and Wilcox boiler.

The Root Boiler.—The general appearance of the Root boiler is shown by Fig. 18, and details of construction are shown by Fig. 19. Pairs of tubes are first expanded into headers at the end, as shown by 1, Fig. 19; then several pairs are assembled, as shown by 2, to form a vertical *section*, by the aid of bends, of which 3 gives further details. The joints between the bends and headers are made tight by aid of a metallic packing-ring shown by 4. The conical bearing on the bend shown by 5 expands the ring into a recess in the header, shown by 6, thus making a steam-tight joint. Each

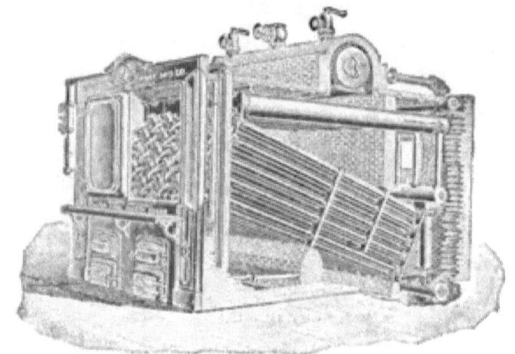

Fig. 18.

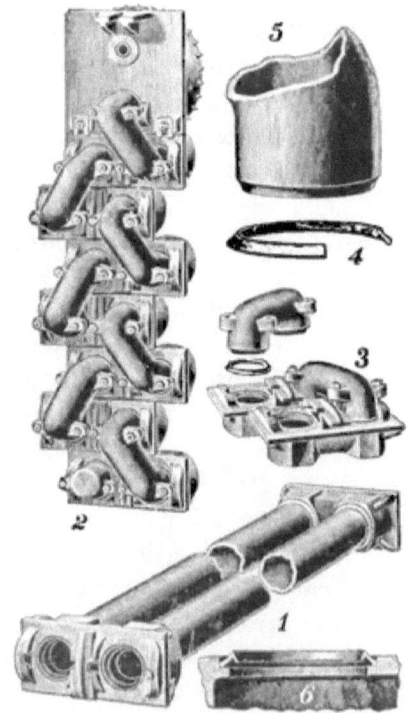

Fig. 19.

section has a steam-drum at the top, as shown in Fig. 18, and at the back end of the steam-drum are pipes leading up into the transverse steam-drum, and downward into a transverse water-pipe at mid-height of the boiler. Near each end of the mid-height water-pipe are vertical pipes communicating with the ends of a transverse mud drum, from which a series of pipes lead to the sections of the boiler. The water circulation is down from the back ends to the mud-drums, then forward through the tubes to the front ends of the steam-drums, in which the steam and water separate, the steam passing into the transverse steam-drum, and the water returning through the back connections to the sections. The products of combustion pass over the tubes three times before escaping to the chimney.

The Stirling Boiler.—This boiler, shown by Fig. 20, has three cylindrical drums at the top and a larger drum at the bottom, connected by tubes having a slight curvature at the ends. The two forward drums at the top have also a connection below the water-line through pipes not indicated. All three upper drums have their steam-spaces connected by piping. The water-line is indicated by a dotted line.

The feed-water is introduced into the rear upper drum, from which it passes down through the rear system of pipes, which act mainly as a feed-water heater, and enter the lower drum, where the water deposits any lime compound that it may contain, from whence it may be blown out at intervals. Fire-brick bridges cause the products of combustion to pass in succession through the three systems of water-tubes as shown by the arrows.

The Cahall Boiler is a vertical water-tube boiler, shown by Fig. 21. It has an annular drum at the top and a cylindrical drum at the bottom, connected by tubes and also by two large circulating pipes outside of the brick setting, one of which is drawn in the figure. The fire is in a brick furnace at one side of the boiler, from which the products of combustion pass back and forth across the tubes to and from the central space between the tubes. For this purpose there are two iron baffle-plates in the central space, as indicated in the figure.

The water-line is carried at about one third the height of the upper drum, and steam is drawn from a nozzle at the top.

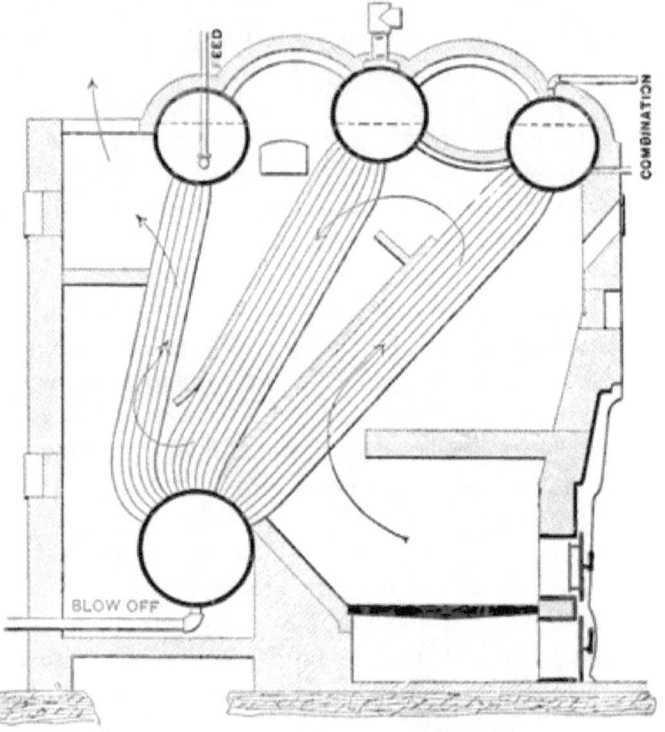

FIG. 20.

The circulation is down the large exterior pipes to the lower drum, and then up the water-tubes to the upper drum. Manholes give access to both drums, and in addition there are eight hand-holes in the top of the upper drum, so that any tube may be cut out and replaced without disturbing the others.

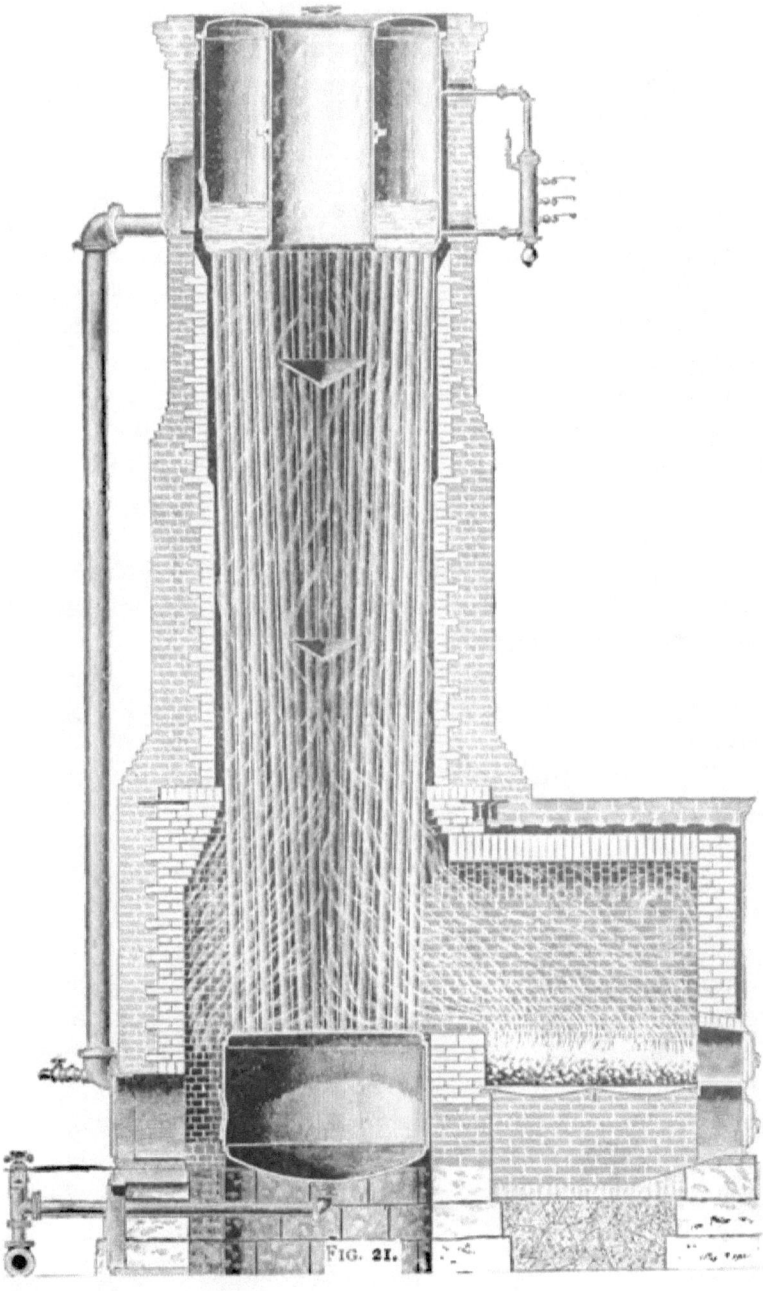

FIG. 21.

Water-tube Marine Boilers.—With the advent of very high steam-pressures on steamships there has been a tendency to replace the Scotch boiler by some form of water-tube boiler. A large number of French merchant steamers and a few French naval vessels have been fitted with Belleville boilers, a type of water-tube boilers that had already found favor for stationary purposes. This type of boiler has also been used to some extent on the Great Lakes. Recently this boiler has been largely introduced in the English Navy. Other water-tube boilers, either designed specially for marine boilers or modified from land boilers, have been used to some extent. In the United States Navy some vessels have been fitted with both shell-boilers and water-tube boilers; the former are intended for use in ordinary service, and the latter when running at high speed.

The objects that are sought in water-tube boilers for steamships are a larger power for the weight and the ability to carry high pressures. It still remains a question whether the water-tube boiler will or can replace the Scotch boiler for ordinary service on steamships. Indeed, it is a question whether there is any real profit in carrying steam at very high pressure.

The Belleville Boiler is represented by Fig. 22; it consists essentially of a series of coils of pipe made up with bends and elbows around which the products of combustion pass on the way to the chimney. At the top there is a steam-drum A, connected by two circulating-pipes B and C, with a drum D at the bottom. From the mud-drum D a rectangular feed-supply runs across the front of the boiler to all the coils or elements of the boiler. Each element is continuous from the feed-supply to the steam-drum, and is made up of slightly inclined pieces of pipe with horizontal bends or connections at the end. The effect is much as though a helical coil were flattened into two vertical tiers of pipes. The amount of water in the boiler is so small that it cannot be run without

TYPES OF BOILERS. 31

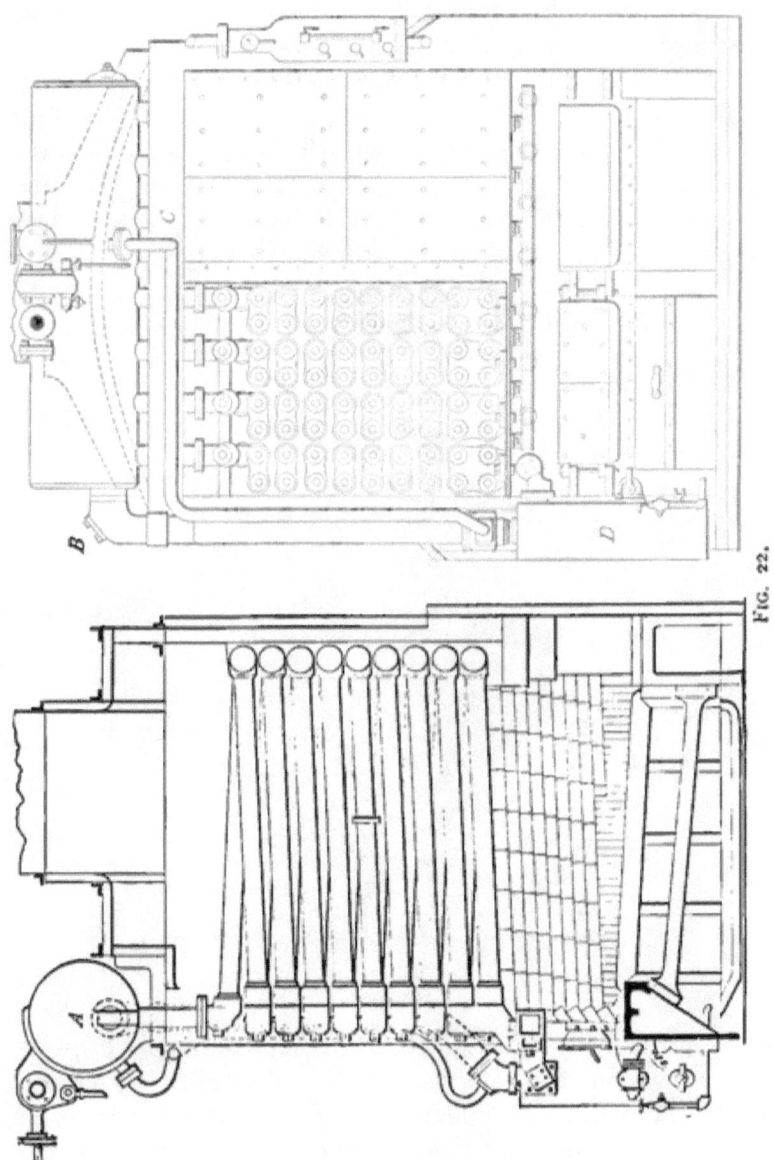

Fig. 22.

an automatic feed-water regulator, which in turn requires the attention of an expert feed-water tender. The several elements deliver a mixture of water and steam to the steam-drum, which does not appear to act efficiently as a separator, as an external separator is placed between the boiler and the engine. The feed-water is supplied to the steam-drum and passes through the external circulating-pipes to the mud-drum, where it deposits much of its impurities,

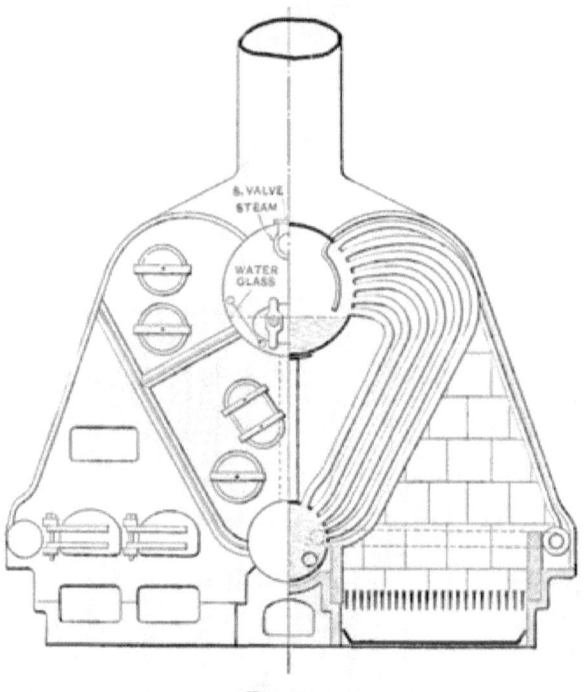

FIG. 23.

Thornycroft Boiler.—The boiler represented by Figs. 23 and 24 was built for the torpedo-boat destroyer, "Daring," by Mr. Thornycroft; boilers of slightly different forms have been fitted by him, in torpedo-boats and steam-launches.

The boiler consists essentially of a large drum or separator at the top and three drums at the bottom, connected by a large number of bent-tubes. There is, inside of the casing, a large tube connecting the top drum to the middle drum at the bottom, and this drum is connected to the side drums by smaller pipes. The circulation is down from the top drum to the middle lower drum, and from that to the side drums, then up through all the bent water-tubes to the upper drum, where mingled water and steam is delivered

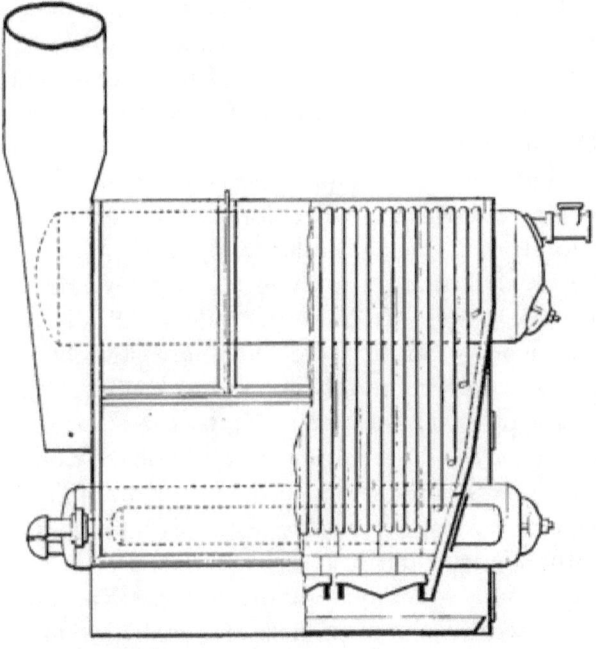

FIG. 24.

against a baffle-plate above the water-line. Steam is drawn from a nozzle at the rear end of the top drum.

The arrangement of grates and fire-doors is shown in elevation and section by Fig. 23. The middle drum divides the grate into two parts; over that drum is a space which is in communication with the uptake, as shown by Fig. 24. The products of combustion pass among the tubes leading from the middle drum; the tubes to the outer drums intercept the radiant heat which would otherwise strike on the boiler-casing.

The boiler-setting is an iron frame, and the casing is thin plate iron lined with incombustible non-conducting material. There are numerous doors through the casing for cleaning the tubes.

This boiler has proved very successful with a forced draught, making steam freely and giving little trouble. The boiler contains so small an amount of water that steam may be raised quickly, and any demand for steam can be quickly met. On the other hand, the feed-supply must be regulated with care and skill, and the pressure is liable to fluctuate.

The Yarrow Boiler.—The form of boiler used by Mr. Yarrow for torpedo-boats, is shown by Fig. 25. It resembles in general arrangement a form used by Mr. Thornycroft with one grate. It, however, differs radically in certain particulars, namely, in that the tubes are straight and that they enter the upper drum below the water-line, and in that there are no pipes outside the casing to carry water from the upper drum to the lower drum or reservoirs. Some of the tubes deliver water and steam to the upper drum, from which steam is drawn; other tubes carry water from the upper drum to the lower drums. A given tube may act sometimes in one way and sometimes in the other. Naturally those tubes which receive the most heat and make the most steam deliver to the upper drum, and tubes that receive less heat carry down water.

The air for the fire is drawn from an iron box or casing outside the boiler-casing, so that the heat escaping from the boiler-casing is largely carried back to the fire, and the fire-room, and also the rest of the vessel, is heated up less.

The Almy Boiler.—This boiler, which is represented by

Fig. 26, is made of short lengths of pipe screwed into return-bends and into twin unions. At the bottom is a large tube or pipe forming three sides of a square at the sides and back

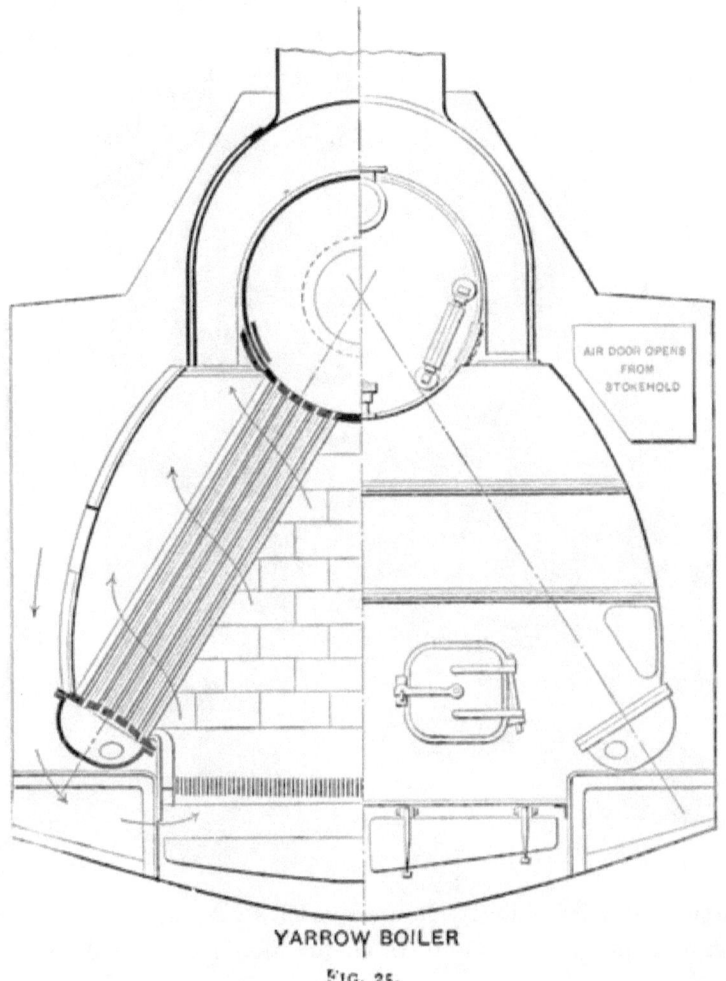

YARROW BOILER

FIG. 25.

of the grate. From this water-space the tubes lead into a similar structure at the top. The steam and water are discharged into a separator in front of the boiler, from which steam is drawn; while the water separated therefrom, together with the feed-water, passes down through circulating-pipes to the bottom of the boiler.

The boiler is provided with a coil feed-water heater

FIG. 26.

above the main boiler. It is enclosed by a casing lined with non-conducting material. It is intended for general marine work.

CHAPTER II.

FUELS AND COMBUSTION.

The fuels used for making steam are coal, coke, wood, charcoal, peat, mineral oil, and natural and artificial gas. Various waste and refuse products, such as straw, sawdust, and bagasse, are burned to make steam.

All coals appear to be derived from vegetable origin, and they owe their differences to the varying conditions under which they were formed or to the geological changes which they have undergone.

Anthracite Coal consists almost entirely of carbon and inorganic matters; it contains little if any hydrocarbon. Some varieties, for example certain coals found in Rhode Island, appear to approach graphite in their characteristics, and are burned with difficulty unless mixed with other coals. Good anthracite is hard, compact, and lustrous, and gives a vitreous fracture when broken. It burns with very little flame unless it is moist, and gives a very intense fire, free from smoke. Even when carefully used, it is liable to break up under the influence of the high temperature of the furnace when freshly fired, and the fine pieces may be lost with the ash.

Semi-anthracite or Semi-bituminous Coal is intermediate in its properties between anthracite coal and bituminous coal; it contains some hydrocarbon, is less dense than anthracite, it breaks with a lamellar fracture, and it burns readily with a short flame.

Bituminous Coals contain a large and varying per cent of hydrocarbons or bituminous matter. Their physical properties and behavior when burning, vary widely and with all intermediate gradations represented. so that classification is difficult. Three kinds may. however, be distinguished, as follows:

Dry bituminous coals, which burn freely and with little smoke and without caking.

Caking bituminous coals, which swell up, become pasty, and cake together in burning. They are advantageously used for gas-making.

Long-flaming bituminous coals, which have a strong tendency to produce smoke; some do and some do not cake while burning.

Coke is made from bituminous and semi-bituminous coal by driving off the hydrocarbons by heat. Coke made as a by-product in gas retorts, is weak and friable, and has little value for making steam. Coke made in coking ovens, by partial combustion of the coal which is coked, is of a dark-gray color, porous, hard, and brittle. It has a metallic lustre, and gives out a slight ringing sound when struck. Sulphur in the coal may be burned out in coking, if the coal is moist or if steam is supplied during coking, so that coke may be comparatively free from this noxious element even when made from a poor coal. Coke burns without flame and makes a fierce fire when forced.

Lignite, or brown coal, is of more recent geological formation than coal, and is in a manner intermediate between coal and peat. It frequently contains much moisture and mineral matter. It is used where good coal is difficult to get, and while the better varieties form a useful fuel, the poorer qualities have little value.

Peat, or turf, is obtained from bogs. It consists of slightly decayed roots of the swamp vegetation mingled with more or less earthy matter. For domestic use it is cut and dried in the air. It is little used for making steam, though when pulverized, dried, and compressed it makes a useful artificial fuel.

Wood is used for making steam either in remote places where coal is hard to get and timber is plenty, or where sawdust or other refuse wood is produced in quantity in manufacturing operations. Wood is also used for kindling coal-fires. One cord of hard wood is equivalent to one ton of anthracite coal; one cord of yellow-pine is equal to half a ton of coal; other soft woods are, as a rule, of less value for fuel.

Charcoal is made by charring wood; it is but little used for making steam.

Mineral Oil, in the form of crude petroleum or the refuse heavy oil left from the distillation of petroleum, is used for making steam, especially in the neighborhood of the Black Sea oil-field, and by steamers carrying oil from those fields. It is customary to throw the oil into the furnace in the form of finely divided spray through special spraying apparatus worked either with compressed air or with superheated steam. The use of superheated steam has its convenience only to recommend it, for it adds to the inert material to be uselessly heated. Special precautions must be taken, when petroleum is burned, to avoid flooding the furnace with oil and to prevent explosions of the vapor and burning of the oil in tanks or receptacles.

Gases.—Natural gas from gas-wells has been used for making steam, usually in a crude and wasteful way. Some attempts have been made to use gas made from poor and smoky coal, in producer-furnaces like those used in metallurgical operations; but the gain to be expected is only the suppression of the smoke nuisance, which is rather a social than an economical problem.

Artificial Fuels.—The small waste from coals and charcoals, sawdust, and other fine combustible material which cannot be sold in such shape, is sometimes made into cakes or briquettes by mixing it with some adhesive material and then compressing it. The adhesive materials have been wood-tar, coal-tar, or else clay. Tar is available in limited quantities only, and clay is disadvantageous since it adds to the inert material, of which fine fuel is liable to have an excess. Artificial fuels have some advantages for special purposes, and can be stored compactly; they are used mostly where good fuel is difficult to get.

Composition of Fuels.—The composition of a number of American coals, together with the total heat of combustion by Dulong's formula, is given in the table on page 41, which has been kindly furnished by Mr. Henry J. Williams. These results are a part of a very extended investigation by Mr. Williams, to be published in full in the near future. Most of the results are the averages of several separate analyses, and all may be depended upon to give a fair representation of the coals named.

Analyses by Mahler of various European coals and of a few American coals, together with the total heat of combustion, are given in the table on page 42.

The following table gives the composition of several representative petroleums:

COMPOSITION OF PETROLEUMS.

	Carbon.	Hydrogen.	Oxygen.	Specific Gravity.
Pennsylvania, crude	84.9	13.7	1.4	0.886
Caucasian, light	86.3	13.6	0.1	0.884
heavy	86.6	12.3	1.1	0.938
Petroleum refuse	87.1	11.7	1.2	0.938

Heat of Combustion.—The number of thermal units developed by the complete combustion of one unit of weight of a fuel is called the heat of combustion. It can be determined by burning the fuel in a properly constructed calorimeter. The most recent and best results are those obtained by the use of Mahler's bomb-calorimeter. This is a strong recep-

COMPOSITION AND HEAT OF COMBUSTION OF AMERICAN COALS (PER CENTS).
By Henry J. Williams.

	Anthracites.			Semi-bituminous.				Bituminous.				Lignite.
	Lehigh.	Lykens Valley.	Drifton, Pa.	Pocahontas.	George's Creek.	Clearfield, Pa.	New River, W. Va.	Connellsville Coking.	Big Muddy, Carterville, Ill.	Dominion, Cape Breton.	Pittsburgh Steaming.	Wyoming.
Proximate analysis:												
Water............	1.97	1.50	1.37	0.54	0.59	0.44	0.77	0.94	5.79	2.43	1.36	5.75
Volatile matter...	4.35	7.84	3.59	16.70	18.52	18.76	21.70	27.39	30.11	30.16	32.00	39.98
Fixed carbon.....	86.49	81.07	89.11	77.28	74.31	73.15	72.69	64.06	55.79	62.25	59.14	47.63
Ash..............	7.19	9.59	5.93	5.48	6.58	7.65	4.84	7.61	8.31	5.16	7.50	6.64
Sulphurs:												
Total sulphur.....	0.64	0.50	0.56	0.79	0.81	1.02	0.77	1.07	1.03	2.00	1.26	
Volatile sulphur...	0.58	0.36	0.45	0.68	0.69	0.91	0.72	0.91	0.90	2.00	1.16	
Sulphur in ash....	0.06	0.14	0.11	0.11	0.12	0.11	0.05	0.17	0.13	0.09	0.10	0.21
Ultimate analysis:												
Carbon...........	85.66	83.20	87.70	83.46	81.05	80.17	83.57	77.81	71.81	77.84	76.47	
Hydrogen.........	2.78	3.29	2.56	4.82	4.91	5.08	4.69	5.24	4.85	4.99	5.19	
Nitrogen..........	0.77	0.95	1.03	1.28	2.15	1.46	1.63	1.56	2.04	1.28	1.45	
Oxygen...........	2.87	2.45	2.26	4.24	4.57	4.69	4.51	6.79	11.52	8.55	8.10	
Ash..............	7.33	9.75	6.00	5.52	6.62	7.69	4.88	7.65	8.82	5.29	7.60	
Volatile sulphur...	0.59	0.36	0.45	0.68	0.70	0.91	0.72	0.92	0.96	2.05	1.19	
Heat of combustion:												
Dulong's formula..	13963	13251	13171	14805	14484	14448	14607	14043	12501	13755	13749	

COMPOSITION AND HEAT OF COMBUSTION OF FUELS (PER CENTS).
By Mahler.

Kind of Coal.	Analysis.					Per cent Vol. Matter exclusive of Water & Ash.	Calorific Power, Observed.	Calorific Power, excluding Water and Ash.	Calorific Power calculated.
	Carbon.	Hydrogen.	Oxygen and Nitrogen.	Hydroscopic Water.	Ash.				
Anthracite from Pennsylvania	86.456	1.995	2.199	3.450	5.900	3.00	7484	8256	8450
Anthracite from Mure	86.564	1.367	2.969	4.700	4.400	2.75	7468	8216	8172
Semi-anthracite from Kébao	85.746	2.733	2.731	2.800	5.450	5.20	7828	8532	8531
Semi-anthracite from Commentry	84.928	2.692	5.005	1.775	5.400	3.19	7850	8456	8360
Semi-anthracite from Blanzy	82.746	2.916	6.278	1.760	6.300	6.00	7773	8203	8215
Semi-bituminous from Anzin	88.473	4.139	4.338	1.350	1.700	14.08	8392	8656	8767
Semi-bituminous from Aniche	85.937	4.198	5.240	0.625	4.000	11.93	8426	8834	8688
Bituminous from Anzin	83.754	4.385	5.761	1.100	4.000	21.51	8051	8574	8689
Bituminous from the collieries of Saint-Étienne	84.546	4.772	5.432	1.250	4.000	20.84	8392	8857	8829
Gas-coal from Béthune	82.418	5.089	7.193	1.200	4.100	30.41	8210	8668	8710
Gas-coal from Commentry	80.182	5.245	8.173	3.000	3.400	39.96	7870	8408	8644
Wigan cannel coal	78.382	5.060	5.058	0.600	10.900	31.64	7761	8768	9011
Lignite from Styria	65.455	4.782	24.303	0.710	4.750	50.34	6284	6646	6610
Lignite from Vaugirard	59.795	4.512	25.799	3.144	6.750	49.95	5536	6076	6270
Norwegian pine, partially dried	47.366	5.581	39.780	6.940	0.333	68.93	4477	4828	4947
Cellulose, $C_{12}H_{10}O_5$	44.440	6.170	49.390	6.920	0.750	68.93	4200	4200	4264
Tourbe from Bohemia	53.183	5.542	34.230	6.125	0.920	63.93	5489	5903	5609
Coke, Commentry coal	92.727	0.414	2.629	0.500	4.200	68.93	7665	8001	7957
Coke, semi-bituminous coal	94.582	0.633	1.585	0.500	3.200	68.93	7787	8044	8130
Coke, Pennsylvania anthracite	91.036	0.685	2.146	0.233	5.900	68.93	7528	8036	8078

tacle of wrought iron or bronze, gold-plated or enamelled inside. The fuel to be tested is placed in a small platinum crucible, with an arrangement for igniting by electricity. The bomb is then filled with oxygen under the pressure of about twenty-five atmospheres, and is placed in a calorimeter-can containing water. There is oxygen in excess, so that the charge when ignited is completely consumed, and the resultant total heat of combustion is absorbed by the metal of the bomb and by the water in the calorimeter. The corrections for the calorimeter are determined by burning in it, some substance like naphthaline, for which the heat of combustion is known. The processes of making combustion determinations are simple and direct; the difficulties are those incident to accurate measurements of temperatures, for which purpose the best physical thermometers are required. The experimenter must be an expert physicist, who has had experience in the use of the apparatus. The table of composition of fuels by Mahler* gives also the total heats of the fuels, determined by the same experimenter by aid of the bomb-calorimeter.

An engineering expert who has had adequate training in a physical laboratory, may learn how to make determinations of the total heat of combustion; an engineer in general practice will find it advantageous to refer such work to an expert physicist. It is not too much to say that all crude forms of apparatus for finding total heat of combustion of fuels are useless and misleading.

The heats of combustion of carbon in various forms as determined by Berthelot † are:

    Diamond .... 7859 calories.
    Diamond bort .... 7860.9 "
    Graphite .... 7901.2 "
    Amorphous from wood .... 8137.4 "

---

* Bulletin de la Soc. d'Encouragement pour Industrie nationale, 1891.
† Comptes rendu, 1889.

The heat of combustion of carbon in fuels may be taken at 8140 calories, a calorie being defined as the heat required to raise one kilogram of water from 15° C to 16° C. This will give in the English system of units 14650 British thermal units, the B. T. U. being defined as the heat required to raise the temperature of a pound of water from 62° F. to 63° F. These definitions are founded on Rowland's determination of the mechanic equivalent of heat; the difference between them and others commonly given are not of practical importance in this connection.

The following table gives the heat of combustion of some elements and simple gases:

Carbon burned to CO_2........ 8,140 calories; 14,650 B. T. U.
Carbon burned to CO....... 4,400 "
Hydrogen............... ...34,500 " 62,100 "
Sulphur....................... 4,032 "
Marsh-gas, CH_4 .... 23,513 "
Olefiant gas, C_2H_4 .... 21,343 "
Carbon monoxide............ 4,393 "

Chemistry of Combustion.—Calculations concerning the heat of combustion of fuels and the amount of air needed for combustion, require a knowledge of the elements of chemistry.

Elementary chemical substances are those that have not been decomposed, such as oxygen, hydrogen, and nitrogen. The elements enter into chemical combination in fixed proportions by weight; these proportions are called the combining weights or the atomic weights of the elements. In the following table are given the most important chemical elements of fuels, their chemical symbols, and their atomic weights. The table gives other useful information which will be referred to later.

A chemical combination, such as water, is represented by a formula consisting of the symbols of the elements entering into the combination, each symbol having a subscript which shows the number of times the combining or atomic weight of

FUELS AND COMBUSTION. 45

	Symbol or Composition.	Atomic or Molecular Weight.	Specific Volumes.	Specific Heat in Gaseous Condition.	Density or Weight of One Cubic Foot.
Carbon.............	C	12	....	......	....
Hydrogen..........	H	1	178.881	3.409	0.005590
Oxygen............	O	16	11.2070	0.2175	0.08928
Nitrogen...........	N	14	12.7561	0.2438	0.07837
Sulphur............	S	12	....	......	....
Carbon dioxide....	CO_2	$12 + 2 \times 16$	8.10324	0.2169	0.12341
Carbon monoxide...	CO	$12 + 16$	12.81	0.2450	0.07806
Water.............	H_2O	$2 + 16$	....	0.4805*	....
Air................	....	....	12.3909	0.2375	0.08071
Ash................	....	....	....	0.2	....

* Superheated steam.

the element occurs in the combination. This water is represented by

$$H_2O,$$

which indicates that water is made up of two portions of hydrogen and one portion of oxygen. It is commonly said that two atoms of hydrogen and one atom of oxygen unite to form one molecule of water. As the atomic weight of hydrogen is 1 and the atomic weight of oxygen is 16, we have water formed of two pounds of hydrogen to 16 pounds of oxygen.

Again, carbon may unite with one portion of oxygen to form carbon monoxide or carbonic oxide, represented by CO; or carbon may unite with two portions of oxygen to form carbon dioxide or carbonic acid, represented by CO_2. Referring to the table on page 44, it appears that the complete combustion to CO_2 gives more than three times the heat obtained from incomplete combustion to CO. But the resulting gas, CO may be burned with one more portion of oxygen, and will finally form CO_2. Assuming that the double process will yield the same amount of heat per pound of coal as is obtained by direct combustion to CO_2, we may calculate the heat of combustion of one pound of carbon monoxide as follows:

In the combustion of carbon to CO, 12 pounds of carbon unite with 16 pounds of oxygen, forming 28 pounds of CO; hence one pound of carbon will form

$$\frac{12 + 16}{12} = 2\tfrac{1}{3} \text{ lbs. of CO.}$$

The heat developed by burning these $2\tfrac{1}{3}$ pounds of carbon monoxide, under our assumption, is

$$14650 - 4400 = 8250 \text{ B. T. U.},$$

so that each pound of carbon monoxide will yield

$$8250 \div 2\tfrac{1}{3} = 4393 \text{ B. T. U.},$$

as given in the table on page 44.

The complete combustion in either case will give

$$\frac{12 + 2 \times 16}{12} = 3\tfrac{2}{3}$$

pounds of carbon dioxide for each pound of carbon.

Calculation of Heat of Combustion.—If a fuel were a mechanical mixture of two chemical elements such as carbon and sulphur, the heat of combustion could obviously be found by calculating the parts separately and adding the results. For example, a mixture of 60 per cent carbon and 40 per cent sulphur would give

$$0.60 \times 14650 = 8790.0$$
$$0.40 \times 4032 = 1612.8$$
$$\overline{10402.8 \text{ B. T. U.}}$$

for each pound of the mixture.

Fuels, as a rule, contain carbon in a free state, and various compounds of carbon and hydrogen, and compounds of carbon, hydrogen, and oxygen. Now the rapid union of chemical elements is usually accompanied by the evolution of heat, as in the combustion of oxygen and hydrogen. Conversely, heat is required to break up a chemical combination. Now the combustion of a fuel is a complex process, involving usually some breaking up of chemical compounds and the union of chemical elements with oxygen; the exact nature of the process is far from certain even when the real chemical compounds and elements of which the fuel is composed are known. As a rule we know only the final analysis of the fuel and do not know the compounds which enter into it. For this reason the only true way of determining total heat of combustion is by experiment. Nevertheless it is customary and convenient to make a calculation of the total heat of combustion by an arbitrary method, when the real heat of combustion of a fuel has not been determined.

Dulong proposed that the heat of combustion should be calculated on the assumption that the oxygen in the fuel and enough hydrogen to unite with it and form water, could be set aside as inert, and that the remainder of the hydrogen and all the carbon could be treated as free elements. From the composition of water and the atomic weights of hydrogen and oxygen it is clear that each pound of oxygen will require

$$\frac{2 \times 1}{16} = \frac{1}{8}$$

of a pound of hydrogen. Dulong's method may therefore be expressed by the equation

$$\text{Total heat} = 14,650\,C + 62,100\,(H - \tfrac{1}{8}O)$$

in which the letters C, H, and O represent the *weights* of carbon, hydrogen, and oxygen in one pound of fuel. No confusion need arise because the letters are used with a different significance from that given them in chemical formulæ. This equation does not give very satisfactory results.

Mahler has proposed an empirical formula for finding heats of combustions which in French units is

$$\text{Total heat} = 8140\,C + 34{,}500\,H - 3000\,(O + N),$$

in which C, H, O, and N represent the weights of the elements carbon, hydrogen, oxygen, and nitrogen in a kilogram of fuel. The result is in calories.

In English units Mahler's equation becomes

$$\text{Total heat} = 14{,}650\,C + 62{,}100\,H - 5400\,(O + N),$$

in which the letters represent the weights of the corresponding elements in one pound of the fuel. The result is in B. T. U. This equation gives results that agree very well with Mahler's experimental determinations, as shown by the table on page 42.

For example, the total heat of combustion of Pittsburg bituminous coal, for which the ultimate analysis in the table on page 41 gives

$$C = 0.7647, \quad H = 0.0519, \quad O = 0.0810, \quad N = 0.0145.$$

appears by Dulong's formula to be

$$14650\,C + 62{,}100\,(H - \tfrac{1}{8}O)$$
$$= 14{,}650 \times 0.7647 + 62{,}100 \left(0.0519 - \frac{0.0810}{8}\right)$$
$$= 13{,}800 \text{ B. T. U.}$$

Mahler's formula for the same coal gives

$$14{,}650\,C + 62{,}100\,H - 54{,}000\,(O + N)$$
$$= 14{,}650 \times 0.7647 + 62{,}100 \times 0.0519$$
$$\qquad - 5400\,(0.0810 + 0.0145)$$
$$= 13{,}910 \text{ B. T. U.}$$

Air required for Combustion.—If the moisture and carbon dioxide in the air be neglected, and if, further, the argon is not distinguished from the nitrogen, then we have for the composition of the atmospheric air,

By weight { Oxygen.................. 0.232
{ Nitrogen................. 0.768
By volume { Oxygen.................. 0.2094
{ Nitrogen................. 0.7906

For rough calculations it is customary to consider that the atmosphere is made up of one volume of oxygen and four volumes of nitrogen. This approximation is sufficient for calculation of air required by fuels, and for similar purposes.

The air required for combustion of a given fuel may be estimated from its composition and from the composition of the air. A few examples will make the process clear.

Thus, carbon burned to CO_2 requires two portions of oxygen, so that one pound of carbon will require

$$\frac{2 \times 16}{12} = 2\tfrac{2}{3}$$

pounds of oxygen. Since air is 0.232 part oxygen by weight, one pound of carbon will require

$$2\tfrac{2}{3} \div 0.232 = 11.5$$

pounds of air for complete combustion.

In like manner one pound of hydrogen will require

$$\frac{16}{2} = 8$$

pounds of oxygen, or

$$8 \div 0.232 = 34.5$$

pounds of air for complete combustion.

Another method of calculation is based on the approximate composition of air, i.e., one volume of oxygen and four of nitrogen. This method depends on the fact that the weights of a cubic foot of different kinds of gases are proportional to their atomic weights; so that if the weight of a cubic foot of hydrogen be taken for the basis of comparison and be called unity, then the weight of a cubic foot of oxygen will be 16, while that of nitrogen will be 14. We shall then have for the approximate composition of air one volume of oxygen having the weight 16, and four volumes of nitrogen having each the weight 14. In order to get one pound of oxygen we must take

$$(16 + 4 \times 14) \div 16 = 4\tfrac{1}{2}$$

pounds of air.

It has already been shown that one pound of carbon will require $2\tfrac{2}{3}$ pounds of oxygen. By the method just stated it appears that a pound of carbon will require

$$2\tfrac{2}{3} \times 4\tfrac{1}{2} = 12$$

pounds of air. This result is often quoted and is easily remembered.

Since a pound of hydrogen requires 8 pounds of oxygen, this method gives

$$8 \times 4\tfrac{1}{2} = 36$$

pounds of air for each pound of hydrogen.

In calculating the air required for a fuel it is customary to use the convention proposed by Dulong for finding heat of combustion, namely, that each pound of oxygen in the fuel renders one eighth of a pound of hydrogen inert, and that the remainder of the hydrogen and all the carbon can be treated as free elements. In using this convention it is customary to take the approximate weights of air just calculated for a pound of carbon and a pound of hydrogen. The convention can then be stated in the form of an equation as follows:

$$\text{Air per pound of fuel} = 12\,C + 36\,(H - \tfrac{1}{8}O),$$

in which the letters C, H, and O represent the weights of carbon, hydrogen, and oxygen in one pound of the fuel.

An application of this equation to Pittsburg coal gives

$$\text{Air} = 12 \times 0.7647 + 36\left(0.0519 - \frac{0.0810}{8}\right) = 10.7 \text{ pounds.}$$

This result is somewhat larger than would be obtained were the more exact composition of the atmosphere given on page 49 used, together with the assumption that the oxygen renders inert its equivalent of hydrogen; but the method is not sufficiently well grounded to warrant much refinement.

As a further illustration of the method the following calculation of the air required for one pound of olefiant gas may be interesting. This gas, having the composition C_2H_4, consists of

$$\frac{2 \times 12}{2 \times 12 + 4 \times 1} = \frac{6}{7} \text{ carbon,}$$

$$\frac{4 \times 1}{2 \times 12 + 4 \times 1} = \frac{1}{7} \text{ hydrogen,}$$

and will require

$$\tfrac{6}{7} \times 12 + \tfrac{1}{7} \times 36 = 15.4 \text{ pounds of air.}$$

Air for Dilution.—In order to secure complete combustion of coal in the furnace of a boiler it is necessary to supply an excess of oxygen, or, what amounts to the same thing, an excess of air. This excess varies from one half the quantity required for combustion to an equal quantity. Thus, roughly, from 18 to 24 pounds of air may be furnished per pound of carbon and from 54 to 72 pounds of air per pound of hydrogen.

Volume of Air for Combustion.—The table on page 45 gives the density or weight of one cubic foot of the several gases mentioned, also the reciprocal of the density, or the volume occupied by one pound of the gas. This is called the specific volume of the gas. The specific volume of air is 12.3909 at the pressure of the atmosphere and at the temperature 32° F. The volume of a pound of gas increases as the temperature rises. At 60° F. one pound of air will occupy about 13 cubic feet. To find the volume of air required per pound of fuel we may simply multiply the weight by 13, for ordinary calculations. Thus we shall have for the air per pound of the principal elements in fuels:

	Without Dilution.	With 50 per cent Dilution.	With 100 per cent Dilution.
Carbon	150	225	300
Hydrogen	450	675	900

These approximate values are sufficient for determining the dimensions of doors or passages through which air is supplied to the fire.

This method applied to Pittsburg coal will give, approximately,

$$10 \times 13 = 130$$

cubic feet of air for each pound of coal without dilution. With dilution of 50 per cent the air required will be about 200 cubic feet for each pound.

Sometimes, in connection with boiler-tests or for other purposes, a more exact estimate of the amount of air is desired. The calculation for this purpose can be best explained by aid of an example.

Example.—Required the weight and volume of air needed for combustion of Pittsburg coal with 50 per cent dilution, the temperature of the atmosphere being 70° F. and the height of the barometer being 29 inches, when reduced to 32° F.

This coal is composed of 76.47 per cent carbon, 5.19 per cent hydrogen, and 8.10 per cent oxygen. Assuming that the oxygen renders inert one eighth of its weight of hydrogen, there will be available

$$5.19 - \frac{8.10}{8} = 4.18 \text{ per cent}$$

of hydrogen and 76.46 per cent of carbon. Since one pound of carbon requires $2\frac{2}{3}$ pounds of oxygen, and one pound of hydrogen requires 8 pounds, the weight of oxygen required per pound of coal is

$$2\tfrac{2}{3} \times 0.7646 + 8 \times 0.0418 = 2.374 \text{ pounds.}$$

But air contains 23.2 per cent of oxygen by weight, so that the air required per pound of coal is

$$2.374 \div 0.232 = 10.2 \text{ pounds.}$$

The specific volume of air is 12.39, so that each pound of coal will require

$$10.2 \times 12.39 = 126$$

cubic feet of air at the normal pressure of the atmosphere and at 32° F.

To find the volume of air required at the actual pressure of the atmosphere and the actual temperature, we have the facts that the volume of a given weight of air is inversely proportional to the absolute pressure and directly proportional to the absolute temperature. Now the absolute pressure of the atmosphere is 29 inches of mercury as given by the barometer, while the normal pressure is 29.92 inches of mercury. To get the absolute temperature we add 460.7 to the temperature by the thermometer; the absolute temperature of 32° F. is 492.7, and that of 70° F. is 530.7. Under the conditions of the problem the air required per pound of fuel will have the volume, without dilution, of

$$126 \times \frac{530.7}{492.7} \times \frac{29.92}{29.00} = 140$$

cubic feet. With 50 per cent dilution the volume will be 206.5 cubic feet.

Determination of Air per Pound of Coal.—The amount of air supplied per pound of coal may be determined either by measuring the air supplied to the furnace or by an analysis of the products of combustion.

For the first method the following arrangement has been used in boiler-tests at the Massachusetts Institute of Technology: The ash-pit doors are removed and a sheet-iron mouthpiece is fitted over the opening into the ash-pit. The air for combustion is supplied by a cylindrical sheet-iron conduit leading into this mouthpiece. The area of the conduit should be at least equal to the area of the fire-door or fire-doors, and its length should be several times its diameter. The velocity of the air in the conduit is measured by an anemometer, from which the volume of air is readily calculated, and its weight determined from the temperature and pressure of the atmosphere. The joint between the mouthpiece and the furnace front must be luted to avoid leakage, and leaks or admission of air to the furnace otherwise than through the sheet-iron conduit must be stopped or allowed for. Anemometers, even when tested and rated, are liable to be affected by errors of two per cent or more. They are commonly tested by swinging them on a revolving arm through still air—a method that is proper for small or moderate velocities, but difficult to use, and is vitiated by the action of centrifugal force at high speeds. An ideal way of testing an anemometer would be to find its reading in such a conduit when the weight, and consequently the velocity, of the air per second is known. The weight may be determined by causing the supply of air to flow through a well-rounded orifice, to which calculations by the proper thermodynamic equations may be applied. This method for large conduits would involve the use of a very large air-compressor, which makes it hardly practicable.

Orsat's Gas Apparatus.—This apparatus, which is well adapted to the analysis of flue-gases, determines the proportion by volume of the carbon dioxide, carbon monoxide, and oxygen in a mixture of gases. The remainder of the flue-gases is commonly assumed to be nitrogen, but it includes unburned hydrocarbon, if there be any, and steam or vapor of water. In Fig. 27, *A*, *B*, and *C* are pipettes containing, respectively, solutions of caustic potash to absorb carbon dioxide, pyrogallic acid and caustic potash to absorb oxygen, and cuprous chloride in hydrochloric acid to absorb carbon monoxide.

At *W* is a three-way cock to control the admission of gas to the apparatus; at *D* is a graduated burette for measuring the volumes of gas, and at *P* is a pressure-bottle connected with *D* by a rubber tube to control the gases to be analyzed. The pressure-bottle is commonly filled with water, but glyc-

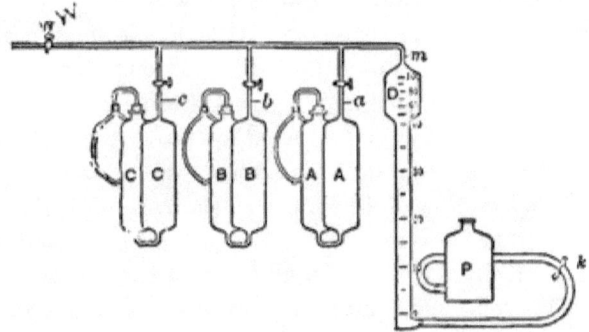

FIG. 27.

erine or some other fluid may be used when, in addition to the gases named, a determination of the moisture or steam in the flue-gases is made.

The several pipettes *A*, *B*, and *C* are filled to the marks *a*, *b*, and *c* with the proper reagents, by aid of the pressure-bottle *P*. With the three-way cock *W* open to the atmosphere, the pressure-bottle *P* is raised till the burette *D* is filled with water to the mark *m*; communication is then made with the flue, and by lowering the pressure-bottle the burette is filled with the gas to be analyzed, and two minutes are allowed for the burette to drain. The pressure-bottle is now raised till the water in the burette reaches the zero-mark and the clamp k is closed. The valve W is now opened momentarily to the atmosphere to relieve the pressure in the burette. Now open the clamp k and bring the level of the water in the pressure-bottle to the level of the water in the burette, and take a reading of the volume of the gas to be analyzed; all readings of volume are to be taken in a similar way. Open the cock a and force the gas into the pipette A by raising the pressure-bottle, so that the water in the burette comes to the mark m. Allow three minutes for absorption of carbon dioxide by the caustic potash in A, and finally bring the reagent to the mark a again. In this last operation, brought about by lowering the pressure-bottle, care should be taken not to suck the caustic reagent into the stop-cock. The gas is again measured in the burette and the diminution of volume is recorded as the volume of carbon dioxide in the given volume of gas. In like manner the gas is passed into the pipette B, where the oxygen is absorbed by the pyrogallic acid and caustic potash; but as the absorption is less rapid than was the case with the carbon monoxide, more time must be allowed, and it is advisable to pass the gas back and forth, in and out of the pipette, several times. The loss of volume is recorded as the volume of oxygen. Finally, the gas is passed into the pipette C, where the carbon monoxide is absorbed by cuprous chloride in hydrochloric acid.

The solutions are as follows:

A. Caustic potash, 1 part; water, 2 parts.
B. Pyrogallic acid, 1 gramme to 25 c.c. caustic potash.
C. Saturated solution of cuprous chloride in hydrochloric acid having a specific gravity of 1.10.

The absorption values per cubic centimetre of the reagents are—

A Caustic potash absorbs 40 c.c. carbon dioxide.
B. Pyrogallate of potassium absorbs 22 c.c. oxygen
C. Cuprous chloride absorbs 6 c.c. carbon dioxide.

FUELS AND COMBUSTION.   57

Samples of gas for analysis by Orsat's apparatus should be taken from the back of the furnace, from the uptake, and from the chimney; the difference in composition of gases at the several points will give the basis for calculations of leakage.

When it is not convenient to draw gases from the flue directly into the measuring burette of the apparatus, samples of gas may be drawn into glass bottles with rubber stoppers, from which gas can be supplied to the burette.

Calculation from a Gas Analysis.—The calculation of the amount of air supplied per pound of carbon and per pound of coal, from the known chemical constituents of the flue-gases, is best shown by an example.

Example.—Let it be assumed that the analysis of the flue-gases resulting from the burning of Pittsburg bituminous coal gives by volume 13 per cent of carbon dioxide, 0.5 per cent of carbon monoxide, and 6 per cent of oxygen. It is convenient to treat the percentages by volume as the number of cubic feet of the several gases in one hundred cubic feet of flue-gas. We will thus have—

Gas.	Volume.	Density. (See page 45.)	Weight.
Carbon dioxide	13	0.12341	1.6043
Carbon monoxide	0.5	0.07806	0.03903
Oxygen	6	0.08928	0.53568

Now one pound of carbon dioxide is composed of

$$\frac{2 \times 16}{12 + 2 \times 16} = \frac{8}{11}$$

of a pound of oxygen and 3/11 of a pound of carbon, and a pound of carbon monoxide is composed of

$$\frac{16}{12 + 16} = \frac{4}{7}$$

of a pound of oxygen and 3/7 of a pound of carbon. Consequently we have

$\tfrac{8}{11} \times 1.6043 = 1.1668$ $\qquad$ $\tfrac{3}{11} \times 1.6043 = 0.4375$

$\tfrac{4}{7} \times 0.03903 = 0.0223$ $\qquad$ $\tfrac{3}{7} \times 0.03903 = 0.0167$

$\phantom{\tfrac{4}{7} \times 0.03903 =\;} 0.5357$

Pounds of oxygen, 1.7248 $\qquad$ Pounds of carbon, 0.4542

And as air consists of 0.232 part by weight of oxygen, the air per pound of carbon from the gas analysis is

$$\frac{1.7248}{0.4542} \div 0.232 = 16.4 \text{ pounds.}$$

The coal in question contains 76.47 per cent of carbon, 5.19 per cent of hydrogen, and 8.10 per cent of oxygen. Of these elements Orsat's apparatus accounts for the carbon only; the oxygen and hydrogen together with unburned volatile matter pass off with the nitrogen.

The analysis shows 16.4 pounds of air for each pound of carbon; consequently the carbon in one pound of coal will require

$$0.7647 \times 16.4 = 12.5$$

pounds of air. Assuming that the oxygen in the coal renders one eighth of its weight of hydrogen inert, and that the remainder will require 36 pounds of air per pound of hydrogen, we shall have

$$36\left(0.0519 - \frac{0.0810}{8}\right) = 1.5$$

of a pound of air required for the hydrogen. So that the total air per pound of coal is about

$$12.5 + 1.5 = 14 \text{ pounds.}$$

The calculation just given, involving the use of the densities of the several gases, is perhaps the most readily understood; there is another method, which gives the same result and is more expeditious, depending on the fact that the weight of a gaseous compound referred to hydrogen as unity, is half its molecular weight. This quantity is called the vapor density of the compound.

Thus the vapor density of carbon dioxide, CO_2, is

$$\tfrac{1}{2}(12 + 2 \times 16) = 22;$$

and the vapor density of carbon monoxide, CO, is

$$\tfrac{1}{2}(12 + 16) = 14.$$

Assuming as before that in each 100 cubic feet of flue-gases there are 13 cubic feet of CO_2, 0.5 of CO and 6.0 of O, we have for the corresponding weights, based upon hydrogen as unity,

$$13 \times 22 = 286 \text{ for } CO_2$$
$$0.5 \times 14 = 7 \text{ for } CO$$
$$6.0 \times 16 = 96 \text{ for } O$$
$$\text{Total, } 389$$

The last result depending on the fact already noted, that the weights of elementary gases are proportional to the atomic weights.

Now each pound of CO_2 contains 3/11 of a pound of carbon, and each pound of CO contains 3/7 of a pound of carbon, so that of the 287 parts by weight of CO_2 we shall have

$$\tfrac{3}{11} \times 286 = 78$$

parts of carbon, and of the 7 parts by weight of CO we shall have

$$\tfrac{3}{7} \times 7 = 3$$

parts of carbon. The total weight of carbon will be

$$78 + 3 = 81.$$

The weight of oxygen is clearly

$$389 - 81 = 308.$$

The oxygen per pound of carbon is therefore

$$308 \div 81 = 3.85,$$

and the air per pound of carbon is

$$\frac{308}{81} \div 0.232 = 16.4$$

pounds, as found by the previous calculation.

Loss from Incomplete Combustion.—The presence of even a small amount of carbon monoxide in flue-gases is evidence of a very appreciable loss of efficiency, as may be seen by the following example, quoted from a test made on a 325-horse-power boiler at Lowell. The coal used was George's Creek Cumberland, fired by hand.

An analysis of flue-gases by Orsat's apparatus showed 12.5 per cent of CO_2, 1.1 per cent of CO, and 4.1 per cent of O, by volume.

Using the method of vapor densities for making the calculation, it appears that the CO_2 contained

$$\tfrac{3}{11} \times 12.5 \times 22 = 75 \text{ parts of carbon,}$$

and the CO contained

$$\tfrac{3}{7} \times 1.1 \times 14 = 6.6 \text{ parts of carbon.}$$

Now 75 pounds of carbon burned to CO_2 gives

$$75 \times 14,650 = 1,098,750 \text{ B. T. U.,}$$

and 6.6 pounds of carbon burned to CO gives

$$6.6 \times 4400 = 29,040 \text{ B. T. U.,}$$

or a total for all the carbon of 1,127,790 B. T. U.

Had all the carbon been burned to CO_2, the heat of combustion would have been

$$(75 + 6.6)\, 14,650 = 1,195,440 \text{ B. T. U.}$$

The loss by incomplete combustion was consequently

$$\frac{1,195,440 - 1,127,790}{1,195,440} \times 100 = 5.6 \text{ per cent.}$$

The actual loss may be placed at a little less figure than 5.6 per cent, since less air is required for burning carbon to CO than for CO_2.

Loss from Excess of Air.—The ideal condition would be to supply just enough air to burn all the carbon in the coal to CO_2 and all the free hydrogen to H_2O; it is necessary to use somewhat more air than required for complete combustion to avoid the formation of CO and the attendant loss of heat. On the other hand, too great an excess of air occasions a loss, as that excess must be heated to the temperature in the chimney.

As an example, suppose that Pittsburg coal can be completely burned with 50 per cent excess of air, but that 100 per cent excess is allowed to pass through the grate.

To simplify the problem we will neglect the effect of sulphur and of the ash, more especially as it is not certain what their effect is; we know only that it cannot be very important.

Each pound of carbon will yield $3\frac{2}{3}$ pounds of CO_2 and each pound of hydrogen will yield 9 pounds of H_2O. There will therefore be

$$3\frac{2}{3} \times 0.7674 = 2.8039 \text{ pounds of } CO_2;$$
$$9 \times 0.0519 = 0.4671 \quad \text{``} \quad \text{``} \quad H_2O.$$

In the calculation for the weight of air (page 45) it has been shown that 2.374 pounds of oxygen and 10.2 pounds of air are required for combustion. There is therefore

$$10.2 - 2.374 = 7.826$$

pounds of nitrogen in the air for combustion. But each pound of coal contains 0.014 of a pound of nitrogen, so that the total nitrogen is 7.840 pounds.

Now the heat required to raise the temperature of one pound of a substance one degree, called the specific heat, is given in the table on page 45. For carbon dioxide the specific heat is 0.2169, and the heat required to raise 2.8039 pounds one degree is

$$2.8039 \times 0.2169 = 0.6082 \text{ B. T. U.}$$

The following are the calculations for the several components of the products of combustion:

	Weight.	Specific Heat.		
Carbon dioxide, CO_2....	2.8039	$\times$ 0.2169 =	0.6082	B. T. U.
Steam, H_2O.	0.4671	$\times$ 0.4805 =	0.2244	"
Nitrogen.	7.840	$\times$ 0.2438 =	1.9114	"
Air for dilution 50%....	5.100	$\times$ 0.2375 =	1.2112	"
Total....................			3.9552	"

If the external air is at 60° F., and the gases in the chimney are at 560° F., then the heat in the chimney-gases above the temperature of the air is

$$500 \times 3.9552 = 1978 \text{ B. T. U.}$$

The total heat of combustion of this coal by Dulong's formula is 13800 B. T. U.; of this about 10 per cent will be lost by conduction and radiation. There will then remain to be transferred to the water in the boiler

$$13800 - (1380 + 1978) = 10442 \text{ B. T. U.}$$

This is about 76 per cent of the heat generated by combustion.

Suppose that the dilution is allowed to be 100 per cent, so that 5 additional pounds of air per pound of coal are admitted to the grate. Then to the above total must be added

1.2112 B. T. U., making in all 5.1664 B. T. U. Multiplying by 500, the difference of temperature assumed

$$500 \times 5.1664 = 2583 \text{ B. T. U.}$$

Assuming, as before, 10 per cent for loss by radiation and conduction leaves

$$13800 - (1380 + 2583) = 9837 \text{ B. T. U.}$$

to be transferred to the water in the boiler. This is about 72 per cent, so that the loss by the excess of dilution is about 4 per cent.

Hypothetical Temperature of Combustion.—A calculation is sometimes made of the temperature of the fire on the assumption that the total heat of combustion is all applied to raising the temperature of the products of combustion, including the ash. In the case of Pittsburg coal it has been found that 3.9552 B. T. U. are required to raise the products of combustion one degree, allowing 50 per cent for dilution. This coal has 7.6 per cent ash, for which a specific heat of 0.2 may be allowed. We must therefore add to the total just quoted

$$.076 \times 0.2 = 0.0152 \text{ B. T. U.},$$

making in all 3.9704 B. T. U. Dividing the total heat by this quantity, we get

$$13800 \div 3.9704 = 3480° \text{ F.}$$

for the elevation of temperature. To this we will add the temperature of the air admitted to the furnace, say 60° F., making 3540° F. for the hypothetical temperature of the fire.

Such a temperature is never reached in the furnace of a boiler, for the combustion is not instantaneous and is not completed in the furnace, as flames commonly extend over the bridge-wall or into the combustion-chamber; meanwhile there is an energetic radiation from the glowing fuel and flame, and a rapid transfer of heat from the hot gases to the heating-surface of the boiler. The better the fuel and the higher the hypothetical temperature of the fire the less chance is there that the actual temperature will approach it.

During a test on a Babcock & Wilcox boiler, at the Massachusetts Institute of Technology, it was found that the temperature immediately over the fire was about 1100° F., while the temperature in the chimney was 400° F.

A test on a boiler of the locomotive type, at the Boston Main Drainage Station, gave for the temperature of the gases escaping from the boiler 439° F., while the steam in the boiler was about 337° F. The gases were afterwards reduced to 194° F. by passing them through a feed-water heater. This boiler was designed for and gave a high efficiency, and the results obtained may be considered to represent first-rate practice.

CHAPTER III.

CORROSION AND INCRUSTATION.

The water supplied to a boiler for forming steam may corrode the iron of the boiler, or it may deposit material that can form a scale or incrustation; both actions may go on at the same time.

Pure water, free from air and carbon dioxide, has little or no solvent action on iron, even though some other metal, such as copper, which may with the iron form the elements of a galvanic couple, be present. On the other hand, iron will not rust if placed in an atmosphere of dry air or dry carbon dioxide. All natural water, rain-water, water from wells, rivers, lakes, or the sea, contains air in solution, and carbon dioxide is not infrequently found in such waters. Iron is rapidly acted upon by water containing air or carbon dioxide, and, on the other hand, iron rusts rapidly in air or carbon dioxide when moisture is present. Again, distilled water, as from the surface condenser of a marine engine containing more or less oil, or the substances resulting from the action of steam on oil, causes corrosion in boilers that are free from scale. To avoid rusting of boilers when not in use they ought to be either quite dry inside or they ought to be entirely filled with water—preferably water that has been freed from air by boiling. In the American Navy it has been the custom to dry out boilers and paint them inside with mineral oil preparatory to laying them up. In the English Navy the boilers are dried out, a pan of glowing charcoal is placed in the boiler to

MINERAL MATTER IN SOLUTION. GRAINS PER U. S. GALLON.

	Charles River.	Long Pond.	Schuylkill River.	Lake Michigan.	Mississippi River.	Missouri River at Council Bluffs.	Mississippi River at Keokuk.	Riverside, Ill. Well.	Downer's Grove, Ill. Well, very bad.	Rockford, Ill. Artesian Well.	Dead-sea Water.
Silica (SiO_2)............	....	....	0.0800	0.306	0.863	1.522	1.190	0.484	0.741	0.624	....
Calcium carbonate ($CaCO_3$)...	0.1610	....	1.8720	4.461	6.870	8.847	4.673	5.237	17.091	8.141	....
Calcium sulphate ($CaSO_4$)....	0.2624	....	....	0.309	0.484	2.251	....	0.776	14.037	....	29.220
Calcium chloride ($CaCl_2$)....	0.0420	0.0308	....	....	....	....	....	....	....	....	....
Magnesium carbonate ($MgCO_3$)..	0.0399	....	0.3510	2.200	4.046	1.866	0.857	4.023	....	7.336	....
Magnesium sulphate ($MgSO_4$)...	....	0.1020	0.0570	....	0.338	3.595	....	....	25.422	....	50.950
Magnesium chloride ($MgCl_2$)...	....	0.0764	....	....	....	....	....	....	....	....	7.950
Magnesium bromide	....	....	....	....	....	....	....	....	....	....	....
Sodium carbonate (Na_2CO_3)...	....	....	....	....	....	....	2.129	....	....	....	....
Sodium sulphate (Na_2SO_4)...	0.3816	....	0.1470	0.225	....	....	0.100	....	....	....	78.050
Sodium chloride (NaCl)...	0.1547	0.0323	....	....	....	....	....	....	....	0.554	....
Potassium carbonate (K_2CO_3).	....	....	....	0.283	....	....	0.430	....	....	0.362	....
Potassium sulphate (K_2SO_4)..	....	0.0380	....	....	....	....	0.489	....	....	0.525	....
Potassium chloride (KCl).	....	....	....	0.029	....	....	2.682	....	....	....	....
Ferrous carbonate ($FeCO_3$)...	....	0.0800	....	....	....	....	....	....	....	....	....
Alumina (with ferric oxide)...	....	....	1.6436	....	0.233	0.233	....	0.146	0.192	0.087	....
Organic matter, etc.....	0.5291	0.5265	....	....	....	....	1.802	....	....	....	....
Suspended mineral matter...	....	....	....	....	....	....	2.155	....	....	....	....
Suspended organic matter...	....	....	....	....	....	....	....	....	....	....	....

consume the oxygen of the air, and quicklime is introduced to absorb moisture.

Mineral Impurities.—The impurities found in water supplied to land boilers are commonly carbonate of lime and sulphate of lime, with more or less organic matter, and sometimes sand or clay held in suspension. The table on page 66 gives the number of grains of various mineral substances held in solution in water from several sources.

Water supplied to land boilers is either hard or soft; the first contains appreciable quantities of lime, and the other usually contains little solid matter of any sort. The first three examples in the table on the preceding page may be taken as typical soft waters, and all the others, except the last two, as typical hard waters. While there is considerable difference in the amounts and the composition of the solids in solution in the several examples of hard water, it will be seen that they are all characterized by a considerable amount of calcium and magnesium carbonates, and (with the exception of Nos. 6 and 9) accompanied by a comparatively small amount of calcium and magnesium sulphates. It will be noticed that Missouri River water is distinctly worse than Mississippi River water, not only in that it contains more of the carbonates, but because it contains a considerable quantity of sulphates. No. 9, from a well at Downer's Grove on the C., B. and Q. R. R., a few miles from Chicago, has been selected as an example of a very bad hard water, especially as it contains so much sulphate. The reason for considering the sulphates of lime and magnesia so deleterious will appear a little later. Note will be made that the water from the Mississippi River at two different places, and presumably at different seasons of the year, vary considerably, especially in the amount of matter held in suspension.

In some places in the western parts of the United States the only available waters for making steam are strongly impregnated with alkalies and borax. Such waters have so deleterious an action on boilers that the advisability of using a surface condenser, as at sea, the distillation of water by a multiple-effect evaporator, or the introduction of a supply of good water even from remote places, is worthy of consideration. If the use of such water cannot be avoided, a competent chemist should be consulted to suggest methods for ameliorating the bad effects so far as possible. As each case is liable to require special treatment, no further discussion appear profitable in this place.

The carbonates of lime and magnesia are held in solution in water by an excess of carbon dioxide and are completely precipitated by boiling. They are thrown down from water supplied to a boiler, in the form of a white or grayish mud, provided there are not other impurities that cement them together and form a hard scale. The customary and sufficient method of treating boilers supplied with water containing carbonates of lime and magnesia is to let the boiler, while full, cool down, and then run out the water and thoroughly wash out the boiler with a strong stream from a hose. If the water is blown out under steam-pressure the deposits are hardened and are removed with difficulty. While pure carbonates are easily treated as just described, the presence of other impurities, such as oil or organic matter, or of sulphate of lime, is likely to make the deposits hard and adhering.

Sulphate of lime is much more soluble in cold than in hot water, and is entirely thrown down from water at a temperature of 280° F., corresponding to 35 pounds pressure of steam above the atmosphere. It forms a hard and adhering scale, and even in comparatively small quantities has a bad effect on scales and deposits composed of carbonates, as has already been suggested. The bad effect of deposits from water containing calcium sulphate is much ameliorated by introducing carbonate of soda or soda-ash into the boiler with the feed-water. The result is to give a deposit of calcium carbonate in the form of a fine white powder, which must be washed or swept out, and sodium sulphate in solution, which must be blown out from time to time.

If the mineral matters in the water are known from a chemical analysis, the quantity of carbonate of soda to be used may be calculated as follows:

Example.—Find the weight of carbonate of soda required per day for a boiler supplied with 1000 gallons of water per day from the well at Downer's Grove.

From the table on page 66 it appears that each gallon of the water contains 14.037 grains of $CaSO_4$ and 25.422 grains of $MgSO_4$. The formula for soda crystals being $Na_2CO_3 + 10H_2O$, the reactions, neglecting the water of crystallization, will be

$$CaSO_4 + Na_2CO_3 = CaCO_3 + Na_2SO_4;$$
$$MgSO_4 + Na_2CO_3 = MgCO_3 + Na_2SO_4.$$

If x_1 is the grains of carbonate of soda to act on the calcium, we have

$$CaSO_4 : Na_2CO_3 + 10H_2O = 14.037 : x_1;$$
$$40 + 32 + 4 \times 16 : 2 \times 23 + 12 + 3 \times 16 + 10(2 + 16)$$
$$= 14.037 : x_1.$$
$$\therefore x_1 = 29.52 \text{ grains.}$$

The magnesium sulphate which is soluble is also changed into the carbonate and thrown down as a white precipitate, adding to the deposit. The number of grains of carbonate of soda required for this reaction is found as follows:

$$MgSO_4 : Na_2CO_3 + 10H_2O = 25.422 : x_2;$$
$$24 + 32 + 4 \times 16 : 2 \times 23 + 12 + 3 \times 16 + 10(2 + 16)$$
$$= 25.422 : x_2.$$
$$\therefore x_2 = 60.59 \text{ grains.}$$

The total weight of carbonate of soda per gallon is therefore

$$29.52 + 60.59 = 90 +,$$

and the weight required for 1000 gallons is

$$\frac{90 \times 1000}{7000} = 12.9 \text{ pounds per day.}$$

It is advisable that soda, or any other chemical for acting on the impurities of feed-water, shall be introduced at regular intervals. Sometimes a weight, or measured portion, is thrown into the feed-water in a tank or reservoir, from which it is pumped. Sometimes the chemical, dissolved in water or diluted with water, is placed in a small tank or receptacle that may be temporarily connected with the suction of the feed-pump. If this method is used care must be taken not to admit air to the pump and so derange its action.

Soda-ash is commonly used instead of carbonate of soda, as it is cheaper and somewhat more efficient, on account of the caustic soda it may contain. Its chemical composition is uncertain, and it is therefore impossible to make satisfactory calculations for the quantity to be used. This, however, is commonly no real objection, for we seldom have a chemical analysis of the water, and cannot determine directly how much soda is required.

An excess of soda in a boiler is liable to cause foaming, and at high temperatures, corresponding to pressures now habitual for steam-boilers, the soda is apt to attack the inside of water-glasses; any indication of either action should raise the question whether too much soda is used, but the absence of such an indication does not show that we are using the right quantity. When a hard scale is formed by a water known to contain lime, we may infer that sulphates are present, and may find by trial the amount of soda to be used. Unfortunately other impurities, such as organic matter, cause the formation of hard scale, and make this method uncertain. Such impurities often produce discoloration, and thus betray their presence. The deposits of lime, whether carbonates or sulphates, are commonly white or grayish, or sometimes fawn color.

It is sometimes proposed to use ammonium chloride, or sal-ammoniac, to break up lime compounds; in the first place, only the carbonates are acted upon by this reagent, and in the second place, the reagent itself, or the resultant chlorides, are liable to be broken up, giving free chlorine, which attacks the boiler.

Tannic acid, either commercial acid or in the crude state, may be used to break up a scale already formed; but as tannic acid does not decompose the sulphates, and as the compound of the acid with lime is not soluble, its use appears to be restricted. Many proprietary boiler compounds depend on tannic acid for their action. Acetic acid may also be used to break up the carbonates, but it likewise has no action on the sulphates; the carbonates are changed into soluble acetates, and can be blown out. Both tannic acid and acetic acid attack iron, but are not so dangerous as sulphuric or hydrochloric acids, which are sometimes recommended for breaking up scale. When a scale is once formed the safer way is to remove it with proper chipping and scaling tools; but this will be found to be impossible for many types of boilers unless they are largely dismembered for that purpose.

When river-water is used in boilers, various earthy impurities are liable to be carried into boilers, such as clay and sand, together with soluble matters. Even waters from ponds or wells may contain considerable matter in suspension. Such substances can sometimes be removed by filtering or by allowing the water to stand so that the insoluble matter may be deposited. Very commonly a systematic blowing out from the surface of the water and the bottom of the boiler will remove such impurities from the boiler. If, however, lime and magnesium carbonates and sulphates are present, suspended matter is carried into the scale, and the scale may be made more troublesome in consequence. The carbonates are more likely to form a hard scale if any binding material, such as clay, is present.

Fig. 28 shows the section of a feed-pipe which was nearly choked with scale from lime-water. Though the deposit of scale in a horizontal piece of feed-pipe where the water may be heated by conduction and otherwise, especially during intervals of feeding, is probably more rapid than in the boiler itself, this may serve to call attention to the extent to which scaling may occur when precautions are not taken.

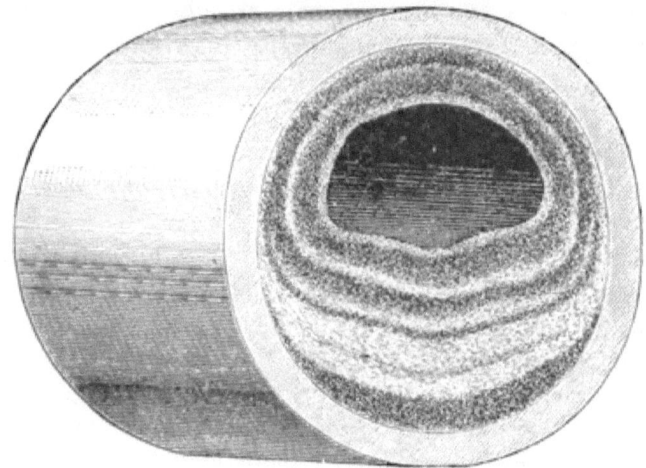

FIG. 28.*

Lime-extracting Feed-water Heater. — It has been pointed out that carbonate of lime can be completely precipitated by boiling to drive off the excess of carbonic acid; carbonate of magnesia if present is thrown down at the same time. Also sulphate of lime is thrown down at 280° F., corresponding to 35 pounds pressure above the atmosphere. It is evident that lime compounds can be removed from feed-water by heating it and removing the precipitated lime before feeding it to the boiler. For this purpose we may use a heater such as the Hoppes heater and purifier shown by Fig. 29, which consists essentially of a series of cylindrical pans

---

* This figure and Figs. 31 to 35 were kindly loaned by the Hartford Steam Boiler Inspection and Insurance Co.

of sheet steel, 1, 2, 3, 4, 5, and 6. The feed-water is pumped into the upper pan, from which it overflows, and, trickling along the bottom, it drops into the pan 2. From 2 the water overflows into 3, and so on.

The capacity of the heater depends on the number of sets of pans, which varies from one to four. The pans are enclosed in a steel shell, from which one end may be removed for cleaning the pans. Feed-water is pumped in at B; steam from the boiler is admitted at A; the feed-water after being heated and purified runs out at D on the way to the boiler:

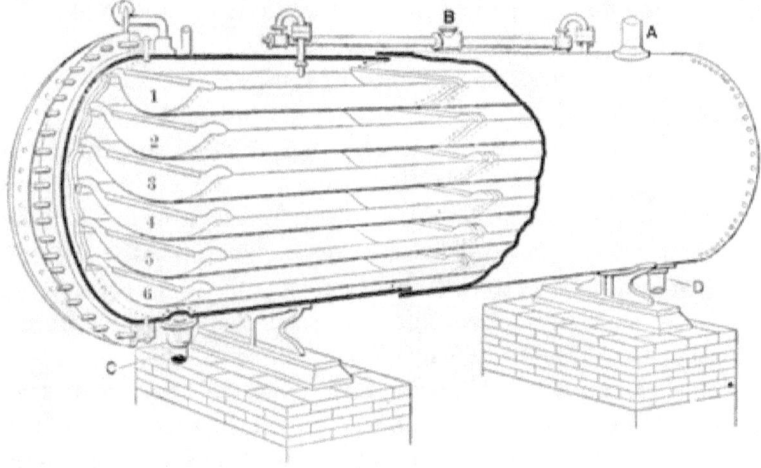

FIG. 29.

at C there is a blow-out, from which air and gases may be blown out when the heater is started, or at other times.

It is desirable that the pipe D shall drop down below the water-level in the boiler before any turns or horizontal pipes are attached. The water runs from the heater to the boiler by gravity only, and the heater must be placed high enough for this purpose. It is also desirable that the feed-pump be supplied with steam from the heater so as to continually remove the carbonic acid, air, or other gases given off from the feed-water.

The feed-water as it trickles along the under sides of the pans in a thin film is heated by the steam, and the lime compounds are deposited in form of a scale or incrustation. Meanwhile mud, sand, and other mechanical impurities settle to the bottom of the pans.

After the heater has been at work a month or so, depending on the amount of lime in the water, the pans must be removed and cleaned. The steam-pipe and the pipe leading to the boiler are shut off by proper valves, and cold water is pumped in and allowed to run to waste at the blow-off. The contraction of the pans cracks off hard scale and makes it easier to remove. When the heater is first opened the scale is usually soft and can be readily removed; it is liable to harden when exposed to the air and allowed to dry.

A heater for use with exhaust-steam, by the same makers, differs from this mainly in that there is a device for extracting oil from the steam before it meets the feed-water, and in that it is run at atmospheric pressure. Such a heater will not remove sulphate of lime; and further, since it is difficult if not impossible to remove oil from exhaust-steam, it is probable that some oil will be carried over into the boiler.

Sea-water.—The following table gives an analysis of sea-water by Professor Lewes of the Royal Naval College, together with an analysis by him of a typical boiler deposit from a marine boiler:

SALTS IN SEA-WATER AND COMPOSITION OF MARINE-BOILER SCALE.*

	Sea-water. Grains per Imperial Gallon.	Marine-boiler Scale. Per Cent.
Calcium carbonate (chalk)	3.9	0.97
Calcium sulphate (gypsum)	93.1	85.53
Magnesium sulphate	124.8	......
Magnesium chloride	220.5	......
Magnesium hydrate	......	3.39
Sodium chloride (salt)	1850.1	2.79
Silicia (sandy matter)	8.4	1.1
Moisture	......	5.9

* Trans. Inst. Naval Arch., vol. xxx. p. 330.

The three principal constituents of the marine scale are calcium sulphate, calcium carbonate, and magnesium hydrate, of which the first forms the greater part of the scale.

The calcium carbonate is kept in solution by the carbonic acid in the sea-water, just as is the case for fresh water containing carbonate of lime, and is deposited when the carbonic acid is driven off by heat. There is, however, a reaction between the calcium carbonate and magnesium chloride at the temperature and pressure in the boiler, giving a deposit of magnesium hydrate and leaving calcium chloride in solution, so that only part of the calcium carbonate appears in the scale; and on the other hand, we may thus account for the presence of the magnesium hydrate in the scale.

The calcium sulphate forms so large a part of the scale, that we will give attention to it only in the further discussion. Calcium sulphate is more soluble in water at 95° F. than at any temperature higher or lower; and the solubility decreases with the rise of temperature, till at about 280° F., which corresponds to 50 pounds pressure absolute to the square inch, or 35 pounds above the atmosphere, the entire amount of calcium sulphate is deposited. In the early history of the marine engine, when low pressures of steam prevailed, we find jet condensers in use, and the boilers, which were fed from the brine in the hot-well, were kept fairly free from scale by blowing out the concentrated brine. It was then customary to supply half again as much feed-water as was evaporated, the excess being compensated by the concentrated brine blown out, and the water in the boiler had three times the degree of concentration found in the sea. As high-pressure steam came into use, surface condensers became indispensable. When surface condensers first came into use the waste of steam from leakage and otherwise was made up from water taken from the sea, with the result that the boilers gradually accumulated a heavy, dense scale. Since it is customary to have an auxiliary boiler, called a donkey-boiler, on steamships, the first device to avoid the scaling from the use of sea-water in the main boilers appears to have been to supply the loss of steam from the donkey-boiler, which was fed from the sea. This of course only transferred the difficulty from one place to another, even though a less objectionable one. At present the loss is made up by vaporizing sea-water in a special boiler, which is heated by steam-coils supplied with steam from the main boilers. The pressure may be low enough in this vaporizer to avoid the total precipitation of the calcium sulphate, and the brine may be kept at any desirable degree of saturation by blowing out, as in the early marine practice; and further, the vaporizer is so made that the steam-coils may be readily cleared from scale.

It should be pointed out that the decomposition of the calcium sulphate in sea-water by the aid of soda is impracticable, on account of the large quantity of magnesium carbonate thrown down by reaction on the magnesium sulphate.

A boiler fed with water condensed in a surface condenser, as is now common in marine practice, is liable to two difficulties: (1) the distilled water is apt to corrode or pit the plates of the boiler, and (2) the cylinder-oil used in the engine is liable to be carried over into the boiler and form oily scales and deposits.

When sea-water is used in the boiler, either as the main boiler-feed or merely to supply the waste, the boiler-plates are protected by the scale of calcium sulphate, and general corrosion or local pitting is seldom troublesome. When care is taken to avoid the use of salt water, supplying the waste with fresh water from a distiller or otherwise, general corrosion and local pitting have both been found to occur to a dangerous degree. A simple remedy appears to be to form a very thin scale by the use of sea-water, and then avoid further use of sea-water. It is, however, found that water from a surface condenser will gradually dissolve off such a scale, and it must be occasionally renewed. There is also an objection to the introduction of any lime compound into a boiler, as will appear in the discussion of the difficulty from the collection of oil in the boiler. In both the United States and the English navies it is customary to use slabs of zinc to protect the boiler-plates from corrosion. The zinc is fastened to or hung from the boiler-stays, with which metallic connection should be made to insure galvanic action. The zinc is gradually consumed, and becomes soft and friable, so that the slabs require renewal. It is recommended to supply 1/4 of a pound of zinc for each square foot of grate-surface.

It is a familiar fact that the cylinders of an engine may be oiled by introducing the oil into the supply-pipe, and that the oil will be carried quite thoroughly over the surface of the cylinder by the steam; and, further, that the oil is carried out of the cylinder by the steam, and will appear in the condensed water in the hot-well. It is evident that any oil is liable to be injurious if it gets into a boiler. It is, consequently, customary to filter the water from a surface-condenser, to remove the oil as far as possible. For this purpose sponges have been used in the navy; they, of course, must be occasionally taken out and washed free from oil. A very simple and efficient filter has been made in the form of a rectangular box, with perforated plates near the ends; the water from the hot-well runs into one end compartment, passes through a mass of hay in the middle compartment, and is drawn from the further end compartment by the feed-pump. When the hay becomes foul it is thrown away, and fresh hay is put in. Professor Lewes advises for a filter a long tube filled with charcoal about the size of a walnut; of course the charcoal should be renewed when necessary. It cannot be expected that any system of filtering will remove all the oil from the water, but the larger part may be removed. It is advisable that no more oil than necessary shall be used in the cylinders of the engine.

Professor Lewes[*] gives the following account of an inves-

---

[*] Trans. Inst. Nav. Arch., XXXII. page 67.

tigation of the collapse of the furnace-flues of a large Atlantic steamer, which made the voyage in twelve days:

The boilers were five and a half years old, and were refilled with fresh water at the end of each voyage, while the waste of the voyage was made up by the use of about 70 tons of fresh water, but during the last voyage sea-water was used for this purpose. Every four hours, while under steam, four pounds of soda crystals were put in the hot-well, making two hundred-weight during the run, the total capacity of the boilers being 81 tons. For lubricating purposes seven pints of valvoline were used in the cylinders every four hours.

When in port the boilers were allowed to cool down, and the water was run off and they were swept down with stiff brushes, and were afterwards sluiced out with a hose shortly before being filled with fresh water. No trouble occurred until five voyages before the final collapse, when some of the furnaces began to creep in; they were stiffened with rings and stays; and on succeeding voyages the whole of the furnaces got out of shape one after the other. Examination showed that they had never been very heavily scaled. On the furnace-crown there was only a slight white scale not more than 1/64 of an inch thick, while on the bottom of the furnaces there was a brown oily deposit 1/16 of an inch thick, which in other parts of the boiler increased to 1/8 or 3/16 of an inch.

The valvoline was a pure mineral oil with a specific gravity of 0.889 and a boiling-point of 371° C.

The composition of scales from several parts of the boiler is shown in the table on the next page.

Careful examination of the organic matter and oil in these deposits showed that half of it was valvoline in an unchanged condition, which had collected around small particles of calcic sulphate.

All the deposits were rich in oily matter except the top of the furnaces, i.e., the place where the collapse occurred. There the scale was not only nearly free from oil, but perfectly harmless both in quantity and quality. It appeared

COMPOSITION OF DEPOSITS IN A MARINE BOILER.

	From Top of Furnace.	From Bottom of Furnace.	Scale on Tubes.	Deposit above Scale on Tubes.	Deposit from Bottom of Boiler.
Calcium sulphate............	84.87	59.11	50.92	11.60	22.52
Calcium carbonate	5.90	6.07	4.18	0.82	....
Magnesium hydrate	2.83	11.29	14.12	22.21	7.09
Iron, alumina, silica.......	2.37	2.85	7.47	9.14	34.85
Organic matter and oil	3.23	19.54	21.06	50.20	27.95
Moisture..................	0.80	1.14	1.17	4.23	5.79
Alkalies...................	....	....	1.08	1.80	1.80

entirely improbable that the scale on the top of the furnaces could be in its original condition.

When oil has entered a boiler the minute globules, if in large quantity, coalesce to form an oily scum on the surface, or if in small quantity remain in separate drops, but show no tendency to sink on account of their low specific gravity. They, however, come in contact with solid particles of calcium sulphate, coat them with oil, and so the light oil becomes loaded till it is easily carried along by convection-currents and adheres to surfaces with which it comes in contact, which are quite as likely to be the under surfaces of tubes as the upper surfaces. Since some brine is liable to find its way to the boiler, from leakage into the condenser or otherwise, even when sea-water is not used directly, this action will occur in a boiler supposed to contain fresh water only.

The deposits thus formed are very poor conductors of heat, and the oily surface interferes with contact with water. On the crown of the furnace this soon leads to overheating of the plates, and the deposit begins to decompose, the lower layer in contact with the plate giving off gases which blow up the greasy layer, ordinarily only 1/64 of an inch thick, to a spongy mass 1/8 of an inch thick, which, because of its porosity, is even a better non-conductor of heat than before, and the plate becomes heated to redness and collapses. During the last stages of this overheating the temperature has risen to such a point that the organic matter, oil, etc., in the deposit burns away, or is distilled off, leaving behind, as an apparently harmless deposit, the solid particles round which it had originally formed.

Such a deposit is more likely to be produced in boilers containing fresh or distilled water, as the low density of the liquid enables the oily matter to settle more quickly, while with a strongly saline solution it is very doubtful if this sinking-point would ever be reached; it is evident also that when oil has found its way into the boiler and is causing a greasy scum on the surface the most fatal thing that can be done is to blow off the boilers without first using the scum-cocks, because as the water sinks the scum clings to the tops of the furnaces and other surfaces with which it comes in contact, and on again filling up with fresh water it still remains there, causing rapid collapse. A very remarkable instance of this is to be found in the case of a large vessel in the Eastern trade, in the boiler of which an oil-scum had formed. The ship having to stop some days in Gibraltar, the engineer took the opportunity of blowing out his boiler and refilling with fresh water, with the result that before he had been ten hours under steam the whole of the furnaces had collapsed. Under some conditions the oil-coated particles coalesce and form a sort of floating pancake, which, sinking, forms a patch on the crown of the furnace at one particular spot, and under these conditions the general result is the formation of a pocket.

A curious fact is that these oily deposits are found to contain a considerable amount of copper. Even mineral oils have a solvent action on copper and its alloys, and it is evident that the copper in the oily deposits has been obtained from the fittings of the cylinder and condenser. Fortunately this copper is protected by oil, otherwise serious galvanic mischief would result.

Professor Lewes found from experiment that a coating, $1/16$ of an inch thick, of the oily deposit found in the bottom of a boiler, applied to the inside of a clean iron vessel, very greatly retarded the transmission of heat from a Bunsen flame, as shown by the time required to heat a known quantity of water to boiling-point. Using an atmospheric blowpipe, he succeeded in raising the outside surface of the vessel, when coated with 1/16 of an inch of the deposit, to the temperature of the melting-point of zinc, and with an oxy-coal-gas flame he fused a hole in the bottom of a thin wrought-iron vessel thus coated and filled with water.

He further says that cylinders should be sparingly lubricated with a pure mineral oil having a high boiling-point, and that animal or vegetable oil should never be used, because they are decomposed by the action of high-pressure steam, producing fatty acids that attack iron, copper, and copper alloys.

Professor Lewes has proposed that marine boilers at sea shall have the water supplied with brine from which the lime compounds have been precipitated in a closed receptacle by the combined action of heat and carbonote of soda. The resulting brine contains mainly sodium and magnesium chlorides and magnesium sulphate, which do not form scale even though the concentration is carried to a higher degree than would occur from the supply of the waste of the boiler in this way for a voyage of some length. This method has not as yet been adopted in practice. Attention is called to the fact that an excess of soda should be avoided, since it would cause a bulky deposit from the action on the magnesium sulphate brought in by leakage of sea-water into the condenser. A description of the apparatus for producing this brine without lime salts is given in the "Transactions of the Institution of Naval Architects" (see the reference, page 74).

Organic Impurities.—Water for feeding boilers, unless taken from a contaminated source, seldom contains much organic matter. Surface water from rivers or ponds may contain some vegetable matter, but if there are no other impurities such organic matter will not cause much trouble unless it is allowed to accumulate. The vegetable and other organic impurities commonly float on the surface of the water when the boiler is making steam, or are carried around by convection-currents, and may be blown out through a surface blow-out, shown by Fig. 30. It consists essentially of a flattened

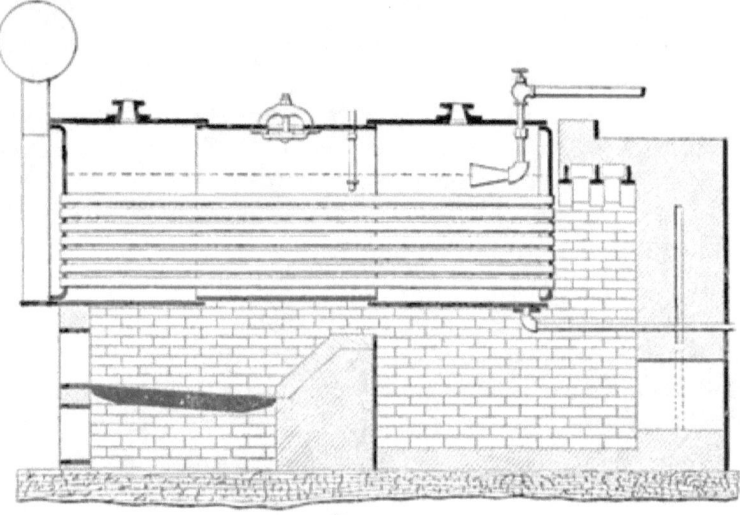

FIG. 30.

bell or cone of sheet metal extending across the boiler at the water-level, and turned so that the convection-currents will carry and lodge floating substances in the mouth of the bell. The valve in the pipe leading from this bell may be opened from time to time to blow out the substances collected in it.

When a boiler has been at rest for some time, overnight for example, the various solids in the boiler, if heavy enough, will settle to the bottom, and may be advantageously blown out before starting the boiler into action again. This may be accomplished by opening the blow-out valve or cock for a short time, until the water-level falls a few inches.

Water from bogs frequently contains vegetable acids that are likely to corrode the plates of the boiler: in such cases carbonate of soda may be used to neutralize the acids; the proper amount must be found by trial.

The oil used in the engine is liable to get into the boiler if surface-condensing is made use of; this subject has already received attention in connection with the discussion of marine-boiler incrustations. Surface condensers are not commonly used in land practice, but very commonly the exhaust-steam from non-condensing engines is used for heating in radiating-coils, and there is an apparent gain from the use of the warm water from the return-pipes. This water is, however, liable to be contaminated by oil, and the oil when it gets into the boiler may cause serious damage, such as was found to occur in marine boilers. If the feed-water has a little vegetable matter in it, the effect of the oil is much worse than if the water-supply is pure. Again, the oil is very troublesome if the water contains lime salts. The bad effect of oil or other impurities on lime-scale has been already noted. Usually it will be found better to reject the water returned from a heating system supplied with exhaust-steam, as the apparent economy is liable to be more than counterbalanced by damage to the boiler. The

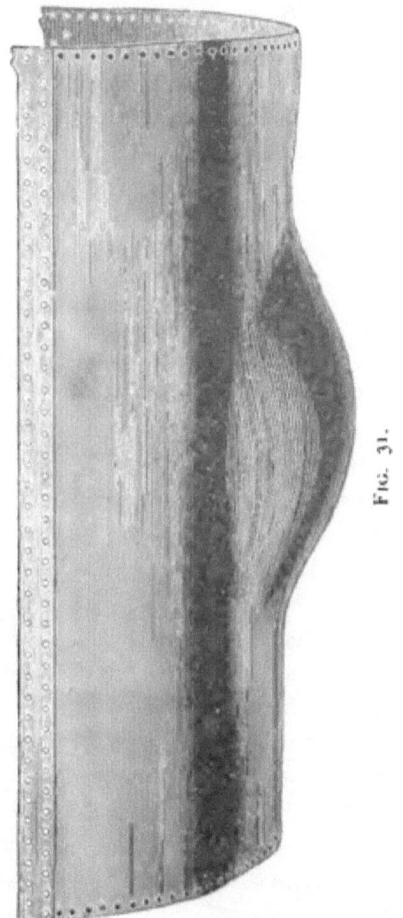

Fig. 31.

externally-fired tubular boilers commonly used in this part of the country are liable to bulge in the sheets over the furnace, as shown in Fig. 31, if oil gets into them. When the plate is cut out a hard deposit of oil, commonly mixed with other impurities, will be found adhering to the plate; this deposit is a very poor conductor of heat, and it causes so much overheating of the plate that it bulges out under the pressure of the steam.

In isolated cases it will be found that water of a stream may be so contaminated with chemicals from some industrial establishment that it acts energetically on the boiler-plates; in such case the water must be abandoned unless the contamination can be stopped.

Kerosene and Petroleum Oils.—Both crude petroleum and refined kerosene have been used in steam-boilers to mitigate the effect of incrustations of calcium carbonate and calcium sulphate. From what is known of the bad effects of the heavier petroleum products, such as the mineral oils used for lubricating steam-engine cylinders, it appears to be unwise to introduce crude petroleum into a steam-boiler. The same objection does not apply to refined kerosene, which is not known to have any bad effect in a boiler. Both oils are said to change the deposits of lime from a hard scale to a friable material, which may be easily removed. It is further said that these oils will soften and loosen scale already formed. In one case 40 gallons of kerosene were used in 24 hours in the boilers of a steamer of about 3000 horse-power. These boilers showed no incrustation, but considerable corrosion.

Corrosion is distinguished as general corrosion or wasting, pitting, and grooving.

General corrosion is difficult to detect, as it acts more or less uniformly over large surfaces, and even at riveted joints the two plates and the rivet-heads waste away equally, so that the thinning of the plates is not easily noticed. Old boilers not infrequently fail from general corrosion, and then are likely to fail in the plate rather than in the riveted joint, where the double thickness of plate gives an advantage. Boilers that have been at work should have the plates below the waterline drilled and the thickness measured; if the effective thickness of the plate is found to be much reduced, the working pressure should be made proportionately lower. Fig. 32

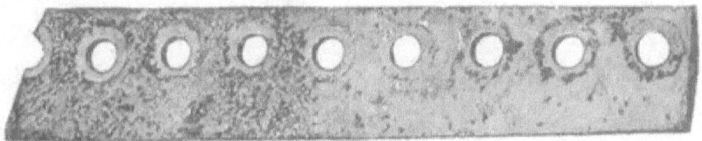

FIG. 32.

shows an example of general corrosion, and Fig. 33 another, but complicated with cracking at the rivet-holes. Both show the protection given to the plate by the rivet-heads, and one may readily see how the wasting of the rivet-heads may be overlooked.

Pitting is likely to occur when the corrosion takes place rapidly. It appears to be due to lack of homogeneity of the metal of the plate, and sometimes appears to indicate galvanic action. Though every precaution to avoid galvanic action should be taken, it is better to assume damage to be due to such action only when there is direct evidence of its existence. Fig. 34 shows pitting over a large surface, and Fig. 35 shows local pitting in the corner of a flanged plate with general corrosion of the flat surface of the plate. It is fair to assume that the disturbance of the metal in the process of flanging may determine the vertical forms of the pitting. The horizontal plate shows irregular pitting.

Grooving is usually due to the combination of springing or buckling of a plate and local corrosion. The buckling may be due to insufficient staying; then the plate springs back and forth as the steam-pressure varies. Or buckling may be due to improper staying or fastenings, which localizes the

86 STEAM-BOILERS.

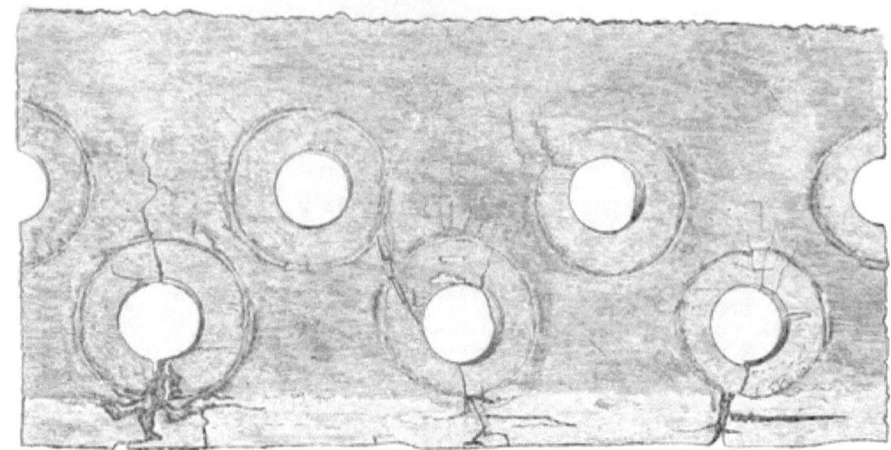

Fig. 33.

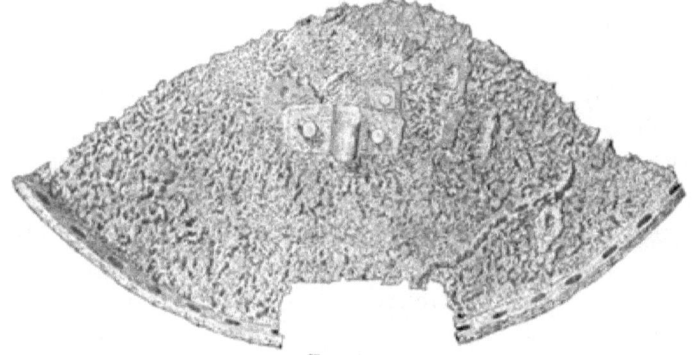

Fig. 34.

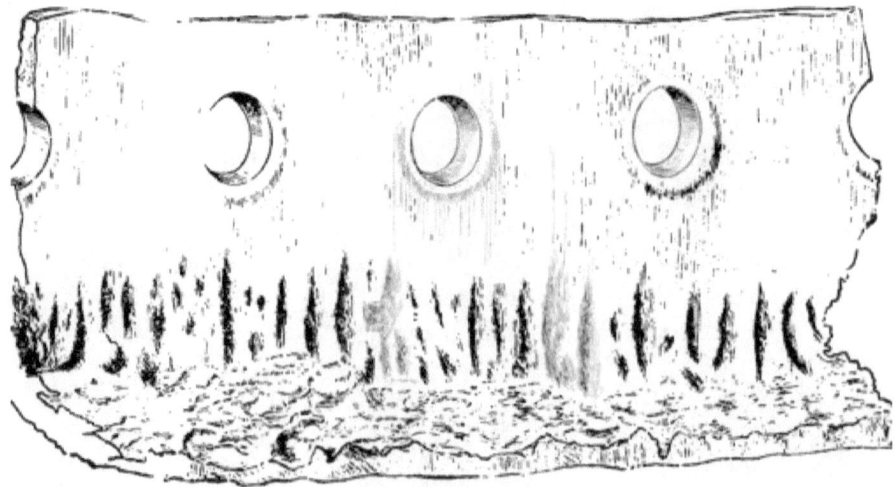

Fig. 35.

change of shape due to expansion. In either case the metal is fretted at the place where the greatest bending takes place, and very much weakened. A crack is liable to be formed, which may grow wider and deeper till the plate shows signs of failure. Such cracks may be very narrow and difficult to find, but usually the fretting of the metal, whether a crack is formed or not, is accompanied by local corrosion, which makes a groove of some width. If the water used forms a scale on the boiler-plates, the working of the metal throws off the scale and exposes the surface to the water so that corrosion takes place there, though elsewhere the plate is protected.

As one example of insufficient staying, we may take the flattened surface in a wagon-top locomotive-boiler (Plate II), where the barrel is expanded to join the shell over the firebox. The surface cannot be stayed from side to side for lack of space between the tubes, and is merely stiffened by riveting three pieces of T iron to the shell. In this case the T irons have through-stays at their upper ends over the tubes. Grooving is liable to occur in this locality even when the plates are stiffened as shown.

Grooving from too great rigidity is liable to occur in the end-plates of Cornish and Lancashire boilers (see pages 6 and 7). The long furnace-flues expand more than the external shell, and expand more at the top than at the bottom, due to the heat of the furnace and of the gases in the flue beyond the furnace; and further, the circulation of water under the flues is likely to be imperfect, so that the bottom of the flue is not so hot as the top. These unequal expansions must be accommodated by the springing of the end-plates, and if the springing is too much localized, grooving is sure to occur. The furnace-flues should be at least nine inches from the shell, and the end-plates should be flanged where they are joined to the flues and shell, instead of using angle-irons. The use of gusset-plates for staying the ends of these boilers is likely to give too much rigidity and to localize the springing of the plates, unless care is taken to avoid it.

Grooving from either too great or too little rigidity can be avoided only by a proper design, which must be guided by experience. If a boiler shows defects of staying, it may be possible to put in additional stays after the boiler is completed and at work; or in some cases too great rigidity may be remedied by rearranging the staying. Such remodelling of a boiler is usually difficult and unsatisfactory.

Loss from Blowing Out Brine.—In the discussion of the use of sea-water in marine boilers, reference was made to the custom of feeding one-and-a-half times as much water as was evaporated. The feed-water was taken from the hot-well of the jet condenser, and was nearly as salt as sea-water, which contains about 1/32 of its weight of salt. The one-half excess of water fed was blown out, and carried with it all the salt of the entire feed-water; it consequently contained 3/32 of its weight of salt, and the brine in the boiler had the same degree of concentration.

In calculating the loss from blowing out hot brine it is customary to assume that the specific heat of sea-water and also of the hot brine is the same as that of fresh water; accuracy in this calculation is not essential.

For example, find the loss from blowing out hot brine to maintain the concentration in the boiler at 3/32, when the boiler-pressure is 30 pounds by the gauge and the temperature in the hot-well is 140° F.

The absolute pressure corresponding to 30 pounds by the gauge is 44.7, found by adding the pressure of the atmosphere. Since no refinement is needed in this calculation we will use instead 45 pounds absolute. A table of the properties of saturated steam (see Appendix) gives for the heat of the liquid at 45 pounds absolute, 243.6 thermal units; this is the heat required to raise one pound of water from 32° F. to 274°.3 F., that is, to the temperature of steam at the pressure of 45 pounds. The same table gives for the heat required to vaporize one pound of steam from water at 274°.3 against a pressure of 45 pounds, 922 thermal units. But it is assumed that the feed-water has a temperature of 140° F. when taken from the hot-well; the corresponding heat of the liquid is 108.2 thermal units. Consequently, to raise a pound of water from 140° F. and vaporize it under the pressure of 45 pounds will require

$$922 + 243.6 - 108.2 = 1057.4$$

thermal units. This is the heat usefully employed.

Meanwhile for each pound of water vaporized half a pound of water is heated from 140° F. to 274°.3 F., and then thrown away. The heat required to raise half a pound of water from 140° F. to 274°.3 F. is

$$\tfrac{1}{2}(243.6 - 108.2) = 67.7$$

thermal units. This is the heat wasted.

The total heat applied to forming steam and heating the brine blown out is

$$1057.4 + 67.7 = 1125.1.$$

The per cent of heat wasted is consequently

$$100 \times \frac{67.7}{1125.1} = 6 \text{ per cent.}$$

A considerable portion of the heat lost in the hot brine may be transferred to the feed-water drawn from the hot-well by the aid of a feed-water heater, and thus saved. A simple form of heater may be made by carrying the hot brine through a small pipe inside the feed-pipe; the currents of water will naturally flow in opposite directions, and thus give the most efficient interchange of heat. If the hot-well is near the boiler, the feed-pipe may not be long enough to allow of this form of heater.

The density of brine in the boiler is ascertained by a salimeter, which is a form of hydrometer graduated to read zero in fresh water, 1/32 in sea-water, and the graduation is extended to give the density of brine in thirty seconds, so far as may be needed. When jet condensers were used at sea it was customary to carry the density to 3/32 only. With surface condensers the density is frequently carried as high as 6/32; no inconvenience is found in this custom, and as less water is taken from the sea the formation of incrustation is less rapid.

CHAPTER IV.

SETTINGS, FURNACES, AND CHIMNEYS.

The Boiler-setting for a stationary boiler consists of the foundation and so much of the flues and furnace as are external to the boiler proper. The entire furnace of externally-fired boilers is in the setting, and in some cases, as with the plain cylindrical boiler, the flues are also formed by the setting. Some internally-fired boilers—for example, the Lancashire boiler—have flues in the setting in addition to the boiler-flues; others, like the upright boiler (Fig. 5, page 10), have only a foundation. Locomotive-boilers rest on the frame of the locomotive; they can scarcely be considered to have any setting. Marine boilers are seated on plates that are built into the framing of the ship.

Cylindrical Tubular Boiler-setting.—The setting for a pair of cylindrical tubular boilers, like the boiler represented on Plate I, is shown by Figs. 36 and 37. The foundation for the boiler-setting is a solid bed of concrete 17 feet 8 inches wide, and 21 feet 8 inches long, and 24 inches thick. On firm soil the foundation may be conveniently made of large rough-stone work, about three feet wide, under the side, middle, and end walls only.

On this foundation there are built the walls that support and enclose the boiler and the furnace. The outer walls at the sides and rear are double, with an air-space to check the conduction of heat. The boilers are each supported by two brackets at each end; the front brackets rest on iron plates which are built into the side walls; the rear brackets have iron rollers interposed to allow for expansion. A brick arch is sprung over the boilers to check the radiation of heat. The space between the side and end walls over the boilers may be filled with sand, for the same purpose. Coal ashes are sometimes used, but they are hygroscopic and liable to harbor moisture when the boilers are not working, and should

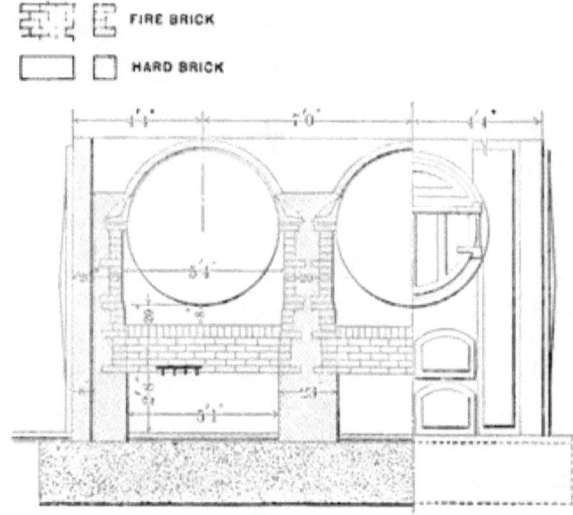

Fig. 36.

not be used. Sometimes the tops of boilers are covered with brick and buried in sand; or the sand may be used without brick. These methods give ready access to the shell for inspection or repairs, but are not so good as a brick arch, as water can more readily get to the boiler if it should drip from leaky valves or fittings. The rear wall is carried a little higher than the top row of fire-tubes, then the space is bridged over from the side walls by a horizontal mass of brickwork, stiffened and supported by T irons. The smoke-box projects over the front wall, and has a rectangular uptake on top, leading to a wrought-iron flue which carries the smoke to the chimney.

The furnaces under the front ends of the boilers are enclosed by the side walls, the front wall, and a bridge just

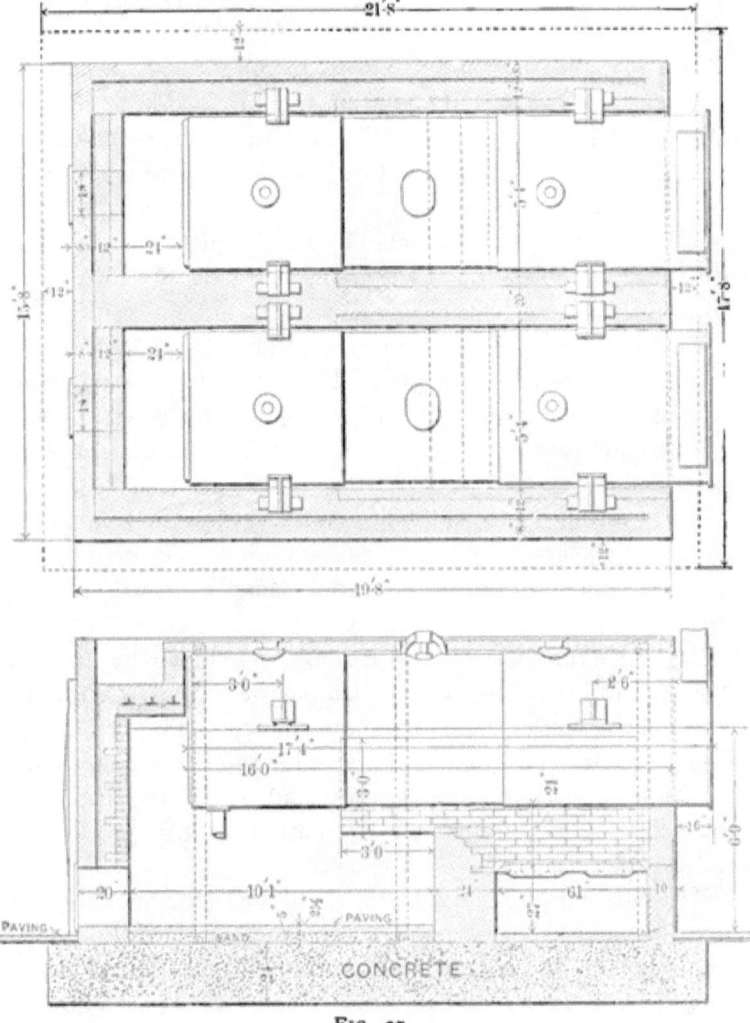

Fig. 37.

beyond the first ring of the boiler-shell. The grates rest on the front wall and the bridge, as shown in vertical section by

Fig. 37 and indicated in black on Fig. 36. There is a clear space of 24 inches between the grate and the boiler, and a clear space of 8 inches over the bridge. The top of the bridge is made of fire-brick, and all the walls of the furnaces and other spaces that are exposed to the fire are lined with fire-brick. All the remainder of the brickwork is of hard, well-burned brick. The ash-pit under the grate is paved with brick. The floor behind the bridge is covered with a layer of sand and paved with brick.

The side walls are braced by three pairs of *buck-staves*, with through-rods under the paving and over the tops of the boilers.

The boiler front is cast iron, with doors opening from the furnaces and from the ash-pits. There are also doors opening from the smoke-boxes to give access to the tubes. Doors through the rear wall give access to the space behind the bridge-wall.

The setting for a two-flue boiler, or for a boiler with several large flues in place of the numerous fire-tubes of the tubular boiler, is substantially the same as those just described.

Settings for Water-tube Boilers, as shown by Fig. 17, page 24, and Fig. 18, page 26, resemble the setting for the cylindrical tubular boiler in external appearance. The furnace and bridge-wall are also similar to those for the cylindrical boiler. Special bridges, extending among and across the tubes, are required to give the proper circulation of the products of combustion.

The Stirling boiler has a setting of special form, shown by Fig. 20, page 28, as required by the design of the boiler. The Cahall boiler has the furnace at one side of the boiler, which is set in a vertical brick casing or stack (Fig. 21, page 29).

Water-tube boilers for marine work, like the Thornycroft boiler, shown by Fig. 23, page 32, and the Almy boiler, Fig.

26, page 36, are enclosed in a sheet-iron casing, lined with blocks of non-conducting material. Asbestos, or a compound of which magnesia is a principal ingredient, is commonly used. Fire-brick and pumice-stone are used with the Thornycroft boiler to intercept heat that would be radiated downward. The spaces in ships under boilers, being more or less inaccessible, and being subject to the influence of heat and moisture, are liable to show excessive corrosion.

Furnaces.—There are certain general conditions to which the construction of furnaces should conform if high efficiency is desired. Some of these depend on the requirements for good combustion, and some depend on the size, strength, and endurance of the human frame, since hand-firing is almost universal. Some of these conditions are violated in the design and arrangement of furnaces in certain types of boilers; deviation from them involves either a demand for greater strength and skill on the part of the fireman, or a loss of efficiency, or both.

These conditions, with examples of good and bad practice, are as follows:

There should be an abundant and uniform supply of air to the under surface of the grate. About the only cases where this condition is not easily fulfilled is in the design of furnace-flues of Lancashire boilers and Scotch marine boilers.

A small supply of air is required over the grate for burning smoky fuels like bituminous coal. This air is very commonly supplied through a circular grid or damper in the fire-door. The fire-door is commonly protected from direct radiation by a perforated wrought-iron plate, which also serves to distribute the air coming through this grid. Since the air thus supplied is cold, it must be small in amount or it will chill the gases and check combustion instead of aiding it.

Leakage of cold air into the furnace, or into the combustion-chamber or flues beyond the furnace, injures the draught and reduces the temperature of the products of combustion, and is a direct source of loss. All externally-fired boilers and water-tube boilers are liable to suffer from leakage of air. Locomotive and Scotch marine boilers are usually free from this defect.

The incandescent fuel on the grate should not come in contact with a cold surface. Furnaces lined with fire-brick, such as are used for externally-fired boilers, conform to this requirement. Vertical boilers, marine boilers, locomotive-boilers, and all other boilers having the furnaces in fire-boxes or flues, violate this condition, as the plates in contact with the fire are kept nearly at the temperature of the water in contact with the other side, and are therefore much colder than the fire.

There should be an abundant opportunity for complete combustion of gases coming from the fuel with hot air drawn through the fuel, before the flame is chilled by contact with cold surfaces. This condition is best fulfilled by having a clear space over the grate. Externally-fired boilers commonly have two feet or more between the grate and the boiler-shell immediately over it, and combustion may continue beyond the bridge. Locomotive-boilers have from four to six feet between the grate and the fire-box crown-sheet, but the flame is quickly drawn into and extinguished by the tubes. To aid combustion and to protect the lower part of the tube-sheet a brick arch is frequently carried across the fire-box, over which the flame must pass on the way to the tubes. The lack of space over the grate of flue-furnaces, as in the Scotch marine boilers, is only partially compensated by the combustion-chamber beyond the furnaces.

Loss from external radiation is almost entirely avoided in internally-fired boilers. Externally-fired boilers are subject to more or less loss from conduction and radiation.

The fire-grate should not be longer nor wider than can be conveniently reached by the fireman in throwing on fuel and in cleaning the grate. A narrow grate should not be so long as a wide grate. In general, a hand-fired grate should not be more than six feet long, and if it is over four feet wide two fire-doors should be provided. These conditions are usually fulfilled by the design of externally-fired boilers, locomotive-boilers, and water-tube boilers. Attention has been called already to the difficulty of getting proper space for the grates in flue-furnaces. With the common diameters of the furnace-flues a length of five feet should not be exceeded. Flues in marine boilers have been made eight feet long; in such case the further end of the grate is sure to be inefficiently fired. To aid in firing, and to use the space below and above the grate to the best advantage for the supply of air and for combustion, the grate is commonly given an inclination downwards of about 3/4 of an inch to the foot.

As an extreme example of deviation from these proportions we may cite the Wooten locomotive fire-box, designed to burn anthracite slack. The grate is made about eight feet wide and twelve feet long.

For convenience in throwing on coal and in cleaning the grates, the floor on which the fireman stands should be about two feet below the grate. This can usually be arranged for stationary boilers. The grate of a locomotive is commonly below the floor of the cab; this facilitates throwing on the coal; some form of rocking grate is used to shake down the ashes. The side furnaces of Scotch marine boilers are commonly too high for convenient firing, and the middle furnaces may be too low for convenience in cleaning the grate.

Excessive heat in the fire-room should be avoided as far as possible; the labor of feeding and cleaning a furnace for rapid combustion is always severe, and when combined with great heat it soon exhausts the fireman. If land boilers are properly clothed to avoid radiation, and if the fire-room is airy and well ventilated, the heat will not be excessive. It is, however, very difficult to avoid excessive heat in the stoke-hole of a steamship. Of course the radiation from the glowing fuel when the fire-doors are open cannot be avoided, but it ought to be possible to clothe the fronts of marine boilers more perfectly than is now the common practice. Moreover, the ventilation of the stoke-hole is commonly defective; the air pours down through the ventilators and makes cold spots immediately beneath them, while other parts of the stoke-hole are hot. Forced draught with closed stoke hole usually gives good ventilation; with closed ash-pit it is liable to give defective ventilation.

Grate-bars are commonly made of cast iron, as it is cheaper and lasts as well as wrought iron. Sometimes wrough-iron bars are used on locomotives and elsewhere, if they are expected to withstand rough usage.

Cast-iron fire-bars are generally 5/8 to one inch thick at the top, and 5/16 to 5/8 of an inch thick at the bottom; they are about two inches deep at the ends, and three to five inches deep at the middle. To provide for wasting of the upper surface, they are made full width for some distance down from the top, thus forming a sort of head; then they are rapidly narrowed down to a web that is tapered gradually toward the bottom. The space between the bars depends on the draught and the nature of the fuel; with ordinary coal and natural draught 3/8 to 1/2 of an inch is allowed. Lugs or projections are cast at the ends and at the middle, so that the bars shall be properly spaced when laid side by side. With forced draught the bars may be 3/8 to 9/16 of an inch wide at the top, and the distance between the bars may be 1/16 to 1/4 of an inch. A dead-plate two inches wide should be fitted to the furnace-tube of marine boilers to prevent admission of air at that place.

The length of fire-bars should not exceed three feet; the length of a fire-grate may be made up of two or three short bars. Bars are commonly cast in pairs, or three or four may be cast together, to resist twisting and warping under heat.

The usual form of grate-bar, cast in pairs with lugs at the side, is shown in Fig. 38. The herring-bone grate shown in the same figure is used for burning fine anthracite coal. The figure also shows a special form of grate-bar for burning sawdust.

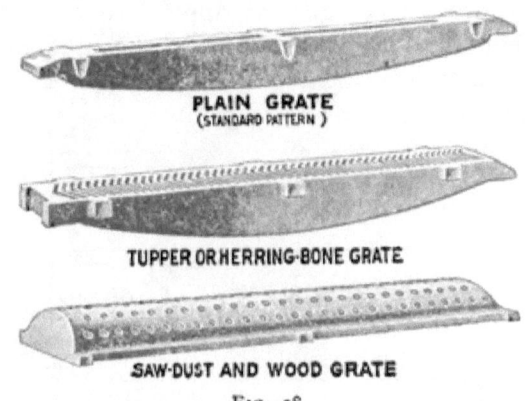

FIG. 38.

Wrought-iron fire-bars are formed with a head and web, but are of uniform depth, as they are cut from a rolled bar; they are bolted together in sets of six, with washers to give the proper spacing. For marine boilers they may be 5/16 of an inch thick at the top, with spaces 3/16 of an inch wide, or less.

Rocking Grates.—The labor of breaking up the clinker which forms on grate-bars, when bituminous coal is used, is very much reduced by employing some form of rocking grate. On locomotives, where the rate of combustion is high and where the fire should always be in good condition, some form of rocking grate is considered essential, in American practice.

In Fig. 39 A and B represent alternate grate-bars which are supported at semicircular notches at the ends. CC' is a cast-iron crank-shaft extending across the furnace at one end of the grate-bars. Shallow bars like A rest on cranks that are above the line CC', and deep bars like B rest on cranks below that line, as shown at a, a', and a'', and at b and b'. The further ends of the grate-bars rest on another crank-shaft like CC'. At the lower right-hand corner of the figure c'' represents the end of the crank-shaft and d represents an upper crank carrying a shallow bar like A. At g is a head to which a lever may be applied to rock the crank-shaft. When the crank-shaft is rocked the alternate bars are thrown back and forth, and grind up the clinker so that it falls through the grate into the ash-pit.

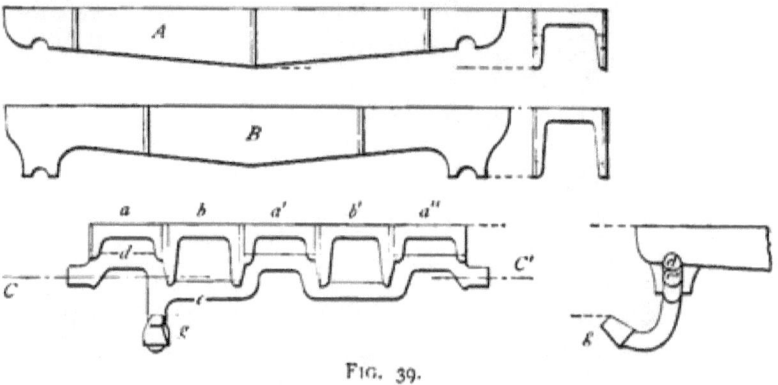

FIG. 39.

Firing.—Care, skill, and intelligence are required to burn coal rapidly and economically. There is a marked difference in the ability of trained firemen to make steam with a given boiler, and probably there is nothing more wasteful and costly than a poor or careless fireman.

The method to be adopted in firing depends on the type of boiler, the kind of coal, and the rate of combustion. Three methods of firing may be distinguished:

Spreading, which consists in distributing small charges of coal evenly over the surface of the fire at short intervals. In this method the object is to deliver the coal just where it is wanted, and then not disturb it. The fire can then be kept in just the right condition at all times, and probably the best results can be thus obtained, both in absolute quantity of steam and in economy, provided the coal used is well adapted to this method. Care must be taken to have the door open as little as possible, or an undue amount of cold air will be admitted through the fire-door.

Anthracite coal should always be fired by spreading, and should be disturbed as little as possible after it is thrown in place. Unless the fire is urged, very little clinker will be formed, and the ashes are readily shaken out by a pick or hook run up between the fire-bars. The thickness of the fire may vary from four to twelve inches, depending on the size of the coal and the strength of the fire.

Dry bituminous coal, and other bituminous coals, if not very smoky and if in small pieces, can be advantageously fired in this way. Each shovelful thrown on will give off volatile matter, which will burn with the excess of air coming through the fuel, and very little smoke will result.

Side firing consists in covering all of one side of the fire with fresh fuel, leaving the other bright. The smoke given off from the fresh fuel can then be burned with the hot air coming through the bright fire. This method of firing is best carried on with two furnaces leading to a common combustion-chamber; the furnaces are fired alternately, at regular intervals, with moderate charges of coal. It is customary to admit air through the grid in the fire-door when the fuel is giving off gas.

Coking the coal on a dead-plate, or on the grate just inside the fire-door, is perhaps the best way of burning a smoky coal. The volatile products driven off from the heap of coal near the furnace-door burn with the hot air, coming through the clear fire at the rear. As soon as the charge is coked it is pushed back and spread over the grate, and a new charge is thrown on.

With bituminous coal the fire should be thicker than with anthracite coal: from 6 to 16 inches gives good results.

The method too often followed by ignorant and indolent

firemen, of throwing on as much coal as the furnace will hold and then sitting down to wait till the steam-pressure falls, needs to be mentioned only to condemn it.

Mechanical Stokers, feeding coal regularly from a hopper, have been invented in a variety of forms from time to time. Since the hopper may be made of considerable size, manual handling of the coal may be entirely avoided, and one man can easily attend to a number of furnaces with little labor and exposure to heat. It would appear also that a more even and better-regulated combustion may be had than with hand-firing. All such devices, which have moving parts inside a confined furnace, quickly get out of order through the combined action of heat and dust.

The Roney stoker, shown by Fig. 40 as applied to a cylindrical tubular boiler, may be taken as an illustration of a mechanical stoker. The grate-bars extend across the furnace and form a series of steps down which the fuel slides, burning on the way down. Each grate-bar is hung on pivots at the ends, near the top, and has a rounded lug at the bottom that rests in a groove in a rocker-bar, as shown by Fig. 41.

The rocker-bar has a slow and regular reciprocation, derived from a small steam-engine, which tips the grate-bars so that the upper surfaces are inclined downward to make the fuel slide, and then rights them to check the motion of the fuel. The coal from the hopper falls onto a horizontal plate, from which it is pushed forward by a " pusher " that is driven by the steam-engine which drives the rocker-bar. The rate of feeding the fuel can be controlled by changing the stroke of the pusher, and by regulating the number of strokes of the pusher and of the rocker-bar per minute. The ashes, clinker, and other unburned refuse collect on a dumping-grate at the foot of the grate-bars. This grate is shown in normal position by heavy lines in Fig. 41, and in the dumping position by light lines.

This grate appears to be well adapted to burn smoky fuel,

as such fuel is well coked at the top of the grate, and the volatile parts driven off by coking can burn with the excess of air coming through the grate at the bottom.

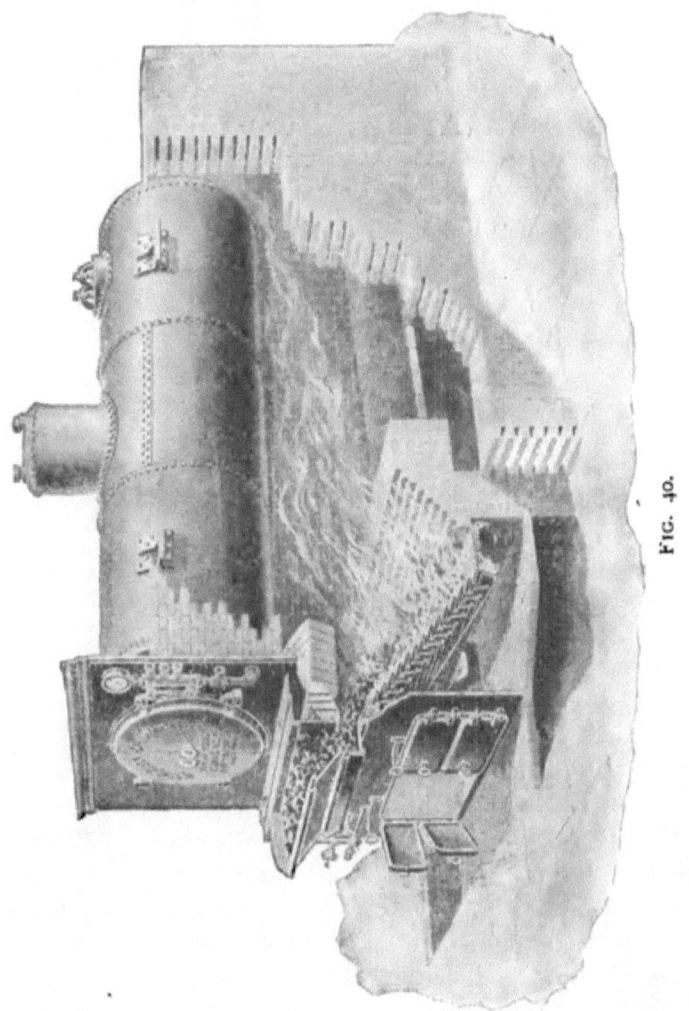

FIG. 40.

If the rate of feed is too fast, it is evident that unburned coal will work down onto the dumping-grate, and will appear in the ashes. If the rate of fuel is regulated so that no coal appears in the ashes, the fire becomes thin at the bottom, and an excess of air is liable to enter there; certain tests on this grate have indicated such an excess of air, which is the side on which the fireman is liable to err, as he may not know how much waste he thus occasions, while he can see the coal in the ashes.

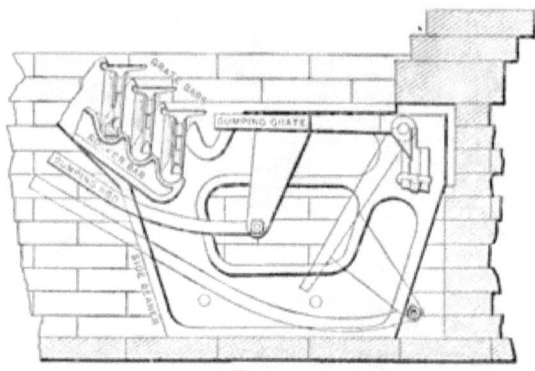

FIG. 41.

Special Furnaces are required for burning various refuse material, such as sawdust, tan-bark, straw, and bagasse; no attempt will be made to describe them here.

When wood is burned on a grate it may be sawn into pieces two feet long, and the grate may have the bars spaced wider than for coal. Cord-wood can be burned on a brick hearth a little longer than the sticks of wood; the wood ashes are small in amount, and are light, so that the draught will sweep a large portion of them into a pit beyond the hearth. Wood is not now burned for making steam except in remote places, unless it be at sawmills and wood-working factories, where a large amount of refuse wood is produced in the form of slabs, sawdust, shavings, etc.

Smoke Prevention has become a matter of great social importance in cities where much smoky coal is used. Though the loss through imperfect combustion of carbon to the form of carbon monoxide may be great, and though there may be an appreciable loss if the volatile parts of coal are driven off unconsumed, it is a fact that the loss in smoke, even when it is dense and black, is not enough to induce coal users to take the trouble to prevent the formation of smoke. Not infrequently it has been found that the methods used to prevent smoke are accompanied by a loss instead of a gain. For example, smoke burning by the alternate firing of two furnaces, leading to a common combustion-chamber, may give a slightly greater efficiency if just enough hot air in excess is admitted through the clear fire, to burn the gases distilled from the fresh charge. If the clear fire must be kept too thin, and thus admit a large amount of air, in order that the smoke may be burned, there will be a loss of efficiency. Though it is not well proved, it is asserted that the mixture of finely divided carbon, in the form of smoke, with carbon dioxide may give a clear gas with the formation of carbon monoxide, and thus with a notable loss. The same difficulties arise when side firing and coking are resorted to with smoky fuels.

One of the most perfect arrangements for smoke prevention which has yet been tried, consisted of a detached furnace with small grate-area and a deficient air-supply, so that the coal was distilled and burned to carbon monoxide; the resulting hot gases were then burned under a steam-boiler. The method was suggested by the producer-furnaces used for making gas for the open-hearth process of steel-making. The objections are the loss of heat by radiation from the detached furnace and the space occupied by that furnace. Though reported to be a success so far as the prevention of smoke was concerned, it does not meet with approval.

It is a common experience, that when laws against making smoke are enforced, users of fuel have chosen to buy anthracite coal or coke, or in some cases have used crude petroleum oil.

Down-draught Furnaces.—In connection with the subject of smoke prevention, attention should be called to down-draught furnaces, which have the connection with the chimney below the grate. The supply of air is through the fire-door to the top of the fire, which has a very attractive appearance, as it burns brightly at the upper surface unless obscured by fresh fuel. A natural inference is, that the combustion is perfect in a down-draught furnace, and that it should give a notable gain in economy of fuel, but a little consideration shows that such a furnace is subject to the same conditions as an

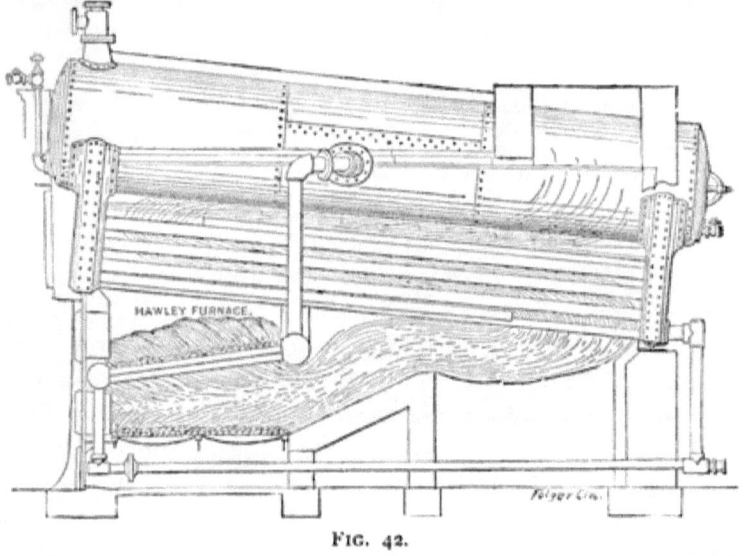

FIG. 42.

ordinary furnace. If there is either an excess or a deficiency of air, the combustion will be imperfect; in the latter case, as with an ordinary furnace, smoke may appear at the top of the chimney. Tests made on a boiler using first an ordinary and then a down-draught grate have commonly shown little if any advantage in favor of the latter.

Down-draught furnaces, if properly arranged and fired, can be made to burn inferior fuels which have a large amount of volatile matter without making much smoke; this may be a matter of great importance in cities where laws against smoke are enforced.

Fig. 42 shows a Hawley down-draught furnace applied to a Heine boiler.

The grate which is shown by Fig. 43 consists of two transverse wrought-iron headers at the front and back of the furnace, between which are two rows of two-inch tubes, acting as grate-bars. The tubes in the lower row are placed under the spaces between the upper rows of tubes. Water is supplied to the front header by a pipe from the back end of the boiler, and steam formed in the tubes and headers is discharged

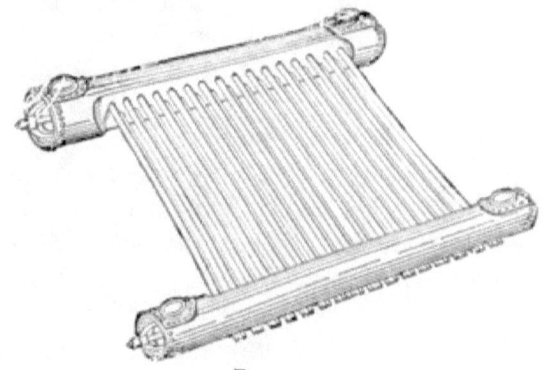

FIG. 43.

through a pipe which enters the drum of the boiler near the water-line. The headers are flattened to receive the double row of tubes, and are provided with hand-holes for cleaning. Opposite each tube of the grate there is a plug in the front header which can be taken out when the tube needs cleaning. A second grate, with solid bars, is placed beneath the water-tube grate, to catch unburned fuel that falls through.

If a boiler has deficient heating-surface or poor circulation there may be a direct gain from the use of a water-tube grate, as in the down-draught furnace. Such a grate is more expensive to install and to repair than the ordinary solid-bar grate. Rocking grates or automatic stoking are of course precluded.

Oil-burning Furnaces have the oil thrown in by sprayers or atomizers, and the oil burns in a flame that is about four inches in diameter and two to four feet long. The sprayer has two conical converging tubes, one inside the other, something like the steam and water nozzles of a steam-injector. Compressed air or superheated steam is supplied to the inner tube, and the oil is drawn through the outer tube and thrown into fine spray mingled with air. Compressed air is the better, considering proper combustion only, but the great convenience of using steam near a steam-boiler has led to its common use. The proportions of air and oil may be nicely regulated, so that perfect combustion may be secured without smoke.

Thus far oil has been used for fuel mainly at the Caspian oil-fields, or on steamers coming from that region. The petroleum obtained there gives a large amount of refuse that cannot be used for other purposes, and coal, which must be brought from a distance, is expensive. In the United States oil has been used for fuel at or near oil-fields, or in cities where laws against smoke are enforced,

The use of oil for fuel on war-ships has received favorable consideration from some authorities, the evident advantages being the great calorific power of oil and the ease with which the fires may be maintained and regulated. The fact that oil in tanks may be set on fire by explosive shells has prevented any extensive adoption of oil for fuel on war-ships.

Oil for fuel should be stored in tanks outside the fire-room, and if possible the tanks should be lower than the burners. The oil is pumped from the tank to the burners as required. This is to avoid accidental flooding of the furnace and the fire-room with oil, and the attendant danger of conflagration. Crude oil is more dangerous than refuse oil, since the former contains all the volatile components that vaporize at ordinary temperatures and form explosive mixtures with air.

Forced Draught.—When a higher rate of combustion is required than can be had with natural draught, resort is had to forced draught, by aid of which 150 pounds of coal can be burned per square foot of grate-surface per hour.

Three systems of forced draught are in common use, namely, with a *closed stoke-hole*, with *closed ash-pits*, and *induced draught*.

Induced draught has long been used on locomotives, by the action of the exhaust-steam thrown through the smoke-stack. The same method is used to some extent on tug-boats. This method is simple and effective, but can be used only with non-condensing engines. Induced draught may be obtained by a centrifugal, or other form of blower, in the chimney. It is essential that an economizer should be used to cool the gases before they come to the blower.

On steamships forced draught has been obtained by the aid of centrifugal fan-blowers. The method with closed ash-pit has been used with success on merchant steamers and some war-ships. With this method air drawn from the fire-room passes through a blower and is delivered to the ash-pit, which has an air-tight door. If the pressure in the ash-pit exceeds the resistance to the passage of air through the fuel, flame comes out around the fire-door unless it is also made air-tight. When the fire-door is opened to throw on coal the blast must be shut off from that furnace and all others having a common combustion-chamber, or flame will shoot out into the fire-room in a dangerous manner. One reason why it has not been used on war-ships is the difficulty of properly ventilating the many small fire-rooms in which boilers are placed.

The closed stoke-hole has been the customary way of getting a forced draught on torpedo-boats and on other naval vessels. The stoke-hole is closed air-tight, admission and egress being through air-locks, and air from without is forced in through a centrifugal blower till the pressure exceeds that of the atmosphere. When a fire-door is opened to attend to the fire, there is a strong inrush of air that is liable to make the tube-plates leak. So great difficulty has been experienced from this cause, when forced draught has been used with the Scotch boiler, that many naval officers doubt its advisability for large ships. The success of forced draught on the locomotive and on torpedo-boats with modified locomotive-boilers may be attributed partly to the type of the boiler and partly to the fact that there is only one boiler and one furnace. When two boilers are used on a torpedo-boat, each has its own chimney.

On locomotives the induced draught is frequently equivalent to a column of water five or seven inches high. Forced draught on torpedo-boats has approached those figures, but is usually less. Large ships usually have the forced draught restricted to two inches of water. On account of the resistance to the entrance of air to the fire-rooms of war-ships, through ventilating shafts, gratings, etc., it has been common to assist the draught by running the blowers without closing the air-locks.

Howden's System.—The temperature of gases in the uptakes of marine boilers is frequently high, especially when forced draught is used. In Howden's system the products of combustion pass among horizontal tranverse tubes placed in an enlargement of the uptake. Air to supply the fire is drawn through these tubes by a fan-blower and is thereby warmed, thus saving heat and giving quicker combustion. Care must be taken in using this system not to go too far, or the fire may become too hot and rapidly burn out the fire-grates and do other injury.

Cleaning Fires.—Three tools are used in clearing the grate: they are a long straight bar known as the slice-bar, a similar bar with the point bent at right angles to make a hook, and a long-handled rake with three or four prongs. The hook may be run along between the grate-bars from below, to clear the spaces from ashes and clinker. The slice-bar is thrust under the fire on top of the grate to break up the cinder; it is used also to stir and break up caking coals. The rake is used to haul the fire forward or to draw out cinder.

To clean a fire the fireman breaks up the cinder with the slice-bar and rattles down the ashes; if necessary, he works the fire back toward the bridge and exposes the grate in front, which may then be thoroughly cleaned. Then he hauls the fire forward and cleans the back end of the furnace. Cinder which will not break up and pass through the grate is pulled out through the fire-door. Some firemen prefer to clean the grate one side at a time. After the grate is cleaned the fuel left is spread evenly over the grate and fresh fuel is thrown on. The fire should be allowed to burn down before cleaning, but a fair amount of glowing coal should be left to start a new fire briskly. Before beginning to clean the fire the draught should be checked by closing dampers or otherwise.

Green's Economizer.—From time to time attempts have been made to get heat from the products of combustion by passing them through a feed-water heater, after they leave the boiler and before they enter the chimney. The earlier attempts to use such feed-water heaters were unsatisfactory, because the pipes forming the feed-water heater were soon covered with soot, and then became inoperative. Green's feed-water heater, or economizer, is made of several sets of vertical cast-iron tubes four inches in diameter, placed in a chamber between the boiler and the chimney. The feed-water is pumped in succession through the several sets of tubes, beginning at the more remote, and, finally, it passes from the nearer tubes to the boiler. This arrangement brings the hottest water in contact with the hottest gases.

The feature which makes the Green economizer successful

is the scrapers, which are arranged in sets, one on each tube. By power, from a small steam-engine or elsewhere, the scrapers are continually moved up and down on the pipes, and so the pipes are kept free from soot and in good condition to take up heat.

As might be expected, this economizer has been found to be most successful with boilers that have deficient heating-surface.

Chimneys.—The present state of our knowledge of chimneys and chimney draught is very unsatisfactory. The theories given in text-books and elsewhere are not strictly logical and are based on insufficient data; they are of little use in proportioning chimneys. On the other hand, there is no systematic statement of ordinary practice that can be used in designing chimneys.

A statement will be given, first of the ordinary theories and their defects, and second of the information at hand.

Chimney Draught.—The draught produced by a chimney is due to the fact that the gases inside the chimney are hotter and consequently lighter than the outside air. Though these gases at a given temperature and pressure have a little greater specific gravity than air at the same temperature and pressure, the difference is not much, and may be neglected in the discussion of chimney draught.

To get an idea of the production of draught by a chimney, we may consider the conditions that would exist if a chimney were filled with hot air and closed at the bottom by a horizontal partition or diaphragm. The pressure of the air at the top of the chimney, due to the atmosphere above that level, is the same on the gases inside the chimney and the air outside. The pressure on the diaphragm at the bottom is the sum of the pressure at the top of the chimney and of the pressure due to the column of hot air in the chimney. At the under side of the diaphragm the pressure will be that at the top of the chimney plus the pressure due to a column of cold air as high as the chimney. This difference of pressure is considered to be the draught, in all theories of the chimney. It may be readily calculated for an assumed set of conditions. For an actual chimney the draught or difference of pressure inside and outside the chimney may be shown by a U tube partially filled with water, and having one end connected to the inside of the chimney and the other open to the air. The water rises in the leg connected with the inside of the chimney; the difference of level measures the draught.

Suppose now that a small hole is opened in the diaphragm at the bottom of the chimney: cold air from without, under the greater pressure existing there, will enter and will force some of the hot air out at the top of the chimney. If the air is heated as it enters, to the temperature in the chimney, we shall have a continuous flow of cold air into and of hot air out of the chimney. Replacing the diaphragm by a grate charged with burning fuel, through which cold air enters and burns with the fuel, we have the actual conditions of chimney draught.

For an example, we will calculate the difference of pressures, or draught, if a chimney 100 feet high is filled with air (or gas) at 600° F., while the temperature outside is 60° F.

The weight of a cubic foot of air at 32° F. and at the average pressure of the atmosphere (14.7 pounds) is about 0.0807 of a pound. Now the weight of air at a given pressure is inversely proportional to the absolute temperature, that is, to a temperature obtained by adding 460°.7 to the temperature given by a Fahrenheit thermometer. Consequently we have for the weights of a cubic foot of hot gas and of a cubic foot of cold air:

$$\text{Hot gas, } 0.0807 \times \frac{460.7 + 32}{460.7 + 600} = 0.0375;$$

$$\text{Cold air, } 0.0807 \times \frac{460.7 + 32}{460.7 + 60} = 0.0764.$$

A column of hot gas 100 feet high and 1 foot square will weigh 3.75 pounds, and would give that pressure in pounds per square foot on a diaphragm at the bottom of the chimney. The cold air outside will give a pressure of 7.64 pounds per square foot. The difference of pressure or draught will be

$$7.64 - 3.75 = 3.89$$

pounds per square foot,

$$\text{or } 3.89 \div 144 = 0.027$$

of a pound per square inch. In this calculation the variation of the pressure of the atmosphere from 14.7 pounds per square inch, and the effect of the reduction of pressure in the chimney, have been neglected, as they are insignificant.

To find the draught in inches of water we may consider that one cubic foot of water weighs 62.4 pounds. Consequently a column of water a foot square and which produces a pressure of 3.89 pounds per square foot will be

$$3.89 \div 62.4 = 0.0623$$

of a foot high, or

$$0.0623 \times 12 = 0.75$$

of an inch high. This is the draught that would be shown by a U tube if the chimney were closed at the bottom.

It is convenient to express the difference of pressure or draught due to the difference of temperatures inside and outside the chimney in algebraic form as follows: Let T_0 be the absolute temperature of the freezing-point. Let T_c be the temperature in the chimney and T_a the temperature of the air. Let w be the weight of a cubic foot of air at freezing-point. Then the weight of a cubic foot of hot air in the chimney will be

$$w \frac{T_0}{T_c},$$

and the weight of a cubic foot of cold air outside will be

$$w\frac{T_o}{T_a}.$$

The weight of a column of hot air H feet high and one foot square will be

$$wH\frac{T_o}{T_c},$$

and that expression will represent the pressure due to such a column of air. The weight of a column of cold air and the pressure due to it will be

$$wH\frac{T_o}{T_a}.$$

The draught due to a chimney H feet high with the absolute temperatures inside and outside T_c and T_a is then

$$wH\left(\frac{T_o}{T_a} - \frac{T_o}{T_c}\right)$$

This expression gives the draught in terms of pounds per square foot.

All theories of chimney draught that have been proposed treat the difference of pressures, inside and outside the chimney, as though it were a head producing a flow of a fluid, as a head of water produces a flow of that liquid.

Flow of a Liquid.—In hydraulics, it is shown that we may express the relation between the velocity of flow of a liquid in a pipe and the head producing the flow, by the following equation:

$$h = \frac{V^2}{2g}\left(1 + k + k_1 + \frac{fl}{m}\right),$$

in which h is the head, in feet, of the liquid producing the flow; V is the velocity in feet per second; g is the acceleration due to gravity; k and k_1 are coefficients used to express the resistance of obstructions like valves, bends, etc.; l is the length of the pipe; m is the ratio of the area to the perimeter of the pipe; and f is the coefficient of friction of the fluid against the sides of the pipe. The several coefficients vary with the velocity of flow, the size of the pipe, and the nature of the resistance. Experiments in hydraulics have been made, and tables prepared, so that the proper coefficients may be selected for use with the equation, under any given conditions.

It is assumed in theories of chimney draught that an equation of the same form may be used to express the relation between the head, or difference of pressure inside and outside the chimney, and the flow of gases through the furnace, flues, and chimney. The resistances to the passage of gases through the grate and the fuel on it, and through the flues and tubes, may be expressed by aid of coefficients G and C, which are like k and k_1; the resistance of friction of the gases against the side of the chimney may be assimilated to the last term in the parenthesis, replacing l by H.

The thermodynamic theory of the flow of gases leads to equations which differ fundamentally from the equation for the flow of liquids. Only when the changes of temperature and pressure are small, can the hydraulic equation be used at all, and then the approximation is not good. Nevertheless it may be possible to base a working theory of chimneys on the hydraulic equation, provided that the proper constants can be derived from experiments, and provided that the application of the theory be restricted within limits that are determined by proper investigation.

Peclet's Theory of Chimney-draught.—The theory of chimney-draught which is commonly given in text-books was proposed by Peclet many years ago. This theory assumes, first, that the flow of gases through the furnace-flues and chimney, may be represented by an equation like the hydraulic equation just quoted: and second, that the head h in that equation may be calculated by dividing the draught, expressed in pounds per square foot, by the weight of a cubic foot of hot gas. The weight of a cubic foot of hot gas is given by the expression $w\frac{T_o}{T_c}$, and the draught is given by the expression $wH\left(\frac{T_o}{T_a}-\frac{T_o}{T_c}\right)$; so that Peclet's assumption gives

$$h = wH\left(\frac{T_o}{T_a}-\frac{T_o}{T_c}\right) \div w\frac{T_o}{T_c}.$$

$$\therefore h = H\left(\frac{T_c}{T_a}-1\right). \qquad \qquad . \quad . \quad (1)$$

Replacing the coefficients k and k_1 in the hydraulic equation by C and G, which represent the resistance to the flow of gas through the tubes and flues, and the resistance to the flow through the grate, we have

$$\frac{V^2}{2g}\left(1+G+C+\frac{fH}{m}\right)=H\left(\frac{T_c}{T_a}-1\right); \quad . \quad . \quad (2)$$

or, solving for V the velocity of the hot gas through the chimney,

$$V = \sqrt{2gH}\left(\frac{T_c}{T_a}-1\right)^{\frac{1}{2}} \frac{1}{\left(1+G+C+\frac{fH}{m}\right)^{\frac{1}{2}}} \quad . \quad . \quad (3)$$

If the area of cross-section of the chimney is A square feet, the volume of hot gas discharged per second is

$$VA,$$

and the weight discharged per second is

$$W = VA \cdot w\frac{T_o}{T_c}$$

$$\therefore W = Aw\ \sqrt{2gH}\frac{T_o}{T_c}\left(\frac{T_c-T_a}{T_a}\right)^{\frac{1}{2}} \frac{1}{\left(1+G+C+\frac{fH}{m}\right)^{\frac{1}{2}}} \cdot \cdot \quad (4)$$

From some experiments on chimneys and boilers Peclet gives, in connection with this theory, the following values for the coefficients,

$$G = 12, \qquad f = 0.012,$$

under the assumption that from 20 to 24 pounds of coal are burned per square foot of grate per hour; the coefficient C does not appear in his equation.

Equation (4) is in the proper form for calculating the weight of gas discharged by a given chimney, for which the height, area, and perimeter of cross-section are known. If the weight of gas to be discharged and the area and perimeter are known, the equation for a given case leads to a quadratic equation for finding the height H, which can readily be solved numerically. If the weight of gases is known, and the height of the chimney is assumed, then the insertion of linear dimensions in place of A and m leads to an equation of the fourth degree; but as $\dfrac{fH}{m}$ is small compared with $1 + G$, a solution by approximations can be readily made.

It is probable that the equation (4) with the given values for G and f represented satisfactorily the performance of the chimneys which were investigated by Peclet. These chimneys provided draught for boilers then in common use in France, which boilers were probably either plain cylindrical boilers or "double-elephant" boilers. For such boilers the resistance is mainly at the grate. On the other hand, the resistance to the passage of gases through the tubes of cylindrical tubular boilers, locomotive-boilers, and marine boilers is about equal to the resistance to the passage of air through the fuel on the grate.

Under the conditions of modern practice in America, the equation deduced by Peclet, using his values for f and G, gives results that do not accord with observations or with common practice. The theory consequently is not valuable for proportioning chimneys.

If the weight of gas discharged by a chimney of given height and cross-section be calculated successively for different values of T_c, the temperature in the chimney, the weight will be found to increase with the temperature, until the temperature becomes about 1000° absolute or about 600° F.; beyond this temperature the weight decreases.

The temperature for maximum discharge, as calculated by the equation, may be readily found by aid of the differential calculus. The factor which increases with the temperature is

$$\frac{(T_c - T_a)^{\frac{1}{2}}}{T_c}.$$

Differentiating with regard to T_c and equating the first differential coefficient to zero gives

$$T_c \tfrac{1}{2}(T_c - T_a)^{-\frac{1}{2}} - (T_c - T_a)^{\frac{1}{2}} = 0.$$
$$\therefore T_c = 2 T_a.$$

Consequently the maximum discharge of gas will occur when the absolute temperature in the chimney is twice the absolute temperature of the air. If the temperature of the air is 70° F., or

$$70 + 460.7 = 530.7$$

degrees absolute, then the temperature to produce the maximum discharge of gas will be, by Peclet's theory,

$$2 \times 530°.7 = 1061°.4 \text{ absolute,}$$
or $$1061°.4 - 460°.7 = 600° + \text{ F.}$$

This is about the temperature of melting lead, and books on chimneys frequently say that the temperature in a chimney should not exceed that of melting lead. A temperature near this is commonly found in chimneys that are doing good work, a fact that seems to give some support to the theory. But the support is entirely fictitious, for the occurrence of a maximum depends on the assumption that the head h in the hydraulic equation may be replaced by $H\left(\dfrac{T_c}{T_a} - 1\right)$, an assumption that can be justified only by observation or experiment. Such observations or experiments are lacking. All we know is that calculations by the equation do not accord with common practice. It is true that 600° F. should be a sufficiently high temperature in the chimney to give all the draught required. It will be still better if a lower temperature will suffice.

Peclet's Second Theory.—It appears that Peclet was not satisfied with his theory as first propounded, for he afterwards advanced another theory, in which the head is calculated by dividing the draught, expressed in pounds per square foot, by the weight of a cubic foot of cold air. It is noteworthy that this later theory does not show a maximum discharge at 600° F.

Neither the first nor the second theory is strictly logical; the value of either as a working theory must consequently depend on its adaptability for designing chimneys under conditions of ordinary practice. Both theories lack connection, through experiment or observation, with practice, and cannot now be used to advantage.

Tests and Observations.—The data from the tests made by Peclet to determine the values of constants in his equation are not now accessible. He gives only the results, namely, $G = 12$ and $f = 0.012$.

Prof. Gale[*] reports the following results of tests made on a chimney and boiler of ordinary construction:

Area of grate	22.5 sq. ft.
Area through tubes	2.74 "
Coal per square foot of grate per hour	13.5 pounds.
Air per pound of fuel	21 "

[*] Trans. Am. Soc. Mech. Engrs., vol. xi. p. 451.

Temperature boiler-room............................ 60° F
Temperature external air.......................... 40° F.
Height of chimney above grate..................... 72 feet.
Area of chimney (round iron stack) .... 4 sq. ft.
Length of horizontal iron flue..................... 24 feet.

PRESSURES IN POUNDS PER SQUARE FOOT.

Required to produce entrance velocity........................ 0.013
Required to overcome resistance of grate..................... 0.91
Required to overcome resistance of combustion-chamber and boiler-tubes 1.23
Required to overcome resistance of horizontal flue........... 0.06
Required to produce velocity of discharge.................... 0.085

    Total effective draught................................. 2.298
Required to overcome resistance of friction.................. 0.19

    Total draught.. 2.488

On these results Prof. Gale based a set of constants, to be used in an equation like that given in Peclet's first theory. It does not appear to us that observations on one chimney are sufficient for this purpose. We will note only that his value for the coefficient of friction is $f = 0.012$ for an unlined iron stack. For a brick chimney he gives $f = 0.016$.

The following table gives the results of a test made at the Massachusetts Institute of Technology on an unlined steel chimney 3 feet in diameter and 100 feet high above the grate.

	Draught. Inches of Water.		Temperature, Centigrade.	
	Max.	Min.	Max.	Min.
Over the grate........................	0.240	0.218		
At the bridge-wall.....................	0.382	0.372		
Half-way between bridge and back end of boiler............................	0.410	0.374		
At back end of boiler..................	0.354	0.334		
In uptake near boiler..................	0.572	0.543	206	195
In stack 34 feet above grate............	0.440	0.414	202	190
In stack 51 feet above grate............	0.334	0.312	193	187
In stack 68 feet above grate............	0.216	0.168	188	179
In stack 85 feet above grate............	0.122	0.086	174	157

This chimney now serves two boilers similar to that shown on Plate I, each of which is rated at 80 horse-power. It is intended to be sufficient for four such boilers. The heating-surface of each boiler is 1113, and each has 25.9 square feet of grate area. At the time of the test one boiler had the fire banked, and the combustion at the grate of the working boiler was at the rate of 19.8 pounds per square foot of grate-surface per hour.

Kent's Table.—Mr. Wm. Kent* has calculated a table of sizes of chimneys on the following assumptions:

1. The draught-power varies as the square root of the height.

2. Allowance for friction against the sides of the chimney may be made by subtracting from the actual area of the section of the chimney, a strip two inches wide and as long as the perimeter of the section.

3. The power of the chimney is directly proportional to the area remaining after the strip is deducted from the actual area.

The first assumption is equivalent to using the hydraulic equation on page 115 in the simplified form

$$H = \frac{V^2}{2g} \quad \text{or} \quad V = \sqrt{2gH},$$

in which H is the height of the chimney in feet and V is the velocity of discharge.

The second assumption is purely arbitrary, and can be used only within limits, or it may lead to absurd results. Thus a flue 4 inches in diameter would give no draught at all by this assumption.

The third assumption follows naturally from the second.

If the side of a square chimney be represented by D, then the area is

$$A = D^2.$$

---

* Trans. Am. Soc. Mech. Engs., vol. XI. p. 81.

and the strip to be subtracted is (nearly)
$$4D \times \tfrac{2}{12} = \tfrac{2}{3} D = 0.6 \sqrt{A}.$$

If the effective area allowing for the strip is E, then for square chimneys
$$E = A - 0.6 \sqrt{A}.$$

The same equation may be used for round chimneys with but little error.

Combining the first and third assumptions, Mr. Kent writes the equation
$$\text{H.P.} = CE \sqrt{H},$$

which he uses for calculating the commercial horse-power of the chimney, or more properly of the boilers to be connected with the chimney.

To obtain a value for the arbitrary constant C he chooses a typical chimney, 80 feet high and 42 inches in diameter, for which the effective area calculated by his method is 9.62 square feet. He says that such a chimney should be capable of carrying a combustion of 120 pounds of coal per hour for each square foot of effective area. If the area of the chimney is one eighth of the area of the grates connected with it, then this is equivalent to a combustion of $120 \div 8 = 15$ pounds of coal per square foot per hour—a very common performance.

This typical chimney should then burn
$$9.62 \times 120 = 1154.4$$

pounds of coal per hour. If it be assumed that five pounds of coal will be burned per horse-power per hour, the chimney may be considered to correspond to
$$1154.4 \div 5 = 231 \text{ H.P.}$$

Substituting in the formula for horse-power
$$231 = C \times 9.62 \sqrt{80}.$$
$$\therefore C = 3.33$$

And the horse-power equation, substituting for E its value, may be written

$$H.P. = 3.33(A - 0.6\sqrt{A})\sqrt{H}.$$

The following table has been calculated by the equation just written:

SIZES OF CHIMNEYS WITH APPROPRIATE HORSE-POWER OF BOILERS.

Dia. in inches.	Height of Chimneys.										Effective Area, square feet.	Actual Area, square feet.	Side of Square of approximate Area, inches.
	50 ft	60 ft	70 ft	80 ft	90 ft	100 ft	110 ft	125 ft	150 ft	175 ft. 200 ft.			
	Commercial Horse-power.												
18	23	25	27								0.97	1.77	16
21	35	38	41								1.47	2.41	19
24	49	54	58	62							2.08	3.14	22
27	65	71	78	83							2.78	3.98	24
30	84	92	100	107	113						3.58	4.91	27
33		115	125	133	141						4.47	5.94	30
36		141	152	163	173	182					5.47	7.07	32
39			183	196	208	219					6.57	8.30	35
42			216	211	245	258	271				7.76	9.62	38
48				311	330	348	365	389			10.44	12.57	43
54				363	427	449	472	503	551		13.51	15.90	48
60				505	505	565	591	632	692	748	16.98	19.64	54
66					658	694	728	776	849	918	981	20.83	23.76	59
72					792	835	876	934	1023	1105	1181	25.08	28.27	64
78						995	1038	1107	1212	1310	1400	29.73	33.18	70
84						1163	1214	1294	1418	1531	1617	34.76	38.48	75
90						1344	1415	1496	1639	1770	1893	40.19	44.18	80
96						1537	1616	1720	1876	2027	2167	46.01	50.27	86

The number of pounds of coal per hour that can be burned with a given chimney can be found by multiplying the horse-power given in the table by five.

Only that part of the table is filled in which corresponds to ordinary proportions, depending on the judgment of the author of the method. In general it will be better to select proportions from the table for a chimney to be used with a given commercial boiler horse-power, rather than to calculate by the formula, as extraordinary proportions will then be avoided.

This table has been used to a considerable extent, and apparently with satisfactory results.

Areas of Chimneys and Flues.—In common practice it is found that satisfactory results are obtained if the area of the section of a chimney is made one eighth of the area of all the grates connected to the chimney. The ratio is sometimes as large as 1/7 and sometimes as small as 1/9, or for tall chimneys 1/10.

Height of Chimney.—Professor Trowbridge* gives the following table of heights of chimney required to give certain rates of combustion, obtained by collecting reliable data and drawing a curve to represent mean results:

HEIGHT OF CHIMNEY.

Heights in Feet.	Pounds of Coal per Square Foot of Section of Chimney per Hour.	Pounds of Coal per Square Foot of Grate per Hour.
20	60	7.5
25	68	8.5
30	76	9.5
35	84	10.5
40	93	11.6
50	105	13.1
60	116	14.5
70	126	15.8
80	135	16.9
90	144	18.0
100	152	19.0
110	160	20.0

The table was made several years ago, but it seems to be conservative and to represent good average practice. By its aid the height of a chimney to give a desired rate of combustion can be determined. This height is then to be used with the ordinary ratio of chimney-area to grate-area as just given.

Forms of Chimneys.—Chimneys are made of brick or of steel plates. Steel chimneys are always round; large brick chimneys are usually round; small ones may be round or square. A round chimney gives a larger draught-area for the same weight of material, and it presents less resistance to the wind.

Plate V gives the general arrangement and some detail of two chimneys: one of brick, 175 feet high, and the other of

---

* Heat and Heat-engines, p. 153.

steel, 200 feet high. The brick chimney is built in two parts: the outer shell, which resists the pressure of the wind; and the lining, which forms the flue proper, and which may expand when the chimney is full of hot gases without bringing any stress on the shell. The shell has a foundation of rough stone and one course of dressed stone at the surface of the ground. The brickwork is splayed out inside to cover the stone foundation, and is drawn in at the top to the same diameter as the inside of the lining. The external form of the top is mainly a matter of appearance. The finish of large tiles at the top sheds rain and keeps water from penetrating the brickwork. The outside of the shell has a straight taper from the base nearly up to the head. A system of internal buttresses, as shown in section at Fig. 3 and Fig. 4, gives the requisite stiffness to the shell without an excessive amount of material. The lining carries its own weight only, being protected from the wind by the external shell; it has a uniform diameter of 6 feet inside, and varies in thickness from 12 inches at the bottom to 4 inches at the top. A rectangular flue with an arched top, leads into the chimney at one side of the foundation.

The shell of the steel chimney is made of vertical half-inch plates at the base, and is splayed out to give additional bearing on the foundation. Above this portion the shell has a straight taper to the top; the plates, each 4 feet wide, vary in thickness from 3/8 of an inch to 1/4 of an inch. At the top an external finish of light plate is given for the sake of appearance. The foundation is of red brick, with a course of stone at the surface of the ground, clamped by a wrought-iron strap. The shell is bolted through a foundation-ring made of cast-iron segments 4 inches thick, and a steel plate $2\frac{1}{2}$ inches thick, by long bolts which take hold of anchor-plates bedded in the foundation. The lining of fire-brick varies in thickness from 18 inches at the bottom to $4\frac{1}{2}$ inches at the top. It lies against and is carried by the steel shell. The internal diameter of the chimney is intended to be 10 feet; at places the size is a little larger on account of the arrangement of the lining. The lining is used to check the escape of heat through the steel shell. It adds nothing to the strength of the chimney; on the contrary, it must be carried by the shell. There is a chance that moisture may be harbored between the lining and the shell and give rise to corrosion. Large steel chimneys are comparatively recent, so that experience does not show whether lined or unlined chimneys are the more durable.

Stability of Chimneys.—On account of the concentration of weight on a small area, and the disastrous results that would follow from defective work, the foundations of an important chimney should be carefully laid by an experienced engineer. A natural foundation is to be preferred, but piling and other artificial methods of preparing the earth for the foundation can be used when necessary. Good natural earth should carry from 2000 to 4000 pounds to the square foot. The base of the chimney should be spread out so that this pressure, or whatever the earth can safely bear, may not be exceeded.

In calculating the stability of a chimney it is customary to assume the maximum pressure of the wind as 55 pounds per square foot on a flat surface. The pressure of the wind on a round chimney is assumed to be half that on a square chimney having the same width. This method has long been in use, and it has been shown to give abundant stability. Experiments on wind-pressure are difficult and uncertain, and, curiously, the pressure determined by small gauges is commonly in excess of that shown by large gauges. Thus, certain experiments made during the construction of the Forth Bridge, gave a maximum wind-pressure of 35 pounds per square foot on a large gauge 20 feet long and 15 feet wide, while a small gauge showed a pressure of 41 pounds at the same time. The highest recorded pressure during violent gales, at the Forth Bridge, was that just quoted, namely, 35 pounds to the square foot. Small wind-gauges have shown a pressure of 80 to 100 pounds to the square foot; but such results are discredited, both because it is known that small gauges give too large results, and because buildings were not destroyed as they would have been if exposed to such wind-pressures.

To determine whether a chimney is stable, treat it as a cantilever uniformly loaded with 55 pounds to the square foot and find the bending-moments and resultant stresses. The stress will be a tension at the windward side and a compression at the leeward side. Calculate the direct stress due to the weight of the chimney, which will be a compression at either side of the chimney. For a brick chimney, subtract the tension due to wind-pressure at the windward side from the compression due to weight: if there is a positive remainder showing a resultant compression the chimney will be stable; otherwise not, because masonry cannot withstand tension. Again, add the compression due to wind-pressure to the compression due to weight, to find the total compression at the leeward side: if the result is not greater than the safe load on masonry, the chimney is strong enough. The safe load may be taken at 8 tons per square foot. A steel chimney may be calculated for compression only, since steel is at least as strong in tension as in compression. The compression should be limited to 10,000 or 12,000 pounds per square inch on the net effective section between rivets. The assumption that rivet-holes are completely filled by the rivets, and that the total compressive strength is not reduced by cutting the rivet-holes, is erroneous. The shearing-resistance of the rivets in the ring-seam should be made equal to the compressive strength of the net section between rivets, in a manner analogous to that used for determining the proportions of boiler-joints.

A calculation like that just described must be made for the section of the chimney at the base, for each section where there is a change of thickness or of construction, and for any other section where there is reason to suspect weakness or instability.

The lining of a brick chimney is to be calculated for compression due to weight, at the base and at each section where there is a reduction of thickness. The lining of a steel chimney must be counted in when the stress due to weight is determined.

A separate calculation must be made for the stability of the foundation of a steel chimney. For this purpose find the total wind-pressure on the chimney and its moment about an axis in the plane of the base of the foundation. Find also the total weight of the entire chimney with its lining, and of the foundation: this will be a vertical force acting through the middle of the foundation. Divide the moment of the wind-pressure by the weight of the chimney and foundation: the result will be the distance from the middle of the foundation to the resultant force due to the combined action of wind-pressure and weight. If this resultant force is inside the middle third of the width of the foundation, the chimney will be stable.

This brief statement is intended to describe the method of calculating the stability of chimneys, and not to give full instructions. The design and calculation for an important chimney should be intrusted only to a competent engineer who has had experience in such work.

CHAPTER V.

POWER OF BOILERS.

The power of a boiler to make steam depends on the amount of heat generated in the furnace, and on the proportion of that heat which is transferred to the water in the boiler. The amount of heat generated depends on the size of the grate, the rate of combustion, and the quality of the coal burned. The transfer of heat to the water in the boiler depends on the amount and arrangement of the heating-surface. In practice it is found that each type of boiler has certain general proportions which give good results; any marked variation from these proportions is likely to give poor economy in the use of coal, or to lead to excessive expense in construction.

The capacity of a boiler is commonly stated in boiler horse-power; the economy of a boiler is given in the pounds of steam made per pound of coal. Neither method is entirely satisfactory, but definite meaning is attached to the terms by definitions and conventions.

Standard Fuel.—A comparison of the composition and of the total heats of the several kinds of coal given in the table on page 41 shows a great difference in the value of a pound of coal, depending on the district and mine from which it comes. In order to introduce some system into the comparison of the performance of boilers in different localities it has been proposed that some coal or coals be selected as standards, and that all boiler-tests intended for comparison be made with a standard coal. For this purpose it has been proposed to select Lehigh Valley anthracite, Pocahontas semi-bituminous, and Pittsburg bituminous coal. More definite comparisons would result if only one coal, such as Pocahontas, were selected. The objections are, first, that some trouble and expense might be incurred in localities where this coal is not regularly on the market; and second, that a furnace designed for a given coal may not give its best results with a different kind of coal. There is a notable difference between furnaces designed for anthracite coal and those designed for bituminous coal; for the rest it appears that the use of a standard coal is a question merely of expediency.

In making a boiler-test it is not difficult to make an approximate determination of the per cent of ash in the coal used. When that is done, the economy is usually stated in terms of water evaporated per pound of combustible, as well as per pound of coal. This gives somewhat more definiteness to the statement; but as no account is taken of the volatile matter in the coal, nor of the oxygen, this method also is indefinite.

Value of Coal.—The actual value of a coal for making steam can be determined only by accurate tests with a furnace and boiler which are adapted to develop and use the heat that the coal can produce. While many boiler-tests have been made, and there is a good deal of material that could be used for the purpose, there has not yet been made a satisfactory statement of the value of the fuel in common use.

It appears probable that the real value of a coal for making steam is proportional to the total heat of combustion. If this can be shown to be true, then coals should be sold on the basis of heat of combustion, just as steel is required to have certain physical properties which are determined by making proper tests.

Quality of Steam.—When the economy of a boiler is stated in terms of water evaporated per pound of coal, it is assumed that all the water is evaporated into dry saturated steam. But the steam which leaves the boiler may contain some water, or it may be superheated.

The moisture carried along by steam is called priming. The steam from a properly designed boiler, working within its capacity, seldom carries more than three per cent of priming. Under favorable circumstances steam from a boiler will be nearly dry.

If steam, after it passes away from the water in the boiler, passes over hot surfaces it will be superheated; that is, raised to a temperature higher than that of saturated steam at the same pressure. Vertical boilers with tubes through the steam-space give superheated steam. If steam is to be superheated to any considerable extent, it must be passed through a superheater, which usually is in the form of a coil of pipe subjected to hot gases outside. Now a boiler filled with water will keep the plates and tubes which form the heating-surface somewhere near the temperature of the water; such heating-surface will endure service for a long time. But superheating-surface is likely to be at a temperature about half-way between that of the steam inside and the gases outside, and is liable to be rapidly destroyed. For this reason superheated steam, though it gives a notable gain in economy when used in a steam-engine, is not looked upon with favor.

Steam-space.—The steam-space and the free surface for the disengagement of steam should be sufficient to provide for the efficient separation of the steam from the water. Cylindrical tubular boilers frequently have the steam-space equal to one third of the volume of the boiler-shell. Marine return-tube boilers usually have a smaller ratio of steam-space to water-space.

The more logical way appears to be to proportion the steam-space to the rate of steam-consumption by the engine. Thus the ratio of the volume of the steam-space of cylindrical boilers to that of the high-pressure cylinder of multiple-expansion engines varies from 50 : 1 to 140 : 1. The ratio of the steam-space of a simple locomotive-engine to the volume of the two cylinders is about $6\frac{1}{2}$: 1.

The capacity of the steam-space is sometimes equal to the volume of steam consumed by the engine in 20 seconds. It was found in some experiments with marine boilers having a working-pressure less than 50 pounds per square inch, that a considerable quantity of water was carried away by the steam when the steam-space was equal to the volume of steam consumed in 12 seconds, but that no water was carried into the cylinders when the steam-space was equal to the volume of steam used in 15 seconds and that no trouble from water was ever experienced when the steam-space was proportioned for 20 seconds.

All the preceding discussion refers to engines that run at a considerable speed of rotation—not less than 60 revolutions per minute. Engines that make but few revolutions per minute and take steam for only a portion of the stroke require a larger proportion of steam-space. As an example we may cite the walking-beam engines for paddle-steamers.

Equivalent Evaporation.—The heat required to evaporate a pound of water depends on the temperature of the feed-water, the pressure of the steam, and the per cent of priming.

For example, if water is supplied to a boiler at 140° F., and is evaporated under the pressure of 80 pounds by the gauge, with 2 per cent of priming, the heat required will be calculated as follows:

The heat of the liquid at 140° F., or the heat required to raise a pound of water from 32° F. to that temperature, is 108.2 B. T. U. The heat of the liquid at 94.7 pounds absolute, corresponding to 80 pounds by the gauge, is 293.8 B. T. U. Consequently the heat required to raise the feed-water up to the temperature in the boiler is

$$293.8 - 108.2 = 185.6 \text{ B. T. U.}$$

The heat of vaporization, or the heat required to change

a pound of water into steam, at 94.7 pounds absolute, is 886.9 B. T. U. But 2 per cent of water is found in the steam which comes from the boiler, leaving 98 per cent of steam; consequently the heat required is

$$0.98 \times 886.9 = 869.2 \text{ B. T. U.}$$

The total amount of heat is therefore

$$185.6 + 869.2 = 1054.8 \text{ B. T. U.}$$

Suppose that each pound of coal evaporates 9 pounds of water, then the heat per pound of coal tranferred to the boiler is

$$9 \times 1054.8 = 9493.2 \text{ B. T. U.}$$

Now the heat required to vaporize a pound of water at 212° F., under the pressure of the atmosphere, is 965.8 B. T. U. Dividing the thermal units per pound of coal by this quantity gives

$$9493.2 \div 965.8 = 9.83,$$

which is called the equivalent evaporation from and at 212 F.

This method of stating the economy of a boiler is equivalent to using a special thermal unit 965.8 as large as the thermal unit defined on page 44.

In making calculations involving quantities of wet steam it is convenient to consider the amount of steam present, rather than the per cent of priming. In the example just considered, there are 0.02 of water or priming, and 0.98 of steam. The part of a pound which is steam is represented by x.

If the heat of vaporization at the pressure of the steam in the boiler is represented by r, the heat of the liquid at that pressure by q, and the heat of the liquid at the temperature of the feed-water by q_0; and if, further, there are w pounds of water evaporated per pound of coal,—then the equivalent evaporation is

$$\frac{w(xr + q - q_0)}{965.8}.$$

The highest equivalent evaporation per pound of coal is about 12 pounds, and to accomplish this result about 80 per cent of the total heat of combustion must be transferred to the water in the boiler. The complete combustion of a pound of carbon develops 14,650 B. T. U.; if all this heat could be applied to vaporizing water at 212° F., then the amount of water evaporated would be

$$14{,}650 \div 965.8 = 15+ \text{ pounds.}$$

Few, if any, coals have a greater heat of combustion, consequently this figure may be considered to be the maximum equivalent evaporative power of coal.

Should any test appear to give a larger evaporative power, or even a power approaching this result, it may be concluded either that there is an error in the test, or that there is a large amount of priming in the steam. Some tests of early forms of water-tube boilers without proper provisions for separating water from the steam, appeared to give extraordinary results; which results were due to the presence of a large amount of priming in the steam. At that time the methods used for determining the amount of priming were difficult and uncertain, and were frequently omitted in making boiler-tests.

Boiler Horse-power.—It has always been the habit to rate and sell boilers by the horse-power. The custom appears to be due to Watt, and at that time the horse-power of a boiler agreed very well with the power of the engine with which it was associated. The traditional method of rating boilers, coming down from that time, was to consider a cubic foot, or $62\frac{1}{2}$ pounds, of water evaporated into steam, as equivalent to one boiler horse-power. This rating is now antiquated, and is seldom or never used.

It is now customary to consider 30 pounds of water evaporated per hour from a temperature of 100° F., under the pressure of 70 pounds by the gauge, as equivalent to one horse-power. This standard was recommended by a committee of the American Society of Mechanical Engineers.*

This standard is equivalent to the vaporization of 34.5 pounds of water per hour from and at 212° F.; it is frequently so quoted. It is also equivalent to 33,320 B. T. U. per hour.

Since the power from steam is developed in the engine, and since the economy in the use of steam depends on the engine only, and may vary widely with the type of engine, it appears illogical to assign horse-power to a boiler. The method appears to be justified by custom and convenience.

Rate of Combustion.—The rate of combustion is stated in pounds of coal burned per square foot of grate-surface per hour. It varies with the draught, the kind of coal, and the skill of the fireman.

In general a slow or moderate rate of combustion gives the best results, both because the combustion is more likely to be complete and because the heating-surface of the boiler can then take up a larger portion of the heat generated. A very slow rate of combustion may be uneconomical, because there is a large excess of air admitted through the grate, and because there is a larger proportionate loss of heat by radiation and conduction. It is claimed that forced draught may be made to give complete combustion with a small amount of air in excess, and that it should give better economy than slower combustion. It will be remembered that a small amount of carbon monoxide due to incomplete combustion will cause more loss than a large amount of air in excess.

Heating-surface.—All the area of the shell, flues, or

---

* Trans., vol. vi, 1881.

tubes of a boiler which is covered by water, and exposed to hot gases, is considered to be heating-surface. Any surface above the water-line and exposed to hot gases is counted as superheating-surface. The upper ends of tubes of vertical boilers are in this condition.

For a cylindrical tubular boiler the heating-surface includes all that part of the cylindrical shell which is below the supports at the side walls, the rear tube-plate up to the brick-arch which guides the gases into the tubes, and all the inside surface of the tubes. The front tube-plate is not counted as heating-surface.

For a vertical boiler like the Manning boiler (page 10) the heating-surface includes the sides and crown of the fire-box and all the inside surface of the tubes up to the water-line. Surface in the tubes above the water-line is superheating-surface. A certain 200-H.P. boiler of this type has 1380 square feet of heating-surface and 470 square feet of super-heating-surface.

The heating-surface of a locomotive-boiler consists of the sides and crown of the fire-box and the inside surface of the tubes.

The heating-surface of a Scotch boiler consists of the surface of the furnace-flues above the grate and beyond the bridge, the inside of the combustion-chamber, and the inside surface of the tubes.

The effective surface of any tube-plate is the surface remaining after the areas of the openings through the tubes is deducted.

Relative Value of Heating-surface.—A review of the kinds and conditions of heating-surface in various kinds of boilers, or even in a particular boiler, shows that the value of heating-surface varies widely. It does not appear possible to assign values to different kinds of heating-surface. We will note only that surfaces like the shell of a cylindrical boiler over the fire, like the inside of a fire-box, or like the flues of a marine boiler, which are exposed to direct radiation from the fire, are the most energetic in their action. Surfaces like combustion-chambers and tube-plates, against which the flames play, are nearly if not quite as good. The inside of small flues and tubes is less favorably situated, more especially as the flame is, under ordinary conditions, rapidly extinguished after it enters such a flue or tube. The length of the flame in small tubes depends on the draught, and with very strong forced draught may extend completely through tubes of some length.

The value of heating-surface in a tube rapidly decreases with the length. It is doubtful if there is any advantage in making the length of a horizontal tube more than fifty times the diameter. Tubes of vertical boilers should have twice that length.

Ordinary Proportions.—The following table gives the ordinary proportions of various types of boilers:

Type of Boiler.	Rate of Combustion.	Square Feet of Heating-surface per Foot of Grate.	Average Equivalent Evaporation.	Square Feet of Grate per Boiler H.P.	Heating-surface per Boiler H.P.
Lancashire	8 to 12	25 to 30	8 to 10	0.36	7.0
Cylindrical multitubular.	8 to 15	35 to 40	9 to 10.5	0.30	11.5
Vertical, Manning	10 to 20	*48 + 16	9 to 10.5	0.23	11.1
Locomotive	50 to 120 average 75	60 to 70	6.7 to 8.5	0.07	4.5
Locomotive type, stationary	8 to 15	40 to 45	9 to 10.5	0.30	12.6
Scotch marine	35 to 45	30	7 to 9	0.11	3.3
Water-tube with cylinder or drum	9 to 15	35 to 45	9 to 10.5	0.28	11.0
Water-tube with separator	15 to 67 average 20	30 to 40	7 to 9	0.22	7.3

* 48 heating-surface, 16 superheating-surface.

The higher rates of evaporative economy are associated with slower rates of combustion and with larger ratios of heating-surface to grate-surface.

No attempt is made to distinguish the kind or location of heating-surface; it must be understood that the ordinary arrangements and proportions for the several types are followed if this table is to be used in designing boilers. For example, it cannot be expected that heating-surface gained by lengthening the tubes of a locomotive-boiler will add materially to the efficiency of the boiler.

This table has been compiled from a large number of examples, and may be taken to represent current good practice. The last two columns giving the grate-surface and heating-surface have been computed on the basis of one horse-power for 34.5 pounds of water evaporated per hour from and at 212° F.

The tables on pages 140 to 145 give the principal dimensions of notable merchant steamships, of ships in the United States Navy, and of ships in the British Navy. The table on page 146 gives the particulars of boilers on the U. S. S. Brooklyn, the most recent and powerful American cruiser.

DIMENSIONS OF SOME NOTABLE ATLANTIC STEAMERS.

Steamer's Name.	Builders.	Date.	Moulded Dimensions.				Displacement.	Gross Tonnage.	Cylinders.			Boilers.			Indicated H.P.	Speed on Trial.
			Length.	Breadth.	Depth.	Draught.			Diam. of Cylinder, in inches	Stroke		Heating surface.	Grate area.	Working Press.		
Servia........	Messrs. Thomson...	1881	515	52' 0"	46' 6"	23' 3"	9900	7392	one 72 / two 100	78	27483	1014	90	10300	17
Alaska........	Fairfield Co......	1881	500	50 0	39 8	22 0	....	6932	one 68 / two 100	72	....	....	100	10900	18
City of Rome..	Barrow Co.......	1881	542.6	52 0	38 9	22 0	11230	8141	three 46 / three 86	72	29286	1398	90	11900	18.23
Aurania.......	Messrs. Thomson.	1882	470	57 0	39 0	....	7269	7259	one 68 / two 91	72	23284	1004	90	8300	17.5
Oregon........	Fairfield Co......	1883	500	54 0	40 0	23 0	9300	7375	one 70 / two 104	72	....	....	110	13300	18.3
America.......	Messrs. Thomson.	1884	432	51 0	38 0	23 0	9300	6500	one 63 / two 91	66	22750	882	95	7354	17.8
Umbria....... (Etruria like Umbria)	Fairfield Co......	1884	500	57 0	40 0	22 6	10500	7718	one 71 / two 105	72	38817	1606	110	14321	20.18
Lahn..........	Fairfield Co......	1887	448	48 10	36 6	23 0	7700	5661	two 32½ / one 68 / two 85	72	....	....	150	8900	17.78
Paris.......... (New York like Paris)	Messrs. Thomson.	1888	527.6	63 0	41 10	23 0	13000	10499	two 45 / two 71 / two 113	60	50265	1293	150	20600	21.8
Augusta Victoria.	Vulcan Co.; Stettin.	1889	460	55 6	39 0	22 9	9500	7661	two 41 / two 66½ / two 106½	63	36000	1120	150	14110	18.31
Columbia......	Messrs. Laird Bros.	1889	462.6	55 6	39 0	22 9	9500	7558	two 41 / two 66 / two 101	66	34916	1226	150	13680	19.15
Teutonic...... (Majestic like Teutonic)	Messrs. Harland & Wolff	1890	565	57 6	42 2	22 0	12000	9686	two 43 / two 68 / two 110	60	40072	1354	180	19500	21.0
Normannia....	Fairfield Co......	1890	500	57 0	38 0	22 0	10500	8716	two 40 / two 67 / two 106	66	46490	1452	160	16352	20.78
Spree......... (Havel like Spree)	Vulcan Co.; Stettin.	1890	463	51 6	37 0	22 6	8900	6963	one 38 / two 75 / two 100	72	....	....	165	13000	19.6
Fürst Bismarck..	Vulcan Co.......	1891	502.6	57 3	38 0	22 6	10200	8000	two 43½ / two 67 / two 106¼	63	47000	1450	157	16412	20.7
Campania..... (Lucania like Campania)	Fairfield Co......	1893	600	65 0	41 6	25 0	18000	12500	two 37 / two 79 / four 98	69	....	....	165	31050	23.10

POWER OF BOILERS.

Class of Ship.	Indicated H.P. Natural Draught.	Indicated H.P. Forced Draught.	Per Cent Increase Due to Forced Draught.	Inches of Water, Air-pressure. Natural Draught.	Inches of Water, Air-pressure. Forced Draught.	Inches of Water, Air-pressure. Increase	Weight. Total Machinery in Tons.	Weight. Machinery per H.P. Natural Draught in Pounds.	Weight of Boilers. Total Tons.	Weight of Boilers. Per I.H.P. Lbs.	Per Cent of Total Machinery. Engines.	Per Cent of Total Machinery. Boilers.	Steam-pressure.	Grate-surface.	Heating-surface.	H/G
Admiral	8227	11271	37	...	1.7	1.7	1209	329	661.6	180	45	55	90	700	20587	26
Trafalgar	8813	12465	41½	.05	1.9	1.4	1015	258	519.4	132	49	51	135	629	18930	30
Hero	4351	6162	41½	...	2.0	1.0	847	420	483.9	250	49	60	60	507	14377	28.7
Royal Sovereign	9430	11500	22	.03	.8	.5	1163	277	595.3	142	49	51	155	731	20348	27.8
Centurion	9882	13188	33½	.09	.6	.5	737	207	364.5	151	49	51	155	774	22217	28.8
Retted cruisers	5734	8835	35	...	1.6	1.5	1304	308	413.6	170	45	55	130	514	15502	30
Blake	14724	21411	45	.03	2.0	1.7	1015	155	770.6	118	24	76	155	1003	28132	29
First-class cruisers	10517	12851	22	.02	1.1	.5	1161	248	646.6	138	45	55	155	812	24908	30.7
Vulcan	8107	12062	48	.04	1.8	1.8	988	272	495.5	137	50	50	155	564	18861	28
Mersey	4217	6151	46	...	1.8	.6	559	300	311.0	170	45	55	110	388	11565	29.5
Second-class cruisers	7437	9274	85	.03	.9	.7	740	223	434.4	132	41	59	155	575	15621	27
Scout	2201	3365	53	...	1.7	.9	296	302	179.2	184	39	61	120	213	6400	30
Archer	2452	3754	52	...	2.0	1.7	353	323	295.7	188	42	58	130	222	6836	30.8
Medea	6065	9268	53	.05	1.5	.9	627	232	366.3	136	42	58	155	526	13264	35
Barracouta	1951	2997	49	.06	2.0	...	246	283	146.6	168	40	60	155	193	4589	23.7
Australian cruisers	4615	...	...	.07	1.5	...	515	251	285.8	134	45	55	155	365	9987	27.2
Bellona	3592	4561	27	1.1	2.1	1.0	274	172	139.6	87	49	51	155	244	7088	29
Early second-class cruisers	5016	7469	49	.2	1.4	1.2	539	240	313.6	141	42	58	155	410	11025	27
Rattlesnake	...	2696	...	...	2.6	2.6	120	100	75.5	121	37	63	150	172	4426	36.5
Sharpshooter	2672	3717	39	.9	2.4	1.5	171	147	101.4	85	41	59	150	...	5470	31.8
Sharp-shooter with Belleville boilers	2620	3738	23	.1	.1	.1	197	170	124.3	107	37	63	...	269	7695	28.5
Hebe	2631	3665	39	.8	2.2	1.4	211	180	124.2	106	41	59	155	171	6204	26.2
Speedy	3046	4203	55	.5	1.7	1.2	212	156	107.2	79	49	51	210	204	17700	86

Name of vessel	Yorktown.	Charleston.	Baltimore.	San Francisco.
Length between perpendiculars	228' 0"	300' 0"	315' 0"	310' 0"
Beam	36' 0"	46' 0"	48' 6"	49' 1½"
Mean draught	14' 2¼	17' 10½"	19' 10½"	18' 9"
Displacement, tons	1710	3580	4500	4088
Type of engine	Two, horizontal, triple-expansion	Two, horizontal, two-cylinder compound	Two, horizontal, triple-expansion	Two, horizontal, triple-expansion
Diameter in inches:				
High-pressure	22	44	42	42
Int.-pressure	31	....	60	60
Low-pressure	50	85	94	94
Stroke in inches	30	36	42	36
Number and types of boilers	(4) low cylindrical	(6) low cylindrical	(4) double-ended (2) single-ended	(4) double-ended (1) single-ended
Number and diameter of furnaces in each	(3) 37"	(3) {42½" / 44½"}	D. E. (8) 36" S. E. (1) 32"	D. E. (6) 42" S. E. (1) 39"
Length and diameter of boilers	17' 9" × 9' 9"	19' 3" × (3) 11' 0" (3) 11' 6"	D. E. 17' 8" × 14' 7" S. E. 6' 2" × 7' 2"	D. E. 19' 2" × 14' 3" S. E. 8' × 8'
Total grate-surface used on trial sq. ft.	220	Main 422.2 Aux. 15	676	567.6
Total heating-surface used on trial, sq. ft.	8092	Main 15147 Aux. 430	17175	20134
Steam-pressure in boilers (gauge)	150	91	....	135
Air-pressure in fire-rooms or ash-pits in inches of water	1.25	For. 1.6 Aft. 2.0	2.09	2.00
Revs. of main engine per minute	Star. 157.98 Port. 155.94	Star. 115.35 Port. 113.95	Star. 116.42 Port. 115.08	Star. 125.80 Port. 123.83
Vacuum in condenser, inches of mercury	Star. 24.84 Port. 25.02	Star. 26.2 Port. 26.1	Star. 24.5 Port. 24.3	Star. 25.7 Port. 26.1
Indicated horse-power total, mean, of machinery	3398.25	6666.16	10064.42	9912.93
Speed per hour in knots	16.14	18.19	19.84	19.52
Aggregate I. H. P. all main engines	3205	6316	9831	9581
I. H. P. per square foot of grate, based on mean I. H. P.	15.42	15.28	14.89	17.46
Heating-surface per I. H. P., based on mean I. H. P.	2.385	2.337	1.710	2.03
Condensing-surface per I. H. P., based on mean I. H. P.	1.403	2.069	1.26	1.46
Mean I. H. P. per ton of machinery	10.6	9.6	10.53	10.84
Coal per hour per square foot of grate-surface, lbs.	....	....	44.9, estimated	....
Coal per hour per I. H. P. (pounds)	....	....	3.15	....

POWER OF BOILERS

NEWARK.	BENNINGTON.	MONTEREY.	DETROIT.
310′ 10″	228′ 0″	256′ 0″	257′ 0″
49′ 2″	36′ 0″	59′ 0¼″	37′ 0″
18′ 3½″	14′ 0″	14′ 5″	14′ 5½″
3980	1706	4000	2068
Two, horizontal, triple-expansion	Two, horizontal, triple-expansion	Two, vertical, triple-expansion	Two, vertical, triple-expansion
34	22	27	26.5
52	31	41	39
76	50	64	63
40	30	30	26
(4) double-ended (1) single-ended	(4) low cylindrical	(2) cylindrical, S. E. (4) Ward coil	(3) double-ended (2) single-ended
D. E. (6) 43″ S. E. (2) 32″	(3) 37″	Cylind. (2) 42″; Ward, annular grate, 10′ 2″ ext. dia., 3′ 0½″ int. dia	D. E. (4) 42″ S. E. (2) 42″
D. E. 19′ 5″ × 13′ 6″ S. E. 7′ 11″ × 8′ 10½″	17′ 9″ × 9′ 9″	Cylind. 10′ 7″ × 11′ 2″; Ward, 12′ 4″ height, 10′ 8″ diameter	(2) D. E. 18′ 1″ × 11′ 8″ (1) D. E. 18′ 3½″ × 11′ 8″ (2) S. E. 9′ 0½″ × 11′ 8″
540	220	383	368
16737	8210	4785	10978
162	166	155	171
2.25	2.45	3.20	0.8
Star. Port. 127.34 126.66	Star. Port. 150.82 151.03	Star. Port. 162.86 161.17	Star. Port. 170.1 170 1
Star. Port. 26.0 25.7	Star. Port. 24.0 23.6	Star. Port. 26.6 26.2	Star. Port. 25.1 25.5
8868.57	3436.09	5243.92	5227.14
19.00	17.05	13.60	18.71
8582	3333	4987	5154
16.42	15.62	13 68	14.21
1.89	2.39	2.81	2.10
1.45	1.43	1.46	1.46
12.08	10.07	13.00	11.29
39.98, estimated	40.56	....	....
2.43	2.60		

Name of vessel	NEW YORK.	MACHIAS.	MASSACHUSETTS.
Length between perpendiculars	380' 0"	190' 0"	348' 0"
Beam	64' 3"	32' 0"	69' 3"
Mean draught	23' 10¼"	12' 0¼"	24' 1"
Displacement, tons	8480	1067.5	10265
Type of engine	Vertical, triple-expansion, two engines on each shaft	Two, vertical, triple-expansion	Two, vertical, triple-expansion
Diameter in inches:			
High-pressure	32	15.75	34½
Int.-pressure	47	22.5	48
Low-pressure	72	35	75
Stroke in inches	42	24	42
Number and type of boilers	(6) double-ended (2) single-ended	(2) marine locomotive	(4) double-ended (2) single-ended
Number and diameter of furnaces in each	D. E. (8) 39" S. E. (2) 33"	(2) { 6' 9" long { 44" and 54" wide	D.E.(8) 40" out., 36" in. S.E. (2) 37" out.,33" in.
Length and diameter of boilers	D. E. 18' 0" × 15' 0" S. E. 8' 6" × 10' 0"	18' 6" × 9' 3"	D. E. (4) 18' × 15' (2) 8' 6" × 10' .8⅜"
Total grate-surface used on trial, sq. ft.	1052	120	616
Total heating-surface used on trial, sq. ft.	32958	4590	19194.6
Steam-pressure in boilers (gauge)	176	163	163.0
Air-pressure in fire-rooms or ash-pits in inches of water	Main fire-room 2.02 Aux. fire-room .7	0.47	.9935
Revs. of main engine per minute	Star. 134.6 Port. 135.00	Star. 218.55 Port. 214.27	Star. 132.3 Port. 133.06
Vacuum in condenser, inches of mercury	Star. 25.3 Port. 25.5	25.7	Star. 25.5 Port. 25.20
Indicated H.-P., total, mean, of machinery	17401.42	1873.41	10402.66
Speed per hour in knots	21.00	15.46	16.208
Aggregate I. H. P. all main engines	16947	1794	10127
I. H. P. per square foot of grate, based on mean I. H. P.	16.54	15.61	16.9
Heating-surface per I. H. P., based on mean I. H. P., sq. ft.	1.89	2.45	1.84
Condensing-surface per I. H. P., based on mean I. H. P., sq. ft.	1.35	1.20	1.298
Mean I. H. P. per ton of machinery	11.37	11.98	
Coal per hour per square foot of grate-surface, lbs.		38.08	
Coal per hour per I. H. P. (pounds)		2.44	

POWER OF BOILERS.

Brooklyn.	Olympia.	Columbia.	Minneapolis.
400' 6"	340' 0"	411' 7½"	411' 7½"
64' 8½"	53'0¼" at load water-line	58' 2½"	58' 2½"
21' 10¼"	20' 8½"	22' 5"	22' 6"
8150	5586	7350	7387.5
Four, vertical, triple-expansion	Two, vertical, triple-expansion	Three, vertical, triple-expansion	Three, vertical, triple-expansion
31½	42	42	42
46½	59	59	59
72	92	92	92
42	42	42	42
(5) double-ended (2) single-ended	(4) double-ended (2) single-ended	(8) double-ended (2) single-ended	(8) double-ended (2) single-ended
D.E.(8) 44" out., 40" in. S.E.(4) 42½" out., 40" in. Greatest outside and least inside diameters D. E. (4) 18' × 16' 3" (1) 19' 11½" × 16' 3" S. E. (2) 9' 5" × 16' 3"	D.E.(8) 43' out., 39" in. S.E.(4) 43" out., 39" in. D. E. (4) 21' 3" × 15' 3" S. E.(2) 10' 11½" × 15' 3"	D.E.(8) 43" out., 39" in. S.E. (2) 43" out., 39" in. D. E. (6) 18' × 15' 9" D. E. (2) 18' × 15' 3" S. E. (2) 8' 6' × 10' 1⅞"	D.E.(8) 43" out., 40" in. (8) 42" out., 39" in. S.E. (2) 37" out.,33" in. D. E. (6) 20' × 15' 9" (2) 18' × 15' 3" S. E. (2) 8' 6" × 10' 1⅞"
1016.2	824	1408 in first half of trial 1344 in last half of trial	1456.2
33432	28298.6	45221	48194
158.3	166.53	147.2	150.3
2.26	2.04	.73	
Star. 136.2 Port. 136.9	Star. 139.98 Port. 138.53	Star. 134.0 Centre. 127.68 Port. 132.9	Star. 131.95 Centre. 132.19 Port. 133.1
Star. 25.5 Port. 24.9	Star. 24.94 Port. 25.59	Star. 25.1 Centre. 25 Port. 25.6	Star. 25.11 Centre. 25.31 Port. 24.76
18769.62	17313.08	18509.24	20862.3
21.912	21.686	22.8	23.073
18248	16850	17991	
18.47	21.011	13.46	14.29
.56	1.635	2.39	2.315
.81	1.152	Star. 1.43 Centre. 1.63 Port. 1.704	1.431
12.16		9.57	10.01
	44.7		
	2.19		

BOILERS OF THE U. S. STEAMSHIP BROOKLYN.

Five Double-ended Boilers (160 *lbs. pressure*).

Length in feet and inches	(4) 18′ 0″, (1) 19′ 11½″
Outside diameter	16′ 3″
Thickness—Shell and top heads	1 25/64″
Heads, bottom	¾″
Tube-sheets	⅞″
Furnaces	9/16″
Combustion-chambers (4) in each boiler, thickness	9/16″
Depth at top	23″
Width	84″
Furnaces—Greatest internal diameter	3′ 8″
Least internal diameter	3′ 4″
Length of grate	(4) 6′ 4″, (1) 6′ 6″
Number in each boiler, corrugated	8
Tubes—Outside diameter	2¼″
Length between tube-sheets	(4) 6′ 6″, (1) 7′ 5¾″
Number of ordinary (B.W.G. No. 12)	904
Number of stay (B.W.G. No. 6)	296
Spaced vertically a distance of	3¼″
Spaced horizontally a distance of	3½″
Diameter of rivets in shell-sheets	1 5/16″
Of screw-stays between backs of combustion-chambers.	1 17/64″
Number and diameter of through upper braces	(10) of 2¾″
Through lower braces	(2) of 2″
Braces around each lower manhole	(3) of 1¾″
Braces from head to back tube-sheet	(2) of 2⅛″, (6) of 2″
Heating-surface—Tubes (4), smaller boilers each 4594 sq. ft., large boiler 5286	
Furnace and combustion-chambers, each of smaller 842, large boiler 890	
Total each of smaller boilers....5436 sq. ft., large boiler 6176 sq. ft.	
Grate-surface, each of smaller boilers 168.6 sq. ft., large boiler 173.2 sq. ft.	
Area through the tubes	173.2 "
Over bridge-walls	25.85 "
Smoke-pipes (3), diameter	7.25 ft.
Area of all	123.84 sq. ft.
Height above lowest grates	100 ft.
Diameter of boiler main stop-valves	10 inches.
Auxiliary stop-valves	7 "

BOILERS OF THE U. S. STEAMSHIP BROOKLYN.—(Continued.)

Two Single-ended Boilers (160 lbs. pressure).

Length...	9' 5"
Diameter outside...	16' 3"
Thickness of shell and top heads...........................	1$\frac{25}{32}$"
Heads, bottom...	$\frac{3}{4}$"
Tube-sheets...	$\frac{7}{8}$"
Furnaces..	$\frac{9}{16}$"
Combustion-chambers, each boiler, number..................	2
Thickness...	$\frac{9}{16}$"
Depth at top..	23"
Width at top..	85$\frac{1}{2}$"
Furnaces—Greatest internal diameter.......................	42$\frac{7}{8}$"
Least internal diameter...............................	40"
Length of grate.......................................	6' 4"
Number in each boiler (corrugated)....................	4
Tubes—Outside diameter.....................................	2$\frac{1}{4}$"
Length between tube-sheets............................	6' 4$\frac{3}{4}$"
Number of ordinary (B.W.G. No. 12)....................	466
Number of stay (B.W.G. No. 6).........................	152
Spaced vertically.....................................	3$\frac{1}{4}$"
Spaced horizontally...................................	3$\frac{1}{2}$"
Diameter of rivets in shell-sheets.........................	1$\frac{15}{16}$"
Screw-stays...	1$\frac{3}{8}$"
Pitch of screw-stays: horizontally 7$\frac{3}{4}$", vertically..........	7"
Number and diameter of through upper braces..............	(10) of 2$\frac{3}{4}$"
Lower braces..	(3) of 1$\frac{3}{4}$"
Number and diameter of braces from head to back tube-sheet...	(6) of 2"
Heating-surface—Tube.......................................	2335 sq. ft.
Furnaces and combustion-chambers......................	421 "
Total square feet......................................	2756 "
Grate-surface..	84.3 "
Area through tubes...	13.3 "
Over bridge-walls.....................................	6.4 "
Diameter of boiler stop-valves.............................	7$\frac{1}{2}$"

Total for all Boilers.

Heating-surface—Tube.......................................	28,332 sq. ft.
Furnaces and combustion-chambers......................	5,100 "
Total...	33,432 "
Grate-surface..	1016.2 "
Area through tubes...	155.86 "
Ratio total heating-surface to grate-surface...............	32.9 to 1
Ratio total area through tubes to grate-surface............	.153 to 1
Ratio total area through tubes to total area smoke-pipes...	1.26 to 1

CHAPTER VI.

STAYING AND OTHER DETAILS.

ALL plates of a boiler that are not cylindrical or hemispherical require staying to keep them in shape. For example, the cylindrical shell of a cylindrical tubular boiler does not require staying, because the internal pressure tends to keep it cylindrical. On the other hand, the pressure tends to bulge out the flat ends, and they must be held in place against that pressure.

Many different methods of staying will be found in the different types of boilers seen in practice, and there are frequently several ways of staying the same kind of a surface. A few methods will be described in a general way. The placing of stays and arrangement of details is an important part of the design of a boiler, and must be worked out for each special design.

Cylindrical Tubular Boiler.—The parts of the tube-sheets at the ends of a cylindrical tubular boiler, through which the tubes pass, are sufficiently stayed by the tubes themselves. The flat ends above the tubes require staying. Also, if there is a manhole at the bottom of the front end, the space thus left unsupported requires staying, and there is a corresponding space at the back end.

An elaborate set of tests was made by Messrs. Yarrow[*] and Co., to determine the holding-power of tubes expanded into a tube-sheet. It was found that from 15,000 to 22,000 pounds

---

[*] London *Engineering*, Jan. 6, 1893.

were required to pull out a two-inch steel tube; in some cases the tube gave way by tension inside the head into which it was expanded.

The staying of a flat surface consists essentially in holding it against pressure at a series of isolated points, which are arranged in a regular or symmetrical pattern. A simple case of staying is found in the side sheets of a locomotive fire-box. Here the stays, which are arranged in horizontal and vertical rows, are screwed and riveted. If possible, the pitch or distance between the supported points should be the same, but this is possible only when arranged in rows as just mentioned. The allowable pitch depends on the thickness of the plate. For cylindrical tubular boilers the pitch of the supported points of the flat ends above the tubes is 3.5 to 5 inches. The outside fibre-stress in the plate stayed may be from 6000 to 8000 pounds per square inch; the calculation of this stress involves a knowledge of the theory of elasticity, and will be referred to later.

It is not advisable, for this type of boiler, to assign a separate stay to each supported point of the flat surface under discussion, consequently the points are grouped, each point of the group being riveted to some support inside the boiler, and then the supports are held by proper stays.

A good method of staying the flat end of a cylindrical boiler is shown by Plate I, and also, with some further details, by Fig. 44. There are two 6-inch channel-bars of proper length, that are riveted to the flat head. The rivets tie the plate to the channel-bars and thus support the plate at isolated points. The channel-bars in their turn are supported by stays that run directly through the boiler and have nuts and washers at each end. The channel-bars act as beams, and must be capable of carrying the load due to the pull on the rivets, and the through-stays must carry the loads on the beams. A short piece of angle-iron is riveted to the upper side of the upper channel-bar; it carries five additional rivets in the flat head, and adds an additional load to the upper channel-bar. The points where the through-stays pass through the head are supported directly by the stays through the washers and nuts.

The lower channel-bar is a continuous girder with four spans and five supports. The stays form three supports and the other two are at the inner edge of the flange of the head. The upper channel-bar is a girder with three spans and four

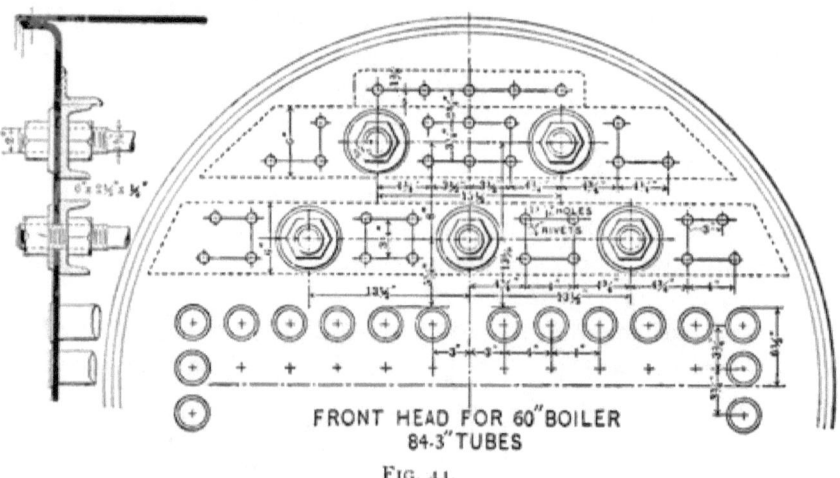

Fig. 44.

supports. The calculation of the stresses in the channel-bars is somewhat unsatisfactory, largely because the support at the flange of the head is uncertain; and this support must be left with some flexibility, and consequently with some uncertainty, as too great rigidity leads to grooving.

In arranging such a staying, we begin by determining the allowable pitch of the points supported by the rivets, assuming them to be in equidistant horizontal and vertical rows. This allowable pitch must not be exceeded, but the pitch may be made less either horizontally or vertically, or in both ways.

A space of at least three inches is left between the top

row of tubes and the lowest row of rivets, and a similar space is left at the sides. This is to avoid grooving.

The two upper through-stays are fifteen and a half inches apart on centres. They must be wide enough apart to allow a man to pass through.

The stay-rods are upset at the ends so that the diameter at the bottom of the threads is greater than the diameter of the body of rod. The washer outside the plate may be made of copper, in which case it is made cup-shaped so as to bear on a narrow ring, and is made tight by calking; or the washer is made of iron, and is bedded in red lead to make a joint. Sometimes cap-nuts are used outside the head to prevent the escape of steam that may leak around the screw-threads. Long stay-rods are sometimes supported at the middle.

A method of staying otherwise similar to that just described, uses two angle-irons in place of a channel-bar. A washer of special form is used to give a proper bearing, for the inner nut on the through-stay, against the angle-irons.

Fig. 45 shows a different method of staying for cylindrical boilers. The left half of the figure represents the end elevation, and the right half represents a section through the manhole; this is a common method for boiler drawings. The supported points are arranged in sets of four, and are tied to forgings known as crowfeet. Fig. 46 represents such a crowfoot with four rivets, known as a double crowfoot; a single crowfoot with only two rivets is shown by Fig. 47. When crowfeet are used they may be arranged in various patterns, in the example given there is a horizontal row of five double crowfeet just above the tubes, and three other double crowfeet are arranged in a circular arc. At the ends of the arc there are two braces like Fig. 48, which are used instead of single crowfeet. From each crowfoot a diagonal stay is carried to the boiler-shell. These stays are flattened at the farther end and bent to lie against the side of the shell, to which they are riveted with two or three rivets; the arrangement is similar to that of the right-hand end of the brace

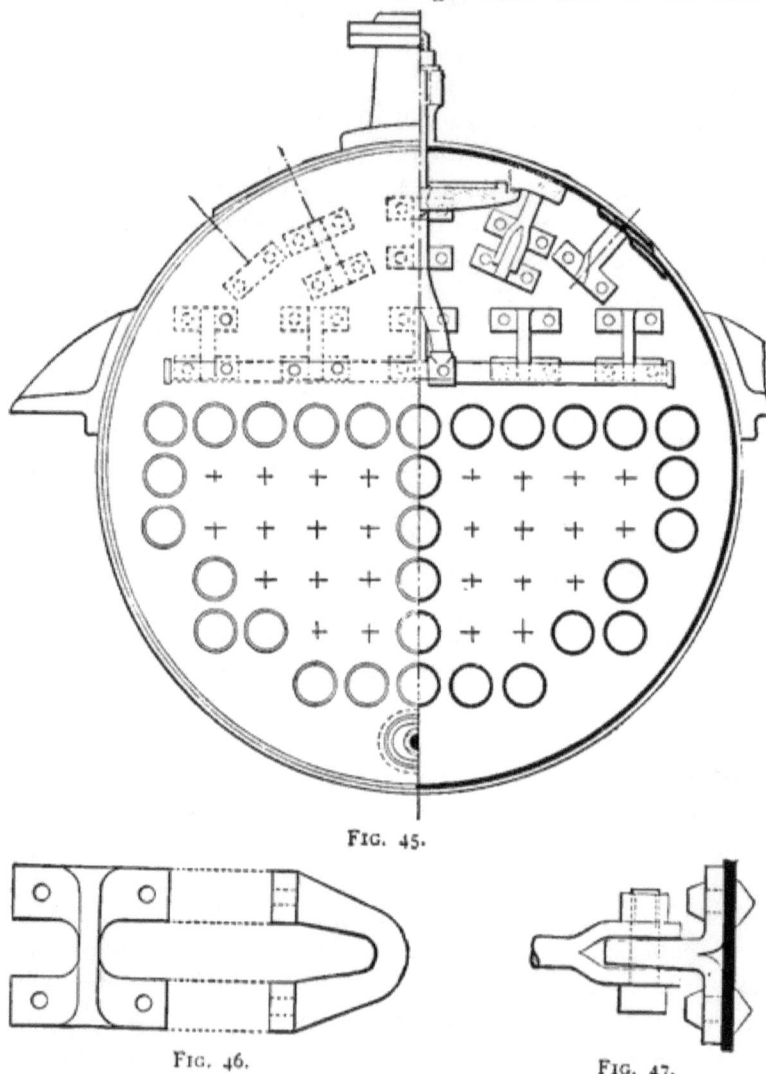

FIG. 45.

FIG. 46.

FIG. 47.

shown by Fig. 48. At the crowfoot the stay has a forked head through which a bolt passes under the arch of the double crowfoot. A nut holds the bolt in place and prevents the head of the stay from spreading.

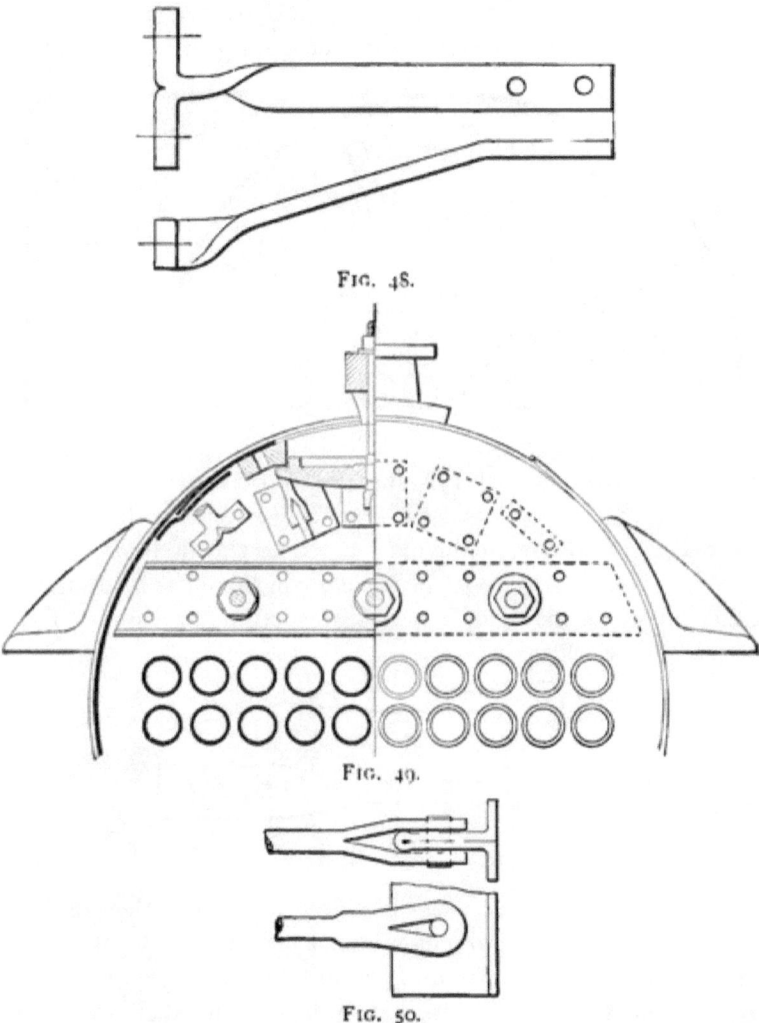

FIG. 48.

FIG. 49.

FIG. 50.

A combination of channel-bar and crowfeet is shown by Fig. 49. The double crowfeet are represented as made of boiler-plate, bent up as shown by Fig. 50.

A method of staying, suitable only for boilers which work under low steam-pressure, is shown by Fig. 51. Short pieces of T iron, arranged radially, are riveted to the head. Each T iron is supported from the cylindrical shell by two

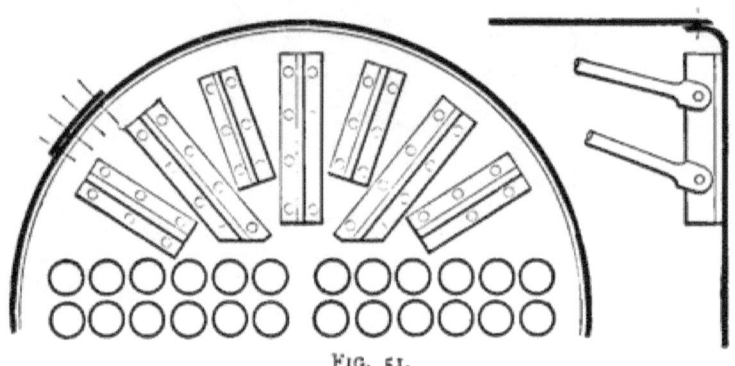

Fig. 51.

diagonal stays; one of the stays is represented by Fig. 52. One end of the stay is split, and is pinned to the T iron; the other end is flattened, and riveted to the shell.

The shell of a cylindrical boiler, whether it is a tubular or a flue boiler, is made of a series of sections or rings. Each

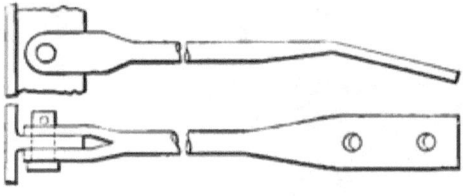

Fig. 52.

ring is made of one or two plates riveted along the edge, or longitudinal seam. This seam has at least two rows of rivets; more complicated joints are commonly used to give more strength to the seam. Alternate rings of the shell are made smaller so that they may be slipped inside the rings at each of their ends. The seams joining adjacent rings are commonly single-riveted. The longitudinal seams are kept above the middle of the boiler, so that they are not exposed to the fire. The first ring at the front end is always an outside ring, so that the first ring-seam has the outside edge pointing away from the fire; there is consequently less liability of injury to the seam from the flames that pass under the boiler toward the back end.

Vertical Boilers.—The tube-sheets of a vertical boiler, as is evident from inspection of Figs. 5 and 6, are usually stayed sufficiently by the tubes. Should the upper tube-sheet be much larger than the crown of the fire-box, it may need staying between the tubes and the shell. Stays like Fig. 48 may be used for this purpose.

The circular fire-box of a vertical boiler is subjected to external pressure, and is prevented from collapsing under that pressure by tying it to the outer shell by screwed stay-bolts, which are put in and set like the stay-bolts for a locomotive-boiler.

Locomotive-boiler.—The parts of a locomotive-boiler that require staying are the fire-box and the flat ends. The tube-sheets are sufficiently stayed by the tubes, but there is a part of the tube-sheet at the smoke-box end and a part of the flat end above the fire-box which requires support. The problems here resemble those met in staying the tube-sheets of a horizontal cylindrical boiler, and similar methods are used. Thus in Plate II there are shown eight through-stays, each $1\frac{1}{2}$ of an inch in diameter. These stays pass through the girder staying of the crown-sheet, and have a simple nut and washer outside the end-plates of the boiler. At the smoke-box end, as shown by Figs. 1 and 3, Plate II, there are two diagonal stays taking hold of single crowfeet and running to the middle of the barrel. At the fire-box end there are four crowfeet or short angle-irons, made by bending up boiler plate; two are shown by the right-hand elevation of Fig. 2 on Plate II. The outer crowfeet have five rivets, and the others six. From the outer crowfeet diagonal stays run to the shell at the ring just in front of the fire-box. From the inner crowfeet stays run to the middle ring of the boiler. There are also two stays like Fig. 48, which run to the shell above the fire-box. Finally, there is a crowfoot and stay at the middle of the row of eight through-stays, this stay fastening to the two end crown-bars.

Below the tubes there is a place in the fire-box tube-sheet which requires support. This is given by three braces like Fig. 48, as shown by Figs. 1 and 2, Plate II. The shell of the boiler, shown by this plate, is higher over the fire-box than it is at the barrel, and a ring of peculiar shape is required to join the two parts together. This ring is cylindrical below and conical on top; at the sides there are flattened spaces which require stiffening to prevent them from springing, and thus start grooving of the plates. For this purpose there are three T irons riveted to the shell at the flattened place mentioned, as shown by Fig. 1, Plate II. The upper ends of the T irons on opposite sides of the boiler are tied together by transverse stays above the tubes.

Coming now to the fire-box of the boiler represented by Plate II, we find that at the front, rear, and sides it is tied to the external shell by screwed stay-bolts set in equidistant horizontal and vertical rows. The holes for these stay-bolts are punched or, better, drilled before the fire-box is in place. After it is in place and riveted to the foundation-ring a long tap is run through both plates, the fire-box plate and the shell, and thus a continuous thread is cut in the plates. A steel bolt is now screwed through the plates, cut to the proper length, and riveted cold at each end. Owing to the screw-thread on the bolts, this riveting is imperfect, and likely to develop cracks at the edge. The thread should be removed from the middle of the bolts, as they are then less liable to crack under the peculiar strains set up by the unequal expansion of the fire-box and outside shell.

The stay-bolts are very likely to be cracked or broken on account of the expansion of the fire-box; to detect such a failure of a bolt, or to show when excessive corrosion has taken place, the stay-bolts are often drilled from the outer end nearly through to the inner end.   In case of failure steam will blow out of the defective stay; serious injury has often been avoided by this method.

The crown-sheet of the fire-box is exposed to intense heat, and is covered with only a few inches of water. The problem of properly staying this flat crown-sheet without interfering with the supply of water to it is one of the most difficult problems in locomotive-boiler construction.   Figs. 1 and 2, Plate II, show the method of staying a crown-sheet with a system of girder-stays. Above the crown-sheet there are fourteen double girders, which are supported at the ends by castings of special form, shown by Figs. 2 and 6; the castings rest on the edges of the side sheets and on the flange of the crown-sheet. In addition the girder-stays are slung to the shell by sling-stays. At intervals of four and a half inches the crown-sheet is supported from the girders by bolts, having each a head inside the fire-box, as shown by Fig. 5, and a nut at the top bearing on a plate above the girder. These plates are turned down at the ends to keep the two halves of the girder from spreading. There is a copper washer under the head of each bolt, inside the fire-box, to make a joint. Between the girder and the crown-sheet each bolt has a conical washer or thimble to maintain the proper distance between the girder and crown-sheet. This thimble is wide above to bear on the girder, and small below to avoid interfering with the flow of water to the crown-sheet, and also so as to cover as little surface as possible on account of the danger of burning the crown-sheet wherever the metal is thickened.   The whole system of girders is tied together, and the girder nearest the fire-door is tied to the outside shell to keep the girders from tipping over. It is evident that such a system of staying is heavy, cumbersome, and complicated. It is also uncertain in its action, since the equalization of stresses depends on a nice adjustment of the form bers of the system, which adjustment is liable to derangement from expansion of the fire-box. The girders or crown-bars are sometimes run lengthwise instead of transversely, but as the fire-box is longer than wide such an arrangement is inferior.

To avoid the cumbersome method of staying the crown-sheet, which has just been described, the fire-box end of the boiler has been made flat on top, as shown by Fig. 53. The

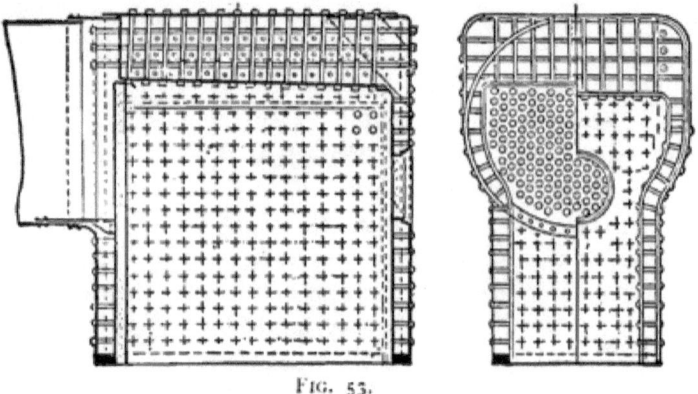

FIG. 53.

crown-sheet can now be stayed to the outside shell by through-stays having nuts and copper washers at each end. The flat side sheets of the shell above the fire-box are also stayed by through-stays, and there are also three longitudinal through-stays in the corners of the shell over the fire-box where it protrudes beyond the barrel. This forms what is known as the Belpair fire-box, from the inventor.

Fig. 54 shows an attempt to combine the use of through-stays, like those of the Belpair fire-box, with a cylindrical top above the crown-sheet. It will be noted that the stays are neither perpendicular to the crown-sheet nor radial when they pierce the shell, and they must be subjected to an awkward side pull at both places.

The locomotive-boiler represented by Plate III has a Belpair fire-box, and shows in addition some peculiarities of staying. Thus the flat end-plate above the fire-box has four T irons riveted to it. Each T iron is tied to the shell by two diagonal stays. Each stay has the usual double head at the T iron; the other end lies between, and is pinned to the flanges of pieces of plate that are riveted to the shell of the boiler. This arrangement is shown by the transverse

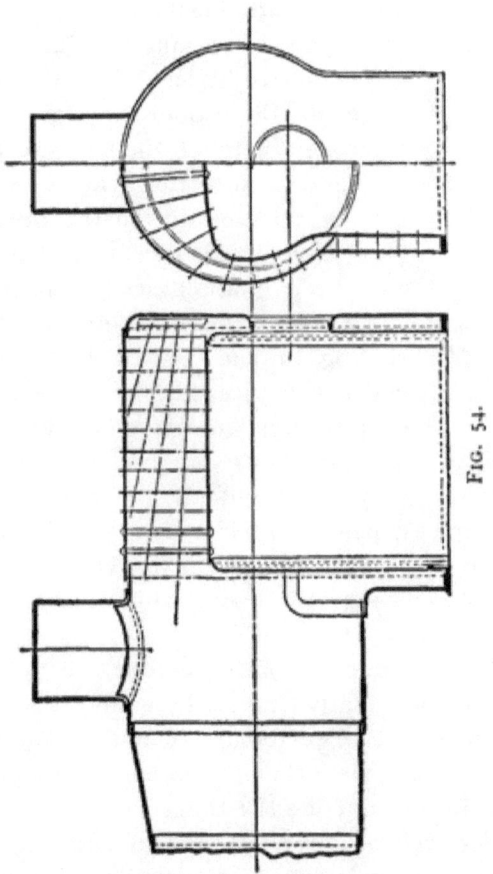

FIG. 54.

and longitudinal sections through the fire-box. It will be noticed that the lower diagonal stays from the end-plate interfere with four transverse through-stays. These stays are cut off and carry short vertical yokes, which are connected by two smaller rods, one above and one below the diagonal stays.

The rings forming the barrel of the locomotive are made progressively smaller from the fire-box to the smoke-box; the slight taper toward the front end of the locomotive is found convenient in the design of the machine.

Fig. 55 shows two ways of making the furnace-mouth of a locomotive-boiler. In one way the end-plate of the boiler-shell and the corresponding plate of the fire-box are flanged in the same direction, and are riveted outside of the boiler. In the other case the two plates are flanged into the water-space and the overlapping edges are riveted.

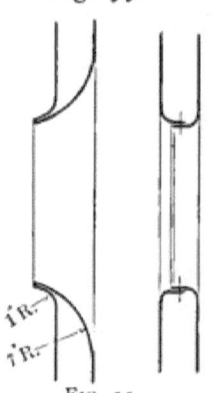

FIG. 55.

Marine Boiler.—The parts requiring staying in the Scotch boiler are the flat ends, the furnaces, and the combustion-chambers. The flat ends above the tubes are stayed by through-stays with nuts inside and with washers and nuts outside the plate. The boiler shown by Fig. 13, page 17, has two rows of through-stays —four in the upper and six in the lower row; two of the upper row pass through the fitting which carries the steam-nozzle.

It is found in practice that the tube-sheets of a marine boiler are not sufficiently stayed by plain tubes expanded into the sheets. It is customary to make a portion of the tubes thicker than the others, and to provide these thick tubes with thin nuts outside the tube-plates, so that they may act more effectively as stays. The thick tubes in Fig. 13 are indicated by heavy circles. Sometimes every other tube of each second row is made a thick tube; that is, something more than one fourth of the tubes are stay-tubes. Usually the number is fewer than this.

Below the tubes the front plate is supported in part by the furnace-flues, and in part by through-stays running to the combustion-chamber. There are two such stays above the furnaces and three below the furnaces in the middle of Fig. 13, each $1\frac{3}{4}$ of an inch in diameter. There are also two stays $2\frac{1}{8}$ inches in diameter, one at each side and above the furnaces. These last stays have one point of attachment to the front end-plate, but each has two points of attachment to the combustion-chamber. For this purpose the rear ends of the stays are bolted to V-shaped forgings, similar to that shown by Fig. 56, page 162.

The furnace-flues are corrugated to stiffen them, and thus maintain their form under the external pressure to which they are subjected. The corrugations in Fig. 13 are made up of alternate convex and concave semicircles; other forms of corrugations and other methods of stiffening flues, together with a discussion of the strength of flues, will be given in the next chapter. The front ends of the furnace-flues in Fig. 13 are made as large as the outside of the corrugations; the rear ends are as small as the inside of the corrugations. Such an arrangement makes it easy to remove the furnaces without disturbing the other parts of the boiler and without destroying the flues.

The combustion-chambers of a Scotch boiler are made up of flat or curved plates subjected to external pressure, and must be stayed at frequent intervals to prevent collapsing. The sides and bottom of the combustion-chamber in Fig. 13 are stayed to the cylindrical shell of the boiler by screwed stay-bolts, spaced seven inches on centres. The back of the combustion-chamber is stayed in like manner to the back end of the boiler, and thus both of these flat surfaces are secured. The plates used for making the combustion-chamber are thicker than those used for a locomotive fire-box, and consequently the stays are spaced wider and are larger in diameter.

The top of the combustion-chamber is stayed by stay-bolts and bridges in a manner that suggests the crown-bar staying of a locomotive fire-box. The space is, however, narrower and the staying is less complicated.

Complex Stays.—Sometimes the points to be connected by stays are so numerous that too many through-stays will be

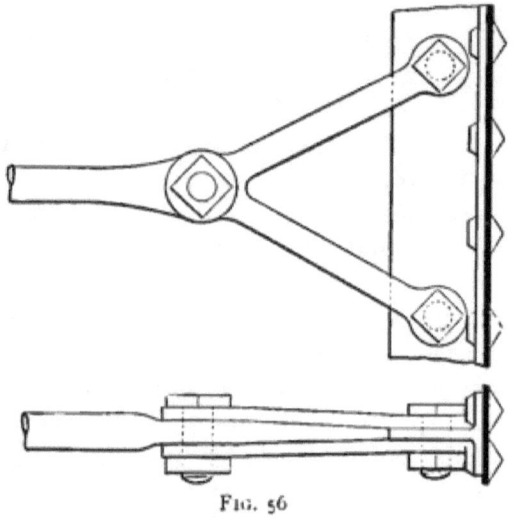

Fig. 56

required if all points are stayed separately. Thus in Fig. 56 there is an angle-iron riveted to a flat plate, and supported at intervals, as indicated by the two bolts passing through it. Instead of using a through-stay for each bolt, the bolts are coupled by two V-shaped forgings, which forgings are bolted to a through-stay at the angle of the V. There is enough freedom of the bolts in their holes to give equal distribution of the pull on the through-stay. By an extension of this method several points may be supported by one stay-rod.

Gusset-stays.—The flat ends of the Lancashire boiler, shown by Fig. 3, page 6, are secured to the cylindrical shell by gusset-stays; such a stay is shown more in detail by Fig. 57. A plate is sheared to the proper form, and is riveted between two angle-irons along the edges that come against the shell and the flat end. The angle-irons in turn are riveted to the shell or to the flat plate. Gusset-stays have the advantages of simplicity and solidity. They interfere less with the accessibility of the boiler than through-stays or diagonal stays. Their chief defect is that they are very rigid and are apt to localize the springing of the flat plates, which

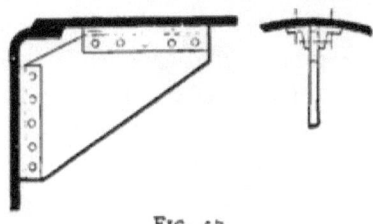

FIG. 57.

is caused by unequal expansion of the furnace-flues and shell. Consequently, grooving near gusset-stays is very likely to be found in Lancashire and Cornish boilers. Gusset-stays are also used to some extent in marine boilers, and in locomotive-boilers.

Spherical Ends.—The ends of cylindrical boilers, or of steam-drums, are commonly curved to form a spherical surface, in which case they retain their form under internal pressure and do not need staying. If the radius of the spherical surface is equal to the diameter of the cylindrical surface, the same thickness of plates may be used for both. If the spherical surface has a longer radius, the thickness may be increased. Such *dished* heads of boilers and steam-drums are struck up between dies while at a flanging heat, and are then flanged to give a convenient riveting edge.

Steam-domes are short, vertical cylinders of boiler-plate fastened on top of the shell of horizontal boilers. Plates II and III show steam-drums on locomotive-boilers. A steam-drum may be used to advantage when the steam-space is so shallow that there is danger that the ebullition may throw water into the pipe leading steam from the boiler. Locomotives usually have steam-domes, for not only is the steam-space shallow, but there is danger of splashing of the water in the boiler, especially if the track is rough or sharply curved.

Stationary boilers ought to have steam-space enough without domes; marine boilers sometimes have domes, but they are less common than formerly. The additional steam-volume in a steam-dome is insignificant, so that a dome should not be added to increase steam-space of a boiler.

The main objection to a steam-dome is that it weakens the boiler-shell, which must be cut away to form a junction with it. The shell may be reinforced, to make partial compensation, by a ring or flange of boiler-plate. Such a flange is clearly shown on Plate III, where the longitudinal seam of the ring carrying the dome is purposely placed at the top of the boiler. A similar arrangement is made for the dome on Plate II.

Dry-pipe.—Any pipe inside of a boiler for the purpose of leading steam from the boiler is known as a dry-pipe; the pressure in such a pipe is frequently less than that of the steam in the boiler, consequently there is a tendency to dry the steam in the pipe. Dry-pipes are found in locomotive and marine boilers and sometimes in stationary boilers.

The dry-pipe of a locomotive opens near the top of the dome. It runs vertically down till it is well below the shell of the barrel, then it runs horizontally through the steam-space and out through the smoke-box tube-sheet. The throttle-valve is at the inlet of the dry-pipe. It is controlled through a bell-crank lever by a rod which enters the head of the boiler from the cab.

The marine boiler shown by Fig. 13 has a dry-pipe which is joined to a steam-nozzle at the front end of the boiler. This dry-pipe is pierced with numerous longitudinal slits on the upper side; the sum of the area of such slits is seven eighths of the area through the stop-valve in the steam-pipe.

Steam-nozzle.—The stationary boiler shown on Plate I has a cast-iron steam-nozzle at each end. The steam-pipe leading steam from the boiler is bolted to the rear nozzle, and the safety-valves are placed above the front nozzle.

Nozzles are often made of cast steel. The best are forged without welds from one piece of steel.

Manholes.—A manhole should be large enough to allow a man to pass easily inside the boiler. That on Plate I is fifteen inches long and eleven inches wide, and has its greatest dimension across the boiler.

The manhole there shown is placed inside the shell of the boiler. Both the ring and the cover are forged from steel without a weld. Fig. 58 shows a form of manhole that is

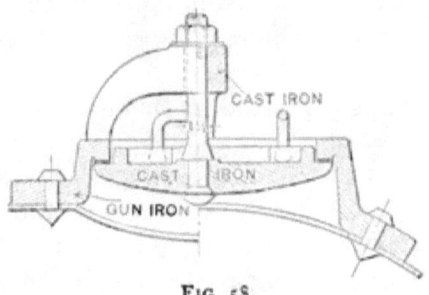

Fig. 58.

placed outside the shell. This form is commonly made of cast iron, but cast steel manholes of similar form are used to some extent.

The manhole-ring should be strong enough to give compensation for the plate cut away from the ring on which it is placed.

The manhole-cover is placed inside the ring so that it is held up to its seat by the steam-pressure. The cover is drawn up to its seat by a bolt and removable yoke. Sometimes there are two bolts each with its yoke. A cast-iron manhole naturally has a cast-iron yoke, and a forged manhole has a wrought-iron or steel yoke.

The manhole-cover is made steam-tight by a rubber gasket; the form of the cover and its seat are such that the gasket cannot be blown out by the pressure of the steam.

Hand-holes are provided at various places on boilers to aid in washing out and cleaning. Thus the boiler on Plate I has a hand-hole near the bottom at each end, and there are several hand-holes near the foundation-ring of the vertical boiler, shown by Fig. 5. The hand-hole covers on Plate I are placed directly against the plate which is not reinforced. Each is held up by a bolt and a small yoke, which has a bearing on the plate completely round the hole. If the yoke has insufficient bearing on the plate, the latter is liable to be damaged and leaks will occur. The hand-holes on the marine boiler shown by Fig. 13 are reinforced by small plates outside the boiler-heads.

Washout Plugs.—Instead of hand-holes, washout plugs, two inches or two inches and a half in diameter, are provided near the corners of the foundation-ring of a locomotive firebox. Such plugs are simply screwed into the outside plate of the boiler. Examples are shown by Plates II and III.

Methods of Supporting Boilers.—Horizontal cylindrical boilers are commonly supported on the side walls of the brick setting, by brackets which are riveted to the shell of the boiler. Thus the boiler shown on Plate I has two such brackets on each side; this boiler is about sixteen feet long. If a boiler is as much as eighteen feet long, three brackets are used. The front brackets rest directly on the brickwork, but the other brackets rest on iron rollers, to provide for the expansion of the boiler. The brackets are set so that the plane of support is a little above the middle of the boiler.

Fig. 59 shows a common form of bracket, made of cast iron, which is riveted to the shell above the flange of the bracket. A better form with rivets both above and below the flange is shown by Fig. 60.

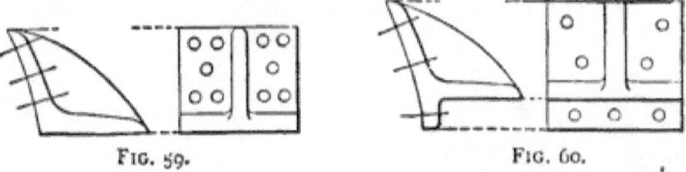

FIG. 59.   FIG. 60.

A detachable bracket, like that shown by Figs. 61 and 62, may be used when the boiler must be put into a building

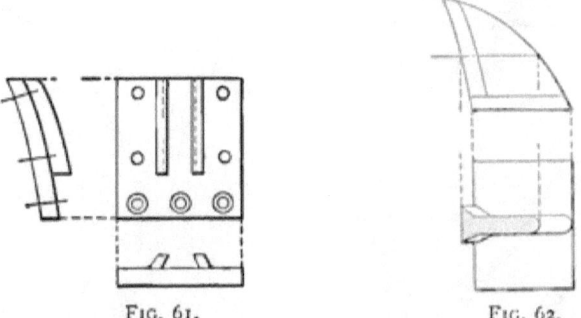

FIG. 61.   FIG. 62.

through a small aperture. Fig. 61 gives an end and side elevation and plan of the body of the bracket; Fig. 62 gives

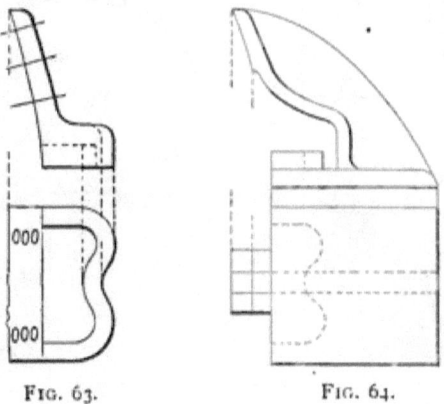

FIG. 63.   FIG. 64.

a side elevation and plan, with section, of the flange. After the boiler is in place the flange is thrust up into the dovetail groove in the body of the bracket. The pressure of the flange against the dovetail groove, intensified by the wedging action of the inclined sides, is liable to be excessive. To overcome this difficulty the bracket shown by Figs. 63 and 64 is often

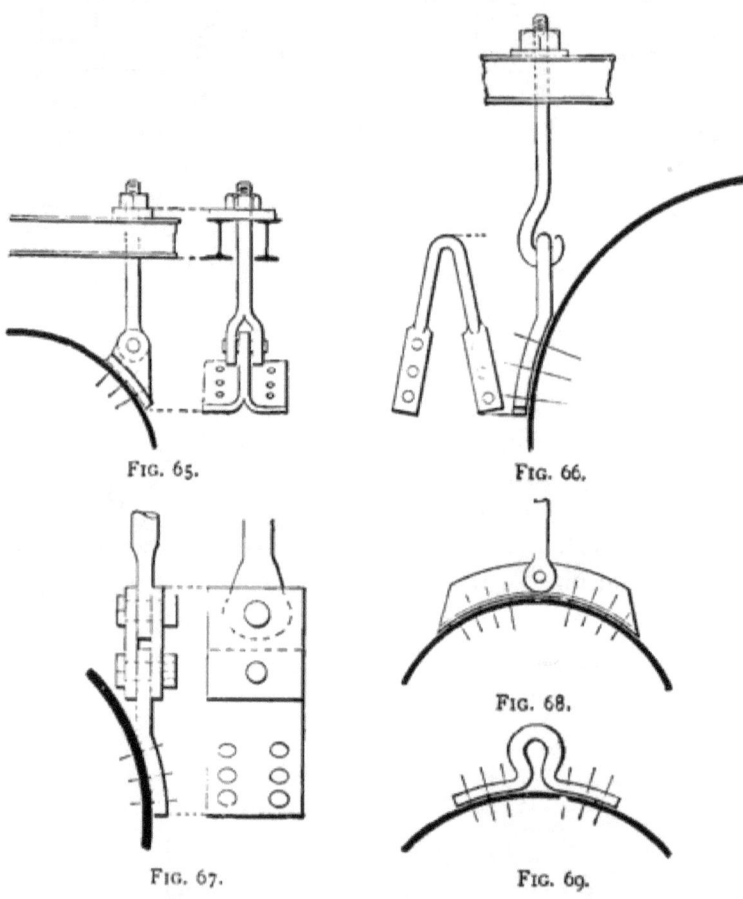

FIG. 65.   FIG. 66.

FIG. 67.   FIG. 68.

FIG. 69.

used. Fig. 63 shows the end elevation and a view from below, of a casting which is riveted to the shell. Fig. 64 shows the same views of a casting which catches into the hollow under Fig. 63 and bears at the top against this same casting, the rivets bolting it to the shell being countersunk.

Horizontal boilers, and especially plain cylindrical boilers, are sometimes hung from a support above the boiler, as shown by Figs. 65, 66, and 67.

Fig. 65 shows a lug, made of boiler-plate, riveted to the shell of the boiler. The lugs are placed in pairs and the boiler is hung from these lugs by bolts that are supported between transverse beams over the boiler. Fig. 66 differs in substituting a loop for the lug.

Fig. 67 shows a method of suspension with two short pieces of plate above the lug, to give some flexibility and provide for expansion.

Figs. 68 and 69 show methods of suspending a boiler from the top. These methods are proper only for boilers which have a small diameter.

CHAPTER VII.

STRENGTH OF BOILERS.

THE determination of the thickness of boiler-plates, the size of stays, and other elements affecting the strength of a boiler, involves a knowledge of the properties of the materials used and a knowledge of the methods of calculating stresses in the several members of the boiler. A brief statement of these subjects, as applied to boilers, will be given here.

Materials Used.—The materials used for making boilers are mild steel, wrought iron, cast iron, malleable iron, copper, bronze, and brass.

In order to insure that materials used for making a boiler shall have the proper qualities, it is customary to require that specimens shall be tested in a testing-machine, and that they shall have certain definite properties, such as ultimate tensile strength, elastic limit, and contraction of area at fracture. In order that these properties shall be properly developed, it is essential that specimens shall be of right size and shape, and that the testing shall proceed in a correct method.

Testing-machines.—The frame of a testing-machine carries two heads, between which the test-piece is placed, and to which it is fastened by wedges or other clamping devices. One head, called the straining-head, is drawn by screws or by a hydraulic piston, and pulls on the test-piece. The other head, called the weighing-head, transmits the pull to some weighing device. Boiler materials are commonly tested in a machine which has the pull applied by screws, driven through gearing by hand or by power; the pull is weighed by a system of levers and knife-edges, arranged like those of a platform scale. Such a machine should be able to exert a pull of fifty or a hundred thousand pounds.

Testing-machines that give a direct tension are commonly arranged to give also a direct compression. There are also machines arranged to give transverse loads, like the load applied to a beam.

Forms of Test-pieces. — A test-piece of boiler-plate should be at least one inch wide, planed on both edges, and should be about two feet long. A piece which is less than eighteen inches long is not fit for testing.

Test-pieces eighteen inches to two feet long may be cut directly from bars or rods for making stays or bolts. If a rod is so large that the available testing-machine will not break it, it is of course possible to turn it down to a smaller diameter, but it would be preferable to send such a rod to a machine that is powerful enough to break it at full size.

Test-pieces of cast metal may be cast in the form of rectangular bars, which should be at least one inch wide and an inch thick. If the bars are rough or irregular it may be necessary to plane the edges, or perhaps to plane them all over.

Test-pieces of boiler-plate should be cut from the edge of at least one plate of each lot of plates. Sometimes specifications require pieces from each plate used for a given boiler. Pieces should be cut from both the side and the end of a plate, for there is a grain developed by rolling either iron or steel boiler-plate, and tests should be made both with the grain and across the grain.

Very hard material may require shoulders on the test-pieces to enable the testing-machine to get a proper hold. But iron or steel that is so hard as to require shoulders is much too hard for boiler-making; consequently there will be no reason for providing test-pieces of boiler iron or steel with shoulders. If test-pieces have shoulders, they should be at least ten inches apart.

Methods of Testing.—A test-piece of proper length is first measured to determine the breadth and thickness or else the diameter, as the case may be. A length of eight inches is laid off near the middle of the test-piece, and clamps for measuring the stretch of the piece are applied at the ends of this eight-inch length, as shown by Fig. 70. The piece is then secured in the machine and a load is applied. The distance between the clamps is now measured by a micrometer caliper with an extension-piece. The method of doing this is to place the head of the micrometer against a point on the flange of the clamp at one end, and adjust the length of the micrometer so that it shall just touch the corresponding point on the other clamp. A little practice will enable the observer to measure to one or two ten-thousandths of an inch. As the load is increased the test-piece stretches, the increase of length being proportional to the increase of the load. The stretch is measured on both sides of the test-piece for each increase of load applied. If the test-piece is not straight or exactly aligned in the machine there may be some irregularity in the stretching at

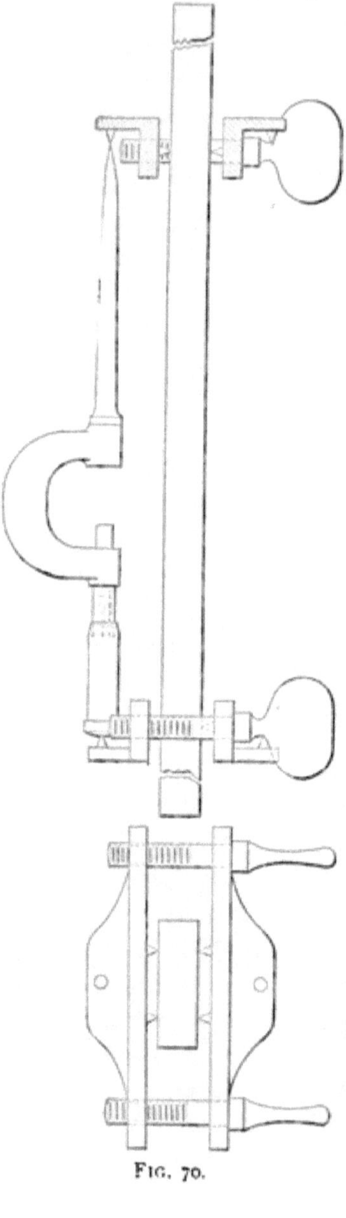

Fig. 70.

first, but after a considerable load is applied the piece stretches uniformly until about half the maximum load that the piece can carry has been applied. During the progress of the test a point is reached beyond which the stretch increases more rapidly than the load; this is known as the elastic limit.

After the elastic limit is reached the clamps are removed and the test proceeds without them, but at about the same rate of loading. A load is soon reached which the piece cannot permanently endure, shown by the fact that the scale-beam will fall though the straining-head remains at rest. This is called the stretch limit. The piece may, however, carry a considerably higher load if the straining-head is kept moving to take up the stretch. Finally, the piece begins to draw down rapidly, somewhere near the middle of its length, and when the piece breaks, the fracture shows about half the area of the piece before testing. Hard materials may draw down little, or not at all; the limit of elasticity may approach the strength of the material.

The jaws or wedges of the testing-machine interfere with the stretching or flow of the material gripped by them. The influence of the wedges may extend two or three inches beyond their edge in the testing of boiler-plate. If a piece has shoulders they will have a like effect. Consequently the points at which a clamp is secured to a test-piece should be two or three inches from a shoulder or from the wedges of the machine. The wedges of a machine of a capacity of fifty or a hundred thousand pounds are four or five inches long. They will grip on three inches at the end of a test-piece, but not on less. The test-piece must have eight inches for measuring stretch, two or three inches at each end for flow, and three to five inches at each end in the wedges. Consequently the piece must be eighteen or twenty-four inches long.

The method just described is slow and laborious, and

requires two observers—one to measure stretch and one to weigh. For commercial work an automatic device is often used which registers loads and corresponding elongations. Such devices commonly record the stretch limit instead of the elastic limit; these two points should never be confused.

Stress.—The number of pounds of force per square inch is called the stress. The stress is uniform on a piece under direct tension, and is equal to the load divided by the area of transverse section. Stress may be expressed in other units, such as tons per square foot or kilograms per square millimeter.

Strain.—The stretch of a piece, under direct tension, per unit of length is called the strain. If the original length is l and the stretch is a, then the strain is $\dfrac{a}{l} = s$.

The Limit of Elasticity is the limiting stress beyond which the strain increases more rapidly than the stress. The limit is not perfectly definite, and can be determined approximately only. A load greater than the elastic limit will produce an appreciable permanent elongation after the load is removed. A stress less than the elastic limit will produce only a slight permanent elongation; such elongation may be inappreciable.

Stretch Limit.—The stress at which the scale-beam of a testing-machine will fall while the straining-head is at rest is called the stretch limit.

Ultimate Strength.—The maximum stress that a piece will endure in a testing-machine is called the ultimate strength of the material. The strength depends somewhat on the rate of testing. The more rapidly the testing proceeds the higher will be the apparent strength. It is desirable that some standard rate of testing may be adopted by engineers so that results may be strictly comparable.

The Modulus of Elasticity is the result obtained by dividing the stress by the strain. If the stress is p pounds per square inch and the strain is s per inch, then the modulus of elasticity is

$$E = \frac{p}{s}.$$

Reduction of Area.—The area of the test-piece of boiler-plate at the rupture is much less than that of the piece before testing. This reduction is important, as it shows the ductility of the metal, and its ability to change shape without too much distress under the influence of unequal expansion of different members of a boiler.

Ultimate Elongation.—After the test-piece is broken the two parts are laid down in a straight line with the broken ends in contact, and the length of the distance between the points of attachments of the measuring clamps is measured. The ratio of the elongation to the original length (eight inches) is called the ultimate elongation. The ultimate elongation is generally given in per cent. This is important, for the same reason given for the contraction of area.

Compression.—The preceding definitions are given for tension only, for sake of simplicity and brevity; they may be applied to pieces in direct compression if the term stretch or elongation is replaced by compression.

Shearing.—Stresses have thus far been considered to be at right angles to the sections of the pieces to which they are applied, and produce either tension or compression at that section. A stress that is not at right angles to a section will tend to produce sliding at that section. A stress that is parallel to a section will tend to produce sliding only, and is called a shearing-stress. If a shearing-stress is uniformly distributed, its intensity may be found by dividing the total force or load by the area of the section.

The rivets of a riveted seam are subjected to a shearing-stress.

Steel.—At the present time boiler-plates are made of mild, open-hearth steel; good wrought-iron plates can be obtained with some difficulty and trouble. Such steel is in reality a tough, ductile, ingot iron, containing about one fourth of one per cent of carbon; it is nearly free from sulphur and phosphorus. The former impurity makes iron hot-short and the latter makes it cold-short, i.e., brittle when hot or cold. Plates of this material can be obtained of all sizes and thicknesses up to eight feet wide and an inch and a quarter thick. There is no limit to length except convenience of handling.

Steel plate for boiler-making should have the following properties:

Tensile strength	55,000 to 60,000
Elastic limit	30,000 to 33,000
Elongation in eight inches	25 per cent or more
Reduction of area at fracture	50 per cent or more.

The plate should be free from blisters, lamination, or blow-holes.

A piece cut from a plate less than three fourths of an inch thick should endure bending double under a heavy hammer, both hot and cold, without showing cracks. Heavy plates should endure bending at a small radius to a large angle without cracking.

The upper end of the ingot into which the molten steel from the open-hearth furnace is cast, is liable to be affected by bubbles and other imperfections when the ingot is poured from the top. Such imperfections, if they are not removed, give rise to lamination in the plates, and therefore when the ingot is rolled into blooms the *crop end* should be cut long enough to remove all the bubbles. There is always a tendency, on account of the reduction of prices through competition, to reduce the length of the crop end, and consequently steel plates, though having the other required physical properties, are liable to show lamination. To guard against this, test-pieces should be cut from the ends of plates and tested in a testing-machine, and also by bending hot and cold. Ingots have been cast from the bottom, in which case bubbles are likely to be distributed throughout the ingot.

Steel plates are sometimes classified as shell-plates and fire-box plates; the latter are supposed to be of special quality to endure the flanging required in the forming of the locomotive fire-box, and to endure the stresses in service due to the action of the fire, draughts of air entering through the fire-door, and from the unequal expansion of the fire-box and the parts of the shell to which it is stayed or otherwise connected. There does not appear to be any difference in the chemical and physical characteristics of these two grades, except the somewhat greater ductility of the fire-box plates, due to greater care in making.

Angle-irons, T irons, bars, and rods used for staying and fastening boilers may be made of steel if welding is not required in forming them.

Blue Heat.—Steel plates, and other forms of mild steel, become brittle at a temperature corresponding, roughly, to a blue heat. A plate that will endure bending double, both hot and cold, is liable to show cracks if bent at a blue heat. In bending, flanging, and forging no work should be done on steel at a blue heat; properly, such work should be done at a bright red heat; work should never be continued after the steel becomes black. After the steel is cold it may be bent as readily as iron at the same temperature.

Wrought Iron.—All the stays and fastenings of boilers that are made by welding should be made of tough, ductile wrought iron. Welds made by a skilful smith may have as great a strength as the bar from which they are made. A ductile bar may break in the clear bar instead of in the weld, on account of the hardening due to the work done on the bar at the weld. It is customary to assume that 25 to 50 per cent of the strength of the bar may be lost by welding.

Wrought-iron plates of a quality suitable for boiler-making are now more expensive than mild-steel plates, which are in every way as well adapted to the purpose, and which have a higher strength. Consequently we find wrought-iron plates used only when specially demanded. Wrought iron does not show cracks when worked at a blue heat, and in general may endure more abuse in working. This caused wrought iron to be preferred by many after reliable steel was produced cheaply, but boiler-makers now understand the working of steel plates and avoid improper handling.

Wrought-iron plates should show a limit of elasticity of 23,000 pounds, and a tensile strength of 45,000 pounds to the square inch.

Wrought-iron rods and bolts should have a strength of 48,000 pounds per square inch.

Rivets.—The rivets used in boiler-making are either iron, or steel similar in quality to steel used for boiler-plates.

A rivet should bend cold around a bar of the same diameter, and it should bend double when hot without fracture. The tail should admit of being hammered down when hot till it forms a disk $2\frac{1}{2}$ times the diameter of the shank, without cracking. The shank should admit of being hammered flat when cold, and then punched with a hole equal in diameter to that of the shank, without cracking.

The rods from which rivets are made should show a tensile strength of about 55,000 pounds for steel and about 48,000 pounds for wrought iron. The other properties, such as ultimate elongation and contraction of area, should be like those for boiler-plate.

The shearing strength of steel rivets is about 45,000 pounds, and of iron rivets about 38,000 pounds; that is, the shearing strength will be about two thirds of the tensile strength.

Cast Iron in different forms will show a tensile strength of 12,000 to 20,000 pounds to the square inch. Gun-iron, which is cast iron made with special care and skill from selected stock, has shown a tensile strength of nearly 30,000 pounds to the square inch. In compression the strength of small pieces may be as high as 80,000 pounds to the square inch, but larger pieces, like columns, fail at 30,000 pounds to the square inch.

Cast iron is used for some or all of the parts of sectional boilers, and for fittings such as manholes, though wrought iron is preferable for such purposes. Flat plates at the ends of cylindrical boilers are sometimes made of cast iron.

In general, cast iron should never be used when it is subjected to severe changes of temperature or to stresses from unequal expansion, and should be replaced by wrought iron or mild steel whenever it is practicable.

Couplings, elbows, and other pipe-fittings are made of cast iron. The brittleness is a convenience when changes are to be made, as joints that cannot be opened are readily broken.

Malleable Iron, which is cast iron toughened by being deprived of part of the carbon, is used for pipe-fittings and for fittings of steam-boilers. It is used in place of cast iron for sectional boilers and for parts of water-tube boilers. Though tougher than cast iron, and though it will endure forging to some extent, its variability in quality and its unreliability prevent much reduction in weight and size when substituted for cast iron.

Copper is largely used in Europe for making fire-boxes of locomotive-boilers and torpedo-boat boilers. Its greater cost is in part offset by the value of the scrap copper after the fire-box is worn out.

Copper for fire-boxes, rivets, and stays should have a tensile strength of 34,000 pounds to the square inch, and should show an elongation of 20 to 25 per cent in 8 inches. It should not contain more than one-half per cent of impurities. The greater ductility of copper, and its greater thermal conductivity, permitting of greater thickness for furnace-plates, recommends it to European engineers.

Copper is largely used on steamships for making piping of all sorts, such as steam-pipes and water-pipes. Such pipes are made of sheet copper, rolled up or hammered to shape, scarfed and brazed at the edges. The pipe is also brazed to brass flanges for coupling lengths of pipe, or for joining to steam-chests or other parts of the engine or boiler. If the brazing is not done with care and skill the brazed joint may lose as much as half the strength of the sheet copper. Several disastrous explosions of such piping have occurred. Consequently wrought-iron piping is finding favor for high-pressure steam.

Bronze and Composition. Brass.—Bronze is properly an alloy of copper and tin; thus gun-metal is 90 parts of copper to 10 of tin. Compositions of various qualities are made of copper and zinc with more or less tin. Brass is an alloy of copper and zinc; for example, brass smoke-tubes are made of 70 parts of copper to 30 parts of zinc. Lead is added to brass and to composition to reduce the cost and to make the metal work easier. It may be considered as an adulteration, as it cheapens the metal at the expense of the quality. There are many special bronzes, such as phosphor-bronze and aluminium-bronze, which are used for special purposes.

Brass is used to some extent for smoke-tubes of locomotive and other boilers, on account of its greater thermal conductivity, by European engineers. In America, brass is used for valves, gauges, and other boiler fittings. Composition or bronze is advantageously used for the valves and seats of safety-valves and wherever the service endured is exceptionally hard. Brass is more commonly used because it is cheaper. In a general way it may be said that the cost and quality of brass and composition is proportional to the copper it contains; thus red brass is better and costs more than yellow brass. Many small brass fittings on the market are sold at a price which precludes the use of proper alloys, and they are consequently soft and worthless.

Stay-bolts are usually arranged in equidistant horizontal and vertical rows; as an example we may take the stay-bolts in the locomotive fire-box on Plate II. These bolts are 7/8 of an inch in diameter outside of the threads, and are spaced 4 inches on centres. The total load on each stay-bolt with a steam-pressure of 170 pounds to the square inch is

$$4 \times 4 \times 170 = 2720 \text{ pounds.}$$

The diameter of the bolt at the bottom of the screw-thread is about 0.7 of an inch, and the area of the section is about 0.4 of a square inch. The stress is consequently

$$2720 \div 0.4 = 6800.$$

Sometimes the area is calculated from the external diameter of the bolt, a proceeding which may lead to a gross error. In the present instance the corresponding area is about 0.6 of a square inch, which gives an apparent stress of about 4500.

Suppose that the thread is turned off from the body of the bolt, and that the diameter is thereby reduced to 5/8 of an inch. The area of the section is then about 0.3 of an inch, and the stress is

$$2720 \div 0.3 = 9000 +.$$

The stress on stay-bolts should always be low to allow for wasting from corrosion, and to allow for unknown additional stresses that may be caused by the unequal expansion of the plates that are tied together by the stay-bolts.

Stay-rods.—Through-stays like those passing through the steam-space of the marine boiler, shown by Fig. 13, page 17, are treated much like stay-bolts. Thus the stays in question are 14 inches apart horizontally and 13 inches apart vertically. If they are each assumed to support a rectangular area 13 inches wide and 14 inches long, the total force from 160 pounds steam-pressure will be

$$14 \times 13 \times 160 = 29120.$$

The diameter of these stays in the body is 2 inches, which gives an area of section of 3.14 square inches. The stress is consequently

$$29120 \div 3.14 = 9300$$

These stay-rods have swaged heads on which the screw-thread is cut, so that the diameter at the bottom of the thread is greater than the diameter of the body.

Stay-rods which are used in connection with girders, as on Plate I, will have to carry loads which depend on the surface supported, the steam-pressure, and the number and arrangement of the stays. The determination of the load may be difficult and uncertain, but the calculation of the stress for a given load is very simple.

Diagonal Stays.—If a stay-rod runs diagonally from a flat plate to the shell of a boiler, it will evidently be subjected

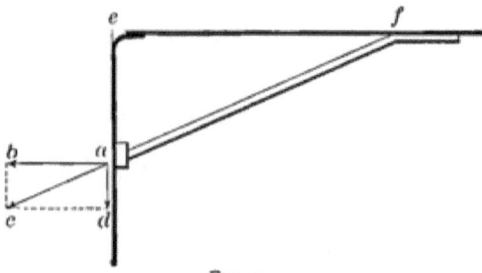

FIG. 71.

to a greater stress than it would be if it were a through-stay. Thus in Fig. 71 we have at the point *a* the parallelogram of

STRENGTH OF BOILERS.

forces *abcd*; *ab* is the total pressure supported by the stay, *ac* is the pull on the stay, and *ad* is a force that must be taken up by the flat plate. But the triangles *abc* and *acf* are similar, so that we have

$$\frac{ac}{ab} = \frac{af}{cf} = \frac{\sqrt{ac^2 + cf^2}}{cf}$$

Suppose, for example, that *ac* is two feet and *cf* is six feet; then

$$\frac{ac}{ab} = \frac{\sqrt{2^2 + 6^2}}{6} = 1.054,$$

or the pull on the stay is more than five per cent in excess of what a through-stay would be required to support.

Gusset-stays are open to the defect that the distribution of stress on the plate forming the stay is uneven and uncertain. It is customary to calculate them on the assumption that the resultant stress acts along a medial line, and is evenly distributed over a section at right angles to that line. A low apparent working-stress should be used.

Thin Hollow Cylinder. — Let Fig. 72 represent a semicircular steam-drum closed at the bottom by a thick flat plate.

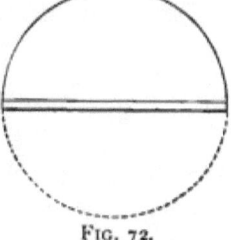

FIG. 72.

If the steam-pressure is *p* pounds per square inch, the radius is *r*, and the length is *l*, then the pressure on the plate is

$$2prl.$$

If the thickness of the cylinder is *t*, and the stress per square inch on the metal of the cylinder is *s*, then the pull of the cylinder at one end of the plate is

$$stl.$$

But this must be equal to half the pressure on the plate, so that

$$stl = prl.$$

$$\therefore s = \frac{pr}{t}.$$

For safety the stress should not exceed the safe working stress for the material of which the cylinder is made; so that we have

$$f = \frac{pr}{t}.$$

It is evident that the pull on the side of the cylinder and the stress per square inch will be the same if another half-cylinder is substituted for this plate, making a complete thin hollow cylinder.

Example 1.—A thin hollow cylinder five feet in diameter and half an inch thick, working at a pressure of 200 pounds, will be subjected to a stress of

$$200 \times \frac{5 \times 12}{2} \div \tfrac{1}{2} = 12{,}000$$

pounds per square inch. If the cylinder is made of one continuous plate of steel without longitudinal joint, this stress will be about one fifth of the ultimate strength.

Example 2.—If it is desired that the stress shall be 9000 pounds in a cylinder 9 feet in diameter when exposed to a pressure of 120 pounds to the square inch, then the thickness of the plate should be

$$t = \frac{pr}{f} = 120 \times \frac{9 \times 12}{2} \div 9000 = 0.72$$

of an inch.

End Tension on a Cylinder.—In the preceding cylinder we have considered the tension on a section at the side of the cylinder. Let us now consider the tension on a transverse section.

If the cylinder is closed by a flat plate at the end, the area of that plate will be

$$3.1416r^2,$$

and the total force due to a pressure of p pounds per square inch will be

$$3.1416r^2p.$$

This force will be resisted by a ring of metal having a circumference $2 \times 3.1416r$, and a thickness t. The resistance of the ring will be

$$2 \times 3.1416rts,$$

representing the stress by s. Consequently we shall have

$$2 \times 3.1416rts = 3.1416r^2p.$$

$$\therefore s = \frac{pr}{2t}.$$

It is evident that the stress from the end pull is half the stress on the section at the side of a cylinder, and consequently a cylinder made of homogeneous material without joints will always be ruptured longitudinally.

It is also evident that the stress from the end pull will be the same if the end of the cylinder is closed by a spherical surface, or by any other figure, instead of a flat plate.

Thin Hollow Sphere.—A section taken through the centre of a sphere is in the same condition as a transverse section of a thin cylinder, and will be subjected to the same stress, if the sphere has the same thickness and is subjected to the same internal pressure.

Formerly the ends of plain cylindrical boilers were made hemispherical, but such ends are difficult to make and are needlessly strong if of the same thickness as the cylindrical shell. It is now the practice to curve such ends to a less radius than that of the cylindrical shell. If the radius of the head is equal to the diameter of the shell, then with the same thickness of plate the stress will be the same per square inch, provided there are no seams in head or shell. The heads usually do not have a seam, and the shells always have a seam; the margin of strength in the head, when the same thickness of plate is used, under this condition may be offset against the possible injury done to the head in shaping it.

The construction known as a *bumped-up head* has the edge flanged into a cylindrical form to make a joint with the shell, and to avoid the awkward stress that would be thrown onto the cylindrical shell if the true cylindrical and spherical surfaces were allowed to intersect.

If it is inconvenient to curve the head to a radius as small as the diameter of the cylinder, then a thicker plate may be used, with a longer radius.

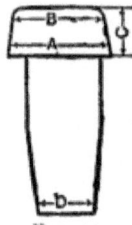

FIG. 73.

Rivets.—The plates of a boiler are joined at the edges by rivets; rivets are also used in stays and other members.

The usual form of rivets is shown by Fig. 73. If the diameter of the rivet is D, then the proportions may be

$$\frac{A}{D} = 1.4;$$

$$\frac{B}{D} = 1.2;$$

$$\frac{C}{D} = 0.7;$$

$$\frac{b}{D} = 3/4.$$

The length of the rivet will depend on the number and thickness of the plates through which it is to pass.

The rivet represented by Fig. 73 has a pan head. Of the rivets shown by Fig. 74, *A*, *B*, and *C* have pan heads, and *D* and *E* have round or hemispherical heads.

The form of the point of a rivet will depend on the way in which the rivet is driven and on the shape of the tools or dies used for forming the point. The rivet *A* has a straight

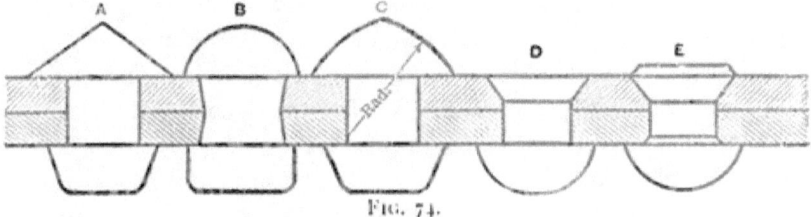

Fig. 74.

conical point; this is the only form that can be made when the rivet is driven by hand with flat-faced hammers.

The rivet *B* has the head formed by a die or snap. The rivet is driven by a few heavy blows of a hammer, and the head is roughly formed; then a die or snap is placed on the point and driven to form the point by a sledge-hammer.

C shows a rounded conical point commonly used for machine-driven rivets. The heads of such rivets may have a similar form.

D represents the usual form of countersunk rivets: the hemispherical head is not a peculiarity of such rivets; it is occasionally used with any form of point. The rivet *E* has some fulness or projection at the point beyond the countersink.

After a rivet is driven, both ends are called heads; the distinction of heads and points is made here for convenience in description.

The straight conical form *A* is liable to be too flat and weak. Its height should be three-fourths the diameter of the rivet.

When rivet-holes are punched in plates they are slightly conical, as shown by *B*, Fig. 74, which shows the two smaller ends of the rivet-holes placed together to facilitate the proper filling of the hole by the rivets. The other rivet-holes are straight, as they would be if drilled.

Riveted Joints.—The proportions of riveted joints, such as diameter and pitch of rivets, are determined in part by practice and in part by a method of calculation to be explained later. In practice it is found necessary to limit the pitch of the rivets, and consequently the diameter, to be used with a given thickness of plate, in order that the joint may be made tight by calking. This limitation frequently makes the joint weaker than it otherwise would be.

The edges of plates are either lapped over and riveted, or brought edge to edge and then joined by a cover-plate which is riveted to each of the two plates. The first method makes a lap-joint and the second a butt-joint.

Fig. 75 shows a single-riveted lap-joint and Figs. 76 and 77 show double-riveted lap-joints. The rivets in Fig. 76 are said to be staggered; the form shown by Fig. 77 is called chain-riveting.

Butt-joints with two cover-plates are shown by Figs. 80 and 81. The outer cover-plate is narrow, with rivets placed close enough together to provide for sound calking. The inner plate is wider, and as its edges are not calked they may have a row of more widely spaced rivets. These joints, and those shown by Figs. 78 and 79, are designed with the view securing more strength than can be had with a plain lap-joint like Fig. 76, or than can be had with a butt-joint with cover-plates of equal width.

Efficiency of a Riveted Joint.—The strength of a riveted joint is always less than that of the solid plate, because some of the plate is cut away by the rivets. This is very evident in the case of a single-riveted joint, such as that shown by Fig. 75; it will be found to be true for more complicated joints, such as those shown by Figs. 80 and 81. The efficiency of a riveted joint is the ratio of the strength of the joint to the strength of the solid plate.

The strength and efficiency of a given riveted joint can be properly determined only by direct test on full-sized specimens, which have considerable width. Tests on narrow specimens are liable to be misleading. Tests on boiler-joints are expensive, and can be made only on large and powerful testing-machines. Tests have been made on behalf of the United States Navy Department at the Watertown Arsenal on a large number of single-riveted joints, on a considerable number of double-riveted joints, and on a few special joints. A few tests have been made elsewhere on full-sized joints. These tests give us important information that can be used in designing joints for boilers, but we cannot in general select a joint directly from the tests.

Methods of Failure.—A riveted joint may fail in one of several methods, depending on the proportions, such as thickness of plate and the diameter and pitch of the rivets. This can be clearly seen in case of a single-riveted joint like that shown by Fig. 75. Such a joint may fail:

(1) By tearing the plate at the reduced section between the rivets. If the rivets have the diameter d and the pitch p, then the ratio of the area of the reduced section to that of the whole plate is

$$\frac{-d}{p}.$$

(2) By shearing the rivets.

(3) By crushing the plate or the rivets at the surface where they are in contact.

(4) By cracking the plate between the rivet-hole and the edge of the plate, or by some method of failure due to insufficient lap. A riveted joint never fails by this method in practice, because the lap can always be made sufficient.

The failure of more complicated joints may occur in various methods, which will be considered in connection with the calculation of some special joints.

Drilled or Punched Plates.—In the better class of boiler-shops it is now the practice to drill rivet-holes in plates after the plates are in place, so that the holes are sure to be fair. Sometimes the holes are punched to a smaller diameter and then drilled out to the final size after the plates are in place. The result is the same as though the holes were drilled in the first place, as the metal near the hole, which was injured in punching, is all removed. The metal remaining between drilled holes does not have its properties changed by the drilling. On the contrary, the metal between punched holes is always injured more or less. In general, soft ductile metal is injured less than hard metal, and further, soft-steel plates are injured less than wrought-iron plates.

When boiler-plates are punched and then rolled to form cylindrical shells, some of the holes are liable to come unfair, so that a rivet cannot be passed through. In such cases the holes should be drilled to a larger size, and a rivet of corresponding diameter should be substituted. Careless or reckless workmen sometimes drive in a drift-pin, and stretch or distort the unfair holes so that a rivet can be forced through. The plate is liable to be severely injured by such treatment, and the rivet cannot properly fill the rivet-holes. Unfortunately it is difficult or impossible to detect bad work of this kind after the rivets are driven.

Tearing.—The metal between the rivet-holes in a riveted joint cannot stretch as a proper test-piece does in the testing-machine, and consequently it shows a greater tensile strength than a test-piece from the same plate. Some tests on single or double riveted joints with small pitches show an excess of strength from this cause, amounting to ten per cent or more. The excess appears to be uncertain and irregular, so that if any allowance is made for it, it should be by a skilled designer after a careful study of all the tests that have been made. Ordinarily it will be safer to use the tensile strength shown by test-pieces in the testing-machine, especially for joints like Fig. 78, which have a large pitch for some of the rivets.

Shearing.—In general it is fair to assume the shearing strength of rivets of iron or steel to be two thirds the tensile strength of the metal from which the rivets are made.

Crushing.—It is customary to assume that the pull on a riveted joint is evenly distributed among the rivets in the joint, and to divide the total pull by the number of rivets to find the shearing or crushing force acting on one rivet. It is further customary to assume that the intensity of the crushing force on the surface where the rivet bears on the plate, may be found by dividing the total force on one rivet, by the product of the diameter of a rivet and the thickness of the plate.

The crushing-stress on rivets in joints that fail by crushing is found by experiment to be high and irregular. In some cases it has amounted to 150,000 pounds per square inch; in a few tests it is less than 85,000 pounds. It is probable that 95,000 pounds may be used with safety in calculating riveted joints for boilers. Now the stress on the bearing-surface will seldom be so much as one third the ultimate strength, even during a hydraulic test of a boiler, and it is not probable that a joint will be injured in this way unless the stress approaches the ultimate strength.

Friction of Riveted Joints.—It is evident that there must be considerable friction between plates that are firmly clamped together by rivets driven hot. It has been proposed to take some account of this friction in calculating riveted joints, or even to allow the friction to be the determining element in proportioning riveted joints. Such a method is shown by experiment to be unsafe, for slipping takes place at all loads, beginning at loads that are much smaller than a safe load, and the effect of friction disappears before a breaking load is reached.

Lap.—The distance from the centre of the rivet-hole to the edge of the plate is called the lap. The lap is usually once and a half the diameter of the rivet, a proportion that appears to be satisfactory.

Diameter of Rivet.—The minimum diameter of punched holes is determined by the consideration that the punch should not be broken. In the ordinary methods of punching boiler-plates the diameter of the punch should at least be as much as the thickness of the plate. It very commonly is once and a half the thickness of the plate.

Drilled rivet-holes may have any diameter. They never have a diameter less than the thickness of the plate. The maximum diameter of rivet to be used with any kind of riveted joint will in general be determined by the consideration that the tendency to crush the plate in front of the rivet should not be greater than the shearing strength of the rivet. The maximum diameter thus found is liable to give too large a pitch.

Pitch.—The maximum pitch for a given plate along a calked edge should be determined by the consideration that the plate should be held up rigidly enough to make a tight joint without excessive calking. The pitch of rivets, like those in the outer row of the joint shown by Fig. 78, need not be governed by this rule. There does not appear to be any explicit rule deduced either from practice or experiment for determining the proper pitch of rivets.

Single-riveted Lap-joint.—In the joint shown by Fig. 75 let the thickness of the plate be t, the diameter of the rivet d, and the pitch p, all in inches. Let the tearing strength of the plate be $f_t = 55{,}000$, the shearing strength be $f_s = 45{,}000$, and the resistance to crushing be $f_c = 95{,}000$, all for mild steel.

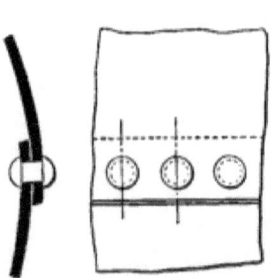

Fig. 75.

Assume the proportions

$$d = 15/16, \quad t = 7/16, \quad p = 2\tfrac{1}{4}.$$

It will be sufficient to consider a portion of the plate having a width equal to the lap. The failure of such a strip may occur in one of three ways:

1st. *Shearing one rivet.* The area to be sheared is $\frac{\pi d^2}{4}$ or $\frac{3.1416 d^2}{4}$. The resistance to shearing is found by multiplying this area by the shearing strength of the rivet:

$$\frac{\pi d^2}{4} f_s = \frac{\pi \times 15 \times 15 \times 45,000}{4 \times 16 \times 16} = 35,340.$$

2d. *Tearing plate between rivets.* The effective width of the strip under consideration, allowing for the rivet-hole, is $p - d$, and the thickness of the plate is t; the resistance to tearing is

$$(p - d) t f = (2\tfrac{1}{4} - \tfrac{15}{16}) \tfrac{7}{16} \times 55,000 = 31,580.$$

3d. *Crushing of rivet or plate.* The conventional method is to assume the effective bearing area to be equivalent to the diameter of the rivet multiplied by the thickness of the plate. The resistance is considered to be

$$d t f_c = \tfrac{15}{16} \times \tfrac{7}{16} \times 95,000 = 38,970.$$

The strength of a strip of the plate $2\tfrac{1}{4}$ inches wide is

$$2\tfrac{1}{4} \times \tfrac{7}{16} \times 55,000 = 54,140.$$

The calculated resistance to tearing is less than the resistance to shearing or compression. The apparent efficiency of the joint is

$$100 \times \frac{31,580}{54,140} = 58.3 \text{ per cent.}$$

If it be assumed that the resistance to tearing of the section between rivets will have an excess of ten per cent over the resistance of a piece in a testing-machine, then the resistance to tearing between rivets will appear to be 34,740. This figure is not far from the resistance to shearing, though still inferior. If it be further assumed that the whole plate outside of the joint will show a tearing strength of only 55,000 pounds per square inch, the efficiency of the joint will appear to be more than five per cent greater than that given above. It is probably wise to ignore the excess of strength due to the fact that the plate between the rivets will not draw down for reasons that have already been stated at length.

Double-riveted Lap-joint.—The rivets in this joint may be staggered as shown by Fig. 76, or chain-riveting may be

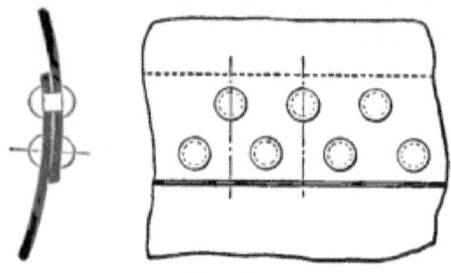

FIG. 76.

used as in Fig. 77. If the rivets are staggered and the two rows are too near together, it is possible that the plate may

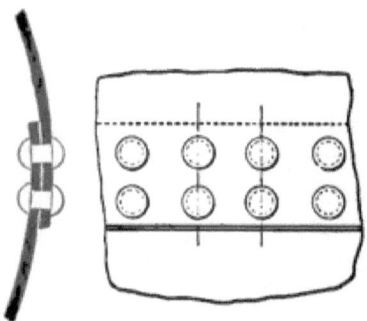

FIG. 77.

tear down from a rivet in one row to the nearest rivet in the next row, and thus have, after tearing, a jagged edge. With the usual proportions such a failure will not occur, but the plate will tear between rivets in the same row, if it fails by tearing. The calculation for efficiency will consequently be the same for both methods of riveting.

Let the dimensions be
$$t = 7/16, \quad d = 13/16, \quad p = 2\tfrac{1}{2}.$$

The joint may fail in one of three ways:

1st. *Shearing two rivets.* The assumed strip having a width equal to the pitch will be held by two rivets; this is apparent at once for chain-riveting. For staggered rivets such a strip will contain one whole rivet and half of two others, so that the same rule holds. The resistance of two rivets to shearing will be
$$\frac{2\pi d^2}{4} f_s = 46,660.$$

2d. *Tearing between two rivets.* The resistance is
$$(p - d)t f_t = 40,600$$

3d. *Crushing in front of rivets.* Just as for shearing, we have here the resistance at two rivets equal to
$$2dt f_c = 67,540.$$

The strength of the plate for a width of the pitch is
$$pt f_t = 60,160.$$

The plate will apparently fail by tearing, and the efficiency of the joint will be
$$100 \times \frac{40,600}{60,160} = 67.5 \text{ per cent.}$$

The increase of efficiency of the double-riveted lap-joint over the single-riveted joint is clearly due to reducing the diameter of the rivet and increasing the pitch. A further increase of efficiency could be obtained by using three rows of rivets; this, however, is practicable only for thick plates, as we are liable to get too wide a pitch for sound calking.

Single-riveted Lap-joint, Inside Cover-plate.—In this joint the plates are lapped and joined by a single row of rivets:

and a plate is worked inside and riveted through the shell with a single row of rivets, which are spaced twice as far apart as the rivets in the lap. In making up the joint all three rows of rivets may be driven at the same time. The lapped joint only is calked; the pitch of rivets through the lap must consequently be small enough to give sound calking. The outer rows of rivets are not controlled by this rule.

We will here consider a strip having the width a, Fig. 78, equal to twice the pitch of the rivets in the lap. Such a strip will be held by two rivets in the lap and by one rivet in an outer row.

Assume the following dimensions:
Thickness of shell and of cover-plate, $t = 5/16$.
Diameter of rivets (iron), $d = 3/4$.
Pitch of rivets in lap, $p = 1\frac{3}{4}$.
Pitch of outer rows of rivets, $P = 3\frac{1}{2}$.
Shearing resistance of iron rivets per square inch or $f_s = 38,000$ lbs.

The joint may fail in one of five ways:

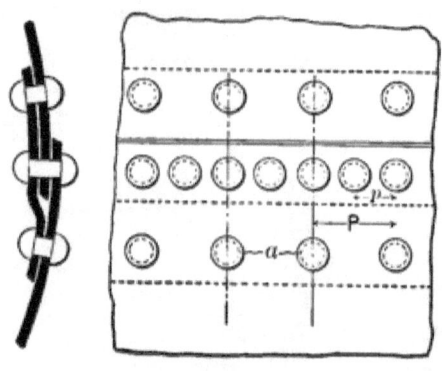

FIG. 78.

1st. *Tearing between outer row of rivets.* The resistance is

$$(P - d)tf_t = 47,270.$$

2d. *Tearing between inner row of rivets, and shearing outer row of rivets.* The resistance is

$$(P - 2d)tf_t + \frac{\pi d^2}{4} f_s = 51,150.$$

Since the rivets are iron, $f_s = 38,000$.

3d. *Shearing three rivets.* The resistance is

$$\frac{3\pi d^2}{4} f_s = 50,350.$$

4th. *Crushing in front of three rivets.* The resistance is

$$3tdf_c = 66,800.$$

5th. *Tearing at inner row of rivets and crushing in front of one rivet in outer row.* The resistance is

$$(P - 2d)f_t + tdf_c = 56,641.$$

The strength of a strip of plate $3\frac{1}{2}$ inches wide is

$$l\,tf_t = 60,160.$$

The least resistance is offered by the first method, giving for the efficiency

$$100 \times \frac{47,270}{60,160} = 78.6 \text{ per cent.}$$

If the inside cover-plate is thinner than the shell, additional complication will be introduced into the calculations for resistance.

Double-riveted Lap-joint with Inside Cover-plate.—The arrangement of this joint is shown by Fig. 79. Assume the dimensions:

Thickness of shell and of cover-plate, $t = 7/16$.
Diameter of rivets (steel), $d = 13/16$.
Pitch of rivets in lap, $2\frac{5}{16}$.
Pitch of outer rows of rivets, $P = 4\frac{5}{8}$.

The methods of failure are:

1st. *Tearing at outer row of rivets.*

$$\text{Resistance } (P-d)tf_t = 91,740.$$

2d. *Shearing four rivets.*

$$\text{Resistance } \frac{4\pi d^2}{4}f_s = 93,310.$$

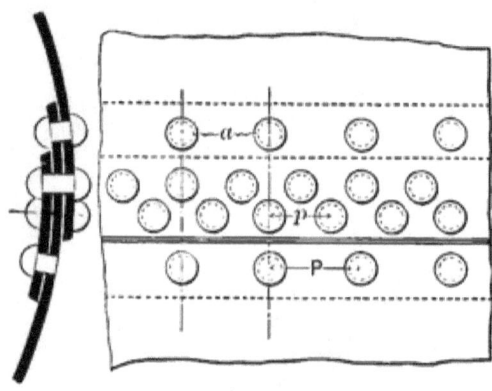

FIG. 79.

3d. *Tearing at inner row and shearing outer row of rivets.* A strip having the width of the pitch of the outer row of rivets will be weakened at the rivets in the lap to the extent of one rivet-hole and half another rivet-hole. The resistance is

$$(P-1\tfrac{1}{2}d)tf_t + \frac{\pi d^2}{4}f_s = 105,285.$$

4th. *Crushing in front of four rivets.*

$$\text{Resistance } 4tdf_c = 135,080.$$

5th. *Tearing at inner row of rivets and crushing in front of one rivet.*

$$\text{Resistance } (P-1\tfrac{1}{2}d)tf_t + tdf_c = 115,730.$$

Strength of strip $4\tfrac{5}{8}$ inches wide,

$$Ptf_t = 111{,}290.$$

$$\text{Efficiency} = 100 \times \frac{91{,}740}{111{,}290} = 82.4 \text{ per cent.}$$

Double-riveted Butt-joint.—The joint shown by Fig. 80 has a cover-plate inside and another, narrower, outside. There are two rows of rivets on each side of the joint. The inner rows are nearer together and pass through both cover-plates.

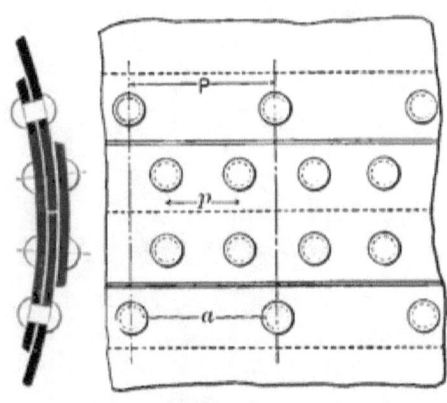

FIG. 80.

The outer row of rivets are wider apart and pass through the inner cover-plate only.

The dimensions assumed are:

Thickness of the plate and of both cover-plates, $t = 7/16$.
Diameter of rivets (iron), 15/16 inch.
Pitch of inner row of rivets, $2\tfrac{5}{8}$.
Pitch of outer row of rivets, $5\tfrac{1}{4}$.

There are five ways in which the joint may fail:
1st. *Tearing at outer row of rivets.* The resistance is

$$(P - d)tf_t = 103{,}770.$$

2d. *Shearing two rivets in double shear and one in single shear.* If the plate pulls out from between the cover-plates shearing off the rivets, then the rivets in the inner row must be sheared through on both sides of the plate, or they are in double shear. The outer row of rivets are sheared at only one place. There are, consequently, five sections of rivets to be sheared for a strip as wide as the larger pitch. The resistance is

$$\frac{5\pi d^2}{4} f_s = 131,100.$$

3d. *Tearing at inner row of rivets and shearing one of the outer row of rivets.* The resistance is

$$(P - 2d)tf_t + \frac{\pi d^2}{4} f_s = 107,430.$$

4th. *Crushing in front of three rivets.* The resistance is

$$3tdf_c = 116,880.$$

5th. *Crushing in front of two rivets and shearing one rivet.* The resistance is

$$2tdf_c + \frac{\pi d^2}{4} f_s = 104,140.$$

The strength of a strip $5\frac{1}{4}$ inches wide is

$$5\frac{1}{4} \times \tfrac{7}{16} \times f_t = 126,560.$$

The efficiency is

$$100 \frac{103,770}{126,560} = 82 \text{ per cent.}$$

Triple-riveted Butt-joint.—The joint shown by Fig. 81 has three rows of rivets on each side. Two rows pass through both cover-plates, and the third or outer row passes through the inner cover-plate only.

The dimensions are:
- Thickness of shell, $t = 7/16$.
- Thickness of both cover-plates, $t_c = 3/8$.
- Diameter of rivets (steel), $d = 15/16$.
- Pitch, inner rows, $p = 3\frac{5}{8}$.
- Pitch, outer row, $P = 7\frac{1}{4}$.

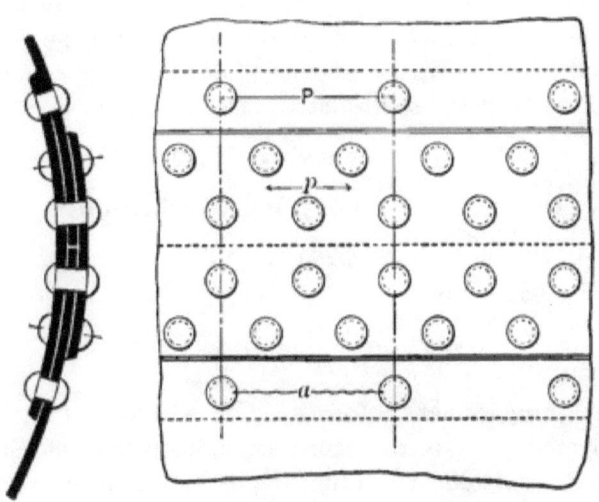

Fig. 81.

The joint may fail in one of five ways:

1st. *Tearing at outer row of rivets.* The resistance is

$$(P - d)t f_t = 151,890.$$

2d. *Shearing four rivets in double shear and one in single shear.* The resistance is

$$\frac{9\pi d^2}{4} f_s = 279,450.$$

3d. *By tearing at middle row of rivets (where the pitch is $3\frac{5}{8}$ inches) and shearing one rivet.* The resistance is

$$(P - 2d)t f_t + \frac{\pi d^2}{4} f_s = 160,340.$$

4th. *By crushing in front of four rivets and shearing one rivet.* The resistance is

$$4dtf_c + \frac{\pi d^2}{4}f = 186,830.$$

5th. *By crushing in front of five rivets.* Four of these rivets pass through both cover-plates and will crush at the shell-plate. The fifth rivet passes through the inner cover-plate only, and will crush at that plate, since the cover-plates are thinner than the shell-plate. The resistance is

$$4dtf + dt_cf_c = 189,170.$$

The strength of a strip of plate $7\frac{1}{4}$ inches wide is

$$Ptf_t = 174,370.$$

The efficiency is

$$100 \times \frac{151,890}{174,370} = 87 \text{ per cent.}$$

Designing Riveted Joints.—One element of the design of a riveted joint is to secure as high an efficiency for the joint as is consistent with other requirements, such as a proper pitch for calking.

A consideration of the example of a single-riveted lap-joint will show that the efficiency can be improved by increasing the diameter of the rivet and by increasing the pitch. In the first place, since the joint will fail by tearing between the rivets, simply increasing the pitch with the same size of rivet will give a greater efficiency. If the pitch is increased till the rivet fails, the failure will be by shearing. Now the resistance to crushing is represented by

$$dtf_c,$$

while the resistance to shearing is represented by

$$\frac{\pi d^2}{4}f_s;$$

that is, the resistance to crushing increases proportionally as the diameter, while the resistance to shearing increases as the square of the diameter. The shearing resistance increases the more rapidly, and can be made equal to the crushing resistance by using a larger rivet. Of course this will demand a further increase of pitch.

In the case of the single-riveted lap-joint now under discussion, the proper proportions for a joint that shall be equally strong against shearing, tearing, and crushing can be calculated directly. The usual way is to determine the diameter of the rivets by making them equally strong against shearing and crushing. Equating the expressions for crushing and shearing resistance, we have

$$dtf_c = \frac{\pi d^2}{4} f_s, \quad \text{or} \quad d = \frac{f_c}{f_s} \frac{4t}{\pi}.$$

For the case in hand with steel plates 7/16 of an inch thick, and steel rivets, the diameter will be

$$d = \frac{95,000}{45,000} \frac{4 \times \frac{7}{16}}{\pi} = 1.17.$$

Having the diameter of the rivets, we may now calculate the pitch by equating the shearing and tearing resistances, which gives

$$\frac{\pi d^2}{4} f_s = (p - d) t f_t, \quad \text{or} \quad p = \frac{f_s}{f_t} \frac{\pi d^2}{4t} + d.$$

For the case in hand we have

$$p = \frac{45,000}{55,000} \frac{\pi \overline{1.17}^2}{4 \times \frac{7}{16}} + 1.17 = 3.2.$$

The efficiency of the joint is the ratio of the resistance to tearing between the rivets to the strength of a strip of plate having a width equal to the pitch, so that the efficiency is

$$\frac{f_t(p-d)t}{f_t pt} = \frac{p-d}{p}.$$

In the case in hand the efficiency is

$$\frac{1}{100}\frac{3.2 - 1.17}{3.2} = 63.4 \text{ per cent.}$$

But the pitch calculated in this method is too great for proper calking with a plate of the given thickness.

The double-riveted lap-joint has three possible ways of failure, which lead to two equations for finding the diameter and pitch of rivets. Equating the shearing and crushing resistance for two rivets, we have

$$2\frac{\pi d^2}{4}f_s = 2dt f_c, \quad \text{or} \quad d = \frac{f_c}{f_s}\frac{4t}{\pi},$$

which will give the same size rivet for a plate of a given thickness as would be found for a single-riveted joint. Now this method has been found to lead to too large a rivet for a single-riveted joint, where a strip having a width equal to the pitch carries one rivet. In the double-riveted joint such a strip carries two rivets, and consequently it is the more certain that the method proposed will give too large a rivet, and of course too large a pitch for proper calking. The advantage of double riveting is that smaller rivets may be used to provide the requisite shearing resistance, and the plate may be less cut away at the section between rivets.

In designing a double-riveted lap-joint it is customary to assume a diameter for the rivets and then determine the pitch by equating the shearing resistance of two rivets to the tearing resistance between the rivets. If the resulting pitch is too large for proper calking, the diameter of the rivets must be reduced. If, on the contrary, the resulting pitch is less than may be allowed, a slightly larger diameter and pitch may be used.

A design of a joint like the single-riveted lap-joint with inside cover-plate, which has a wide and a narrow pitch, involves some difficulty and complexity. The fundamental idea of such a joint is to make the resistance to tearing at the inner row of rivets (when the pitch is small) plus the shearing of the outer row of rivets greater than the resistance to tearing at the outer row of rivets (when the pitch is larger). To insure this condition we may proceed as follows: Equate the resistance to tearing at the outer row of rivets to the resistance to tearing at the inner row plus the resistance to shearing one rivet at the outer row. This gives

$$(P - d)tf_t = (P - 2d)tf_t + \frac{\pi d^2}{4} f_s$$

whence

$$d = \frac{4tf_t}{\pi f_s}.$$

The result is the minimum diameter of rivets allowable. We may now choose a slightly larger diameter of rivets, and then determine the pitch in three different ways, namely, by equating the resistance to tearing at the outer row of rivets, in succession, to the resistance to shearing of three rivets, to the resistance to crushing in front of three rivets, and to the resistance to tearing between the inner rows of rivets and compression before one rivet. The smallest pitch obtained will be the correct one to use with the given diameter of rivet. Should the efficiency of the joint be unsatisfactory, an attempt may be made to raise the efficiency by increasing the diameter of the rivets.

Practical Considerations.—In proportioning a riveted joint, the following considerations, some of which have already been mentioned, must receive attention:

The pitch of rivets near a calked edge must not be too great for proper caulking.

Rivets must not be too near together for convenience in driving.

Punched holes must have a diameter greater than the thickness of the plate.

A riveted seam must contain a whole number of rivets. Again, it is desirable that similar seams, as for example the longitudinal seams for the several rings of a cylindrical boiler, shall have the same pitch.

It is evident that the design of a boiler-joint cannot be considered apart from the general design of the boiler.

Flues.—The tendency of internal pressure in a thin hollow cylinder is to give it a true cylindrical shape; consequently, with fair workmanship, the formulæ for thin hollow cylinders may be applied to the calculation of boiler-shells subjected to internal pressure. But the tendency of external pressure is to exaggerate any imperfection of shape, and cylindrical flues fail by collapsing.

The pressure at which a flue will collapse can be found by direct experiment only.

The earliest and for a long time the only tests on the collapsing of flues were those made by Fairbairn, and published in the Transactions of the Royal Society, in 1858. All of the tubes tested were 0.043 of an inch thick; they varied in diameter from 4 inches to 12 inches, and in length from 20 inches to 60 inches. From these tests he deduced the empirical formula

$$p = \frac{806{,}300 \times t^{2.19}}{l \times d},$$

in which l is the length of the tube in feet and d and t are the diameter and thickness in inches, while p is the collapsing pressure in pounds per square inch. Sometimes the exponent of t is made 2 instead of 2.19, for sake of simplicity. As t is commonly a proper fraction, the use of a smaller exponent will give a higher calculated collapsing pressure.

The tubes in this series were too small, and more especially too thin, to serve as a proper basis for the calculation of boiler-flues. It is quoted because it has been widely used, and is now used by some engineers. It sometimes gives a calculated pressure higher and sometimes lower than that at which flues will collapse, and its use is liable to lead to disappointment if not to disaster.

The following table gives the results of some tests on larger boiler-flues, taken from Hutton's "Steam-boiler Construction." The table gives the dimensions and the actual collapsing pressure, and also the collapsing pressure by Fairbairn's rule and by a rule proposed by Hutton.

EXPERIMENTS ON THE COLLAPSING PRESSURE OF BOILER-FLUES.

Where or by Whom Made.	Dimensions.			Collapsing Pressure in Pounds per Square Inch.		
	External Diameter in Inches.	Length in Inches.	Thickness in 32ds of an Inch.	Found by Experiment.	Calculated by Fairbairn's Rule.	Calculated by Hutton's Rule.
	2	3	4	5	6	7
By Fairbairn................	7.87	276	5	110	109	114
By Fairbairn................	33.5	360	11	99	81	113
By Fairbairn................	42	420	12	97	78	100
By Fairbairn	42	300	12	127	108	119
Engineering Dept., U. S. N.	54	36	8	128	311	120
At Greenock..............	38	86	16	450	740	436
By Knight.................	36	24	8	235	700	218
By Knight.................	36	24	12	468	1568	490
By Knight.................	36	48	12	390	784	350
By J. Howden & Co., Glasgow............	43	23	17	840	2758	842

On the whole the rule proposed by Hutton gives the most concordant results; in most cases Hutton's rule gives a calculated collapsing pressure that is smaller than the actual collapsing pressure; in no case is the calculated result very largely in excess. Fairbairn's rule in some cases shows a very close agreement with experiment, but in others it shows a dangerous excess.

Hutton's rule is

$$p = \frac{Ct^3}{d\sqrt{l}},$$

in which l is the length in inches, d is the diameter in inches, and t is the thickness in *thirty-seconds* of an inch. C is a constant which Hutton makes 600 for iron and 660 for steel.

Mr. Michael Longridge, as a result of an investigation of many boiler-flues, most of which have endured service for years, but some of which failed, gives a rule in the same form but with a constant 540 instead of 600.

For oval tubes and flues it is recommended that the above rules be applied, using for the diameter twice the maximum radius of curvature.

Strengthened Flues.—It is clear from inspection of the preceding table of tests on boiler-flues that the collapsing pressure decreases as the length of the flue increases. Account is taken of this in Hutton's formula, by introducing the square root of the length into the denominator of the expression for calculating the collapsing pressure of a flue. Stating the proposition in the converse manner, the reason why a short flue is the stronger is that the ends of the flue are kept in a circular form by the plates to which the flue is riveted.

It has been customary to strengthen plain flues by the aid of rings placed at regular intervals. The section of a ring made of angle-iron is shown by Fig. 82a. The ring is riveted to the flue at intervals, a thimble being placed over each rivet to give space for circulation of water between the ring and the flue. The rings were sometimes solid, made of one piece of angle-iron bent up and welded. Most frequently the ring was in halves, which were merely belted together at the joint. Such rings could be easily removed when the flue was taken out of the boiler.

A better method of strengthening a flue is to make it of

short pieces so joined at the ends as to make stiffening rings. Fig. 82 shows three ways in which this can be done. At *b* is shown the Adamson ring, formed by flanging the edges of the short lengths of flue outwardly, and riveting through a welded iron ring. At *c* is shown a welded ring of T iron, to which the short lengths can be riveted without flanging. This

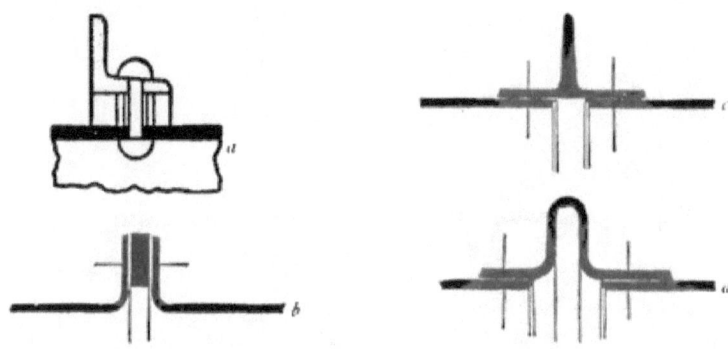

FIG. 82.

method provides for calking both inside and outside. It does not require the flue to be flanged; but flanging by machinery is rapid, and does not give trouble when good iron or steel is used. Material that will not stand flanging should not be used for flues. At *d* is shown the bowling hoop-ring, which has the advantage that it provides for longitudinal expansion of the flue.

Flues for Scotch and other marine boilers with furnace-flues, are stiffened by transverse or helical corrugations, which provide at the same time for longitudinal expansion. A number of methods of corrugating furnace-flues will be illustrated in connection with tests given on the following pages.

Tests on Furnace-flues.—The strength of corrugated and other stiffened flues can be determined only by tests on full-sized specimens. The following tests are taken from a paper by Mr. B. D. Morison, read before the Northeast Coast Institution of Engineers and Shipbuilders.

Furnaces made with Adamson Joints.

Tests made at the Works of Hall, Russell & Co., Aberdeen, in 1882, and of J. Howden & Co. in Glasgow, in 1887.

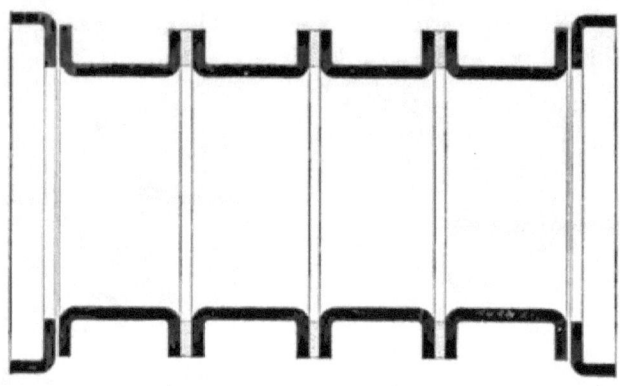

Date of Test.	Length of Furnace.	Number of Rings.	Mean Thickness of Plate.	External Diameter in Inches over Plain Part.	Greatest Diff. in Diameter at any Part.	Collapsing Pressure.	Collapsing Coefficient $P \times D \div T$.	Collapsing Coefficient reduced to Steel of 27 Tons Tensile.
1882	6 ft. 5¾ in. total length. Length of rings: 18¼", 19", 19", and 20"	4	1st ring ½", 2d ring $\frac{15}{32}$", 3d ring $\frac{15}{32}$", 4th ring ½"	43	9/64	3d ring at 700	64,213	61,918
1887	7 ft. ½ in. total length. Length of each ring, 23"	4		43.09		1st ring at 840, 2d ring at 760, 3d ring at 840, 4th ring at 835	64,240	61,945

NOTE.—No record of tensile strength of steel; 28 tons per square inch assumed. The collapsing coefficients are calculated on external diameter of furnace over plane part.

STRENGTH OF BOILERS.   211

Fox's Patent Corrugated Furnace.
Official Tests made at Leeds Forge, Leeds, in 1882, 1890, and 1891.

Date	Length of Furnace	Number of Corrugations	Greatest Length of Flat at End	Thickness of Plate in Inches - Front End	Middle	Back End	Mean	Mean Diam. in Inches	Greatest Diff. in Diameter at any Part	Collapsing Pressure	Collapsing Coefficient $P \times D \div T$	Ultimate Tensile Strength of Steel	Collapsing Coefficient reduced to Steel of 27 Tons	Position of Collapse
Nov., 1882	6′ 9″	13	....	....	....	....	.52	35.875	....	900	62,091	22.7	73,852	....
Nov. 4, '90 / Feb. 11, '91	6′ 3½″	11	5½	.339	.364	.331	.349	34.125	9/32	830	81,157	29.05	75,430	On flat
Feb. 11, 1891	6′ 3½″	11	6	.345	.384	.398	.378	34.25	5/32	800	72,486	29.17	67,094	On flat
Feb. 11, 1891	6′ 7″	12	6	.419	.463	.462	.452	33.625	4/32	1130	84,062	29.26	77,569	Second corrugation
Feb. 11, 1891	6′ 8⅛″	12	6½	.471	.473	.454	.468	34.469	4/32	1090	80,280	29.02	74,692	On flat
Feb. 11, 1891	6′ 2½″	11	6½	.577	.585	.551	.574	33.593	7/32	1400	81,934	29.55	77,485	Eleventh corrugation
Feb. 11, 1891	6′ 6⅛″	11	6¼	.542	.582	.596	.575	34.937	4/32	1410	85,671	29.41	78,650	Front corrugation

NOTE.—The collapsing coefficients are calculated on the mean diameter of furnace. The first, third, fifth, and seventh were annealed.

Farnley Spirally-corrugated Furnace.

Official Tests made at the Works of Farnley Iron Co., Limited, Farnley, near Leeds.

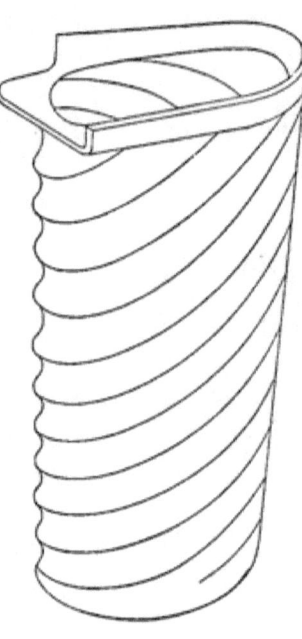

Date of Test.	Length of Furnace.	Mean Thickness of Plate in Inches.	Mean Diameter in Inches.	Collapsing Pressure.	Collapsing Coefficient $P \times D \div T$.	Ultimate Tensile Strength of Steel Assumed.	Collapsing Coefficient reduced to Steel of 27 Tons Tensile.	Position of Collapse.
May, 1888	6' 5¾"	.559	39.231	835	58,601	28	56,508	No record
Do.	6' 6¼"	.548	39.132	850	60,697	28	58,529	
Do.	6' 6¾"	.442	38.928	670	59,008	28	56,900	
Do.	6' 6¾"*	.446	39.394	570	50,346	28	48,548	
Do.	6' 6⅝"	.354	39.296	515	57,168	28	55,126	

* This furnace was not so true as the others.

Note.—The collapsing coefficients are calculated on the mean diameter of the furnaces.

STRENGTH OF BOILERS.   213

Holmes' Corrugated Furnace.
Official Tests made at Chas. P. Holmes & Co.'s, Hull, in 1891.

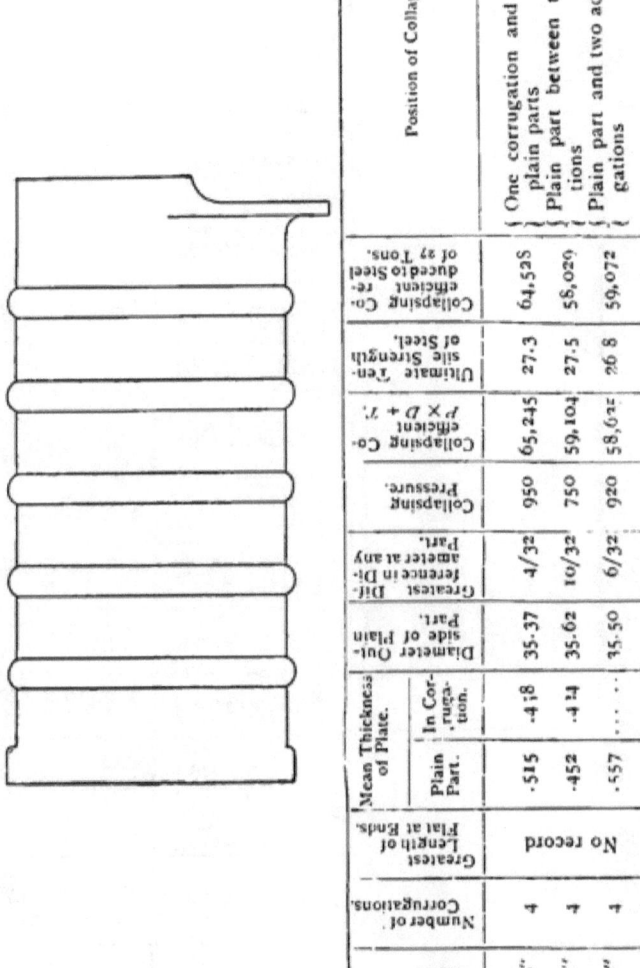

Date of Test.	Length of Furnace.	Number of Corrugations.	Greatest Length of Flat at Ends.	Mean Thickness of Plate. Plain Part.	Mean Thickness of Plate. In Corrugation.	Diameter Outside of Plain Part.	Greatest Difference in Diameter at any Part.	Collapsing Pressure.	Collapsing Coefficient $P \times D \div T$.	Ultimate Tensile Strength of Steel.	Collapsing Coefficient reduced to Steel of 27 Tons.	Position of Collapse.
1891	7' 0"	4	No record	.515	.418	35.37	4/32	950	65,245	27.3	64,528	One corrugation and two adjacent plain parts
1891	7' 0"	4	No record	.452	.414	35.62	10/32	750	59,104	27.5	58,029	Plain part between two corrugations
1891	7' 0"	4	No record	.557	....	35.50	6/32	920	58,625	26.8	59,072	Plain part and two adjacent corrugations

NOTE.—The collapsing coefficients are calculated on the diameter of the furnaces over flats.

Purves's Patent Furnace.

Official Tests made at the Works of Sir John Brown & Co., Sheffield.

Date of Test.	Length of Furnace.	Number of Corrugations.	Greatest Length of Flat End.	Thickness of Plate in Inches.				Diameter over Flats in Inches.	Greatest Difference in Diam. at any Part.	Collapsing Pressure.	Collapsing Coefficient $P \times D \div T$.	Ultimate Tensile Strength of Steel.	Collapsing Coeff. reduced to Steel of 27 tons Tensile.	Position of Collapse.
				Front End.	Middle.	Back End.	Mean.							
Mar. 12, 1887	6′ 6⅝″	7	13 13/16″	·561	·582	·548	·568	38.61	7/32	740	50,302	27.2	49,932	On flat
Mar. 12, 1887	6′ 6⅝″	7	13 13/16″	·544	·554	·532	·546	38.685	11/32	760	53,847	26.6	54,656	" "
1887	6′ 7½″	8	7½	·299	·286	·310	——	38.652	8/32	650	85,165	25.9	88,782	Bet. 1st & 2d ribs
1887	6′ 7½″	8	7½	·327	·302	·313	·311	38.591	6/32	635	78,795	27.9	76,253	Bet. 6th & 7th ribs
1887	6′ 7½″	8	7½	·482	·535	·483	·509	38.83	16/32	800	61,029	26.1	63,133	From end to end
1887	6′ 7½″	8	7½	·529	·589	·529	·559	38.868	8/32	800	55,625	27.9	53,831	" " "
1887	6′ 7½″	8	7½	·583	·632	·597	·611	38.647	10/32	875	55,622	27.6	54,422	" " "
1887	6′ 7½″	8	7½	·576	·638	·60	·613	38.788	11/32	873	55,239	28.3	52,701	" " "

NOTE.—The collapsing coefficients are calculated on the diameter of the furnace over flats.

Purves's Patent Furnaces.

Official Tests made at Sir John Brown & Co.'s Works at Sheffield in 1889.

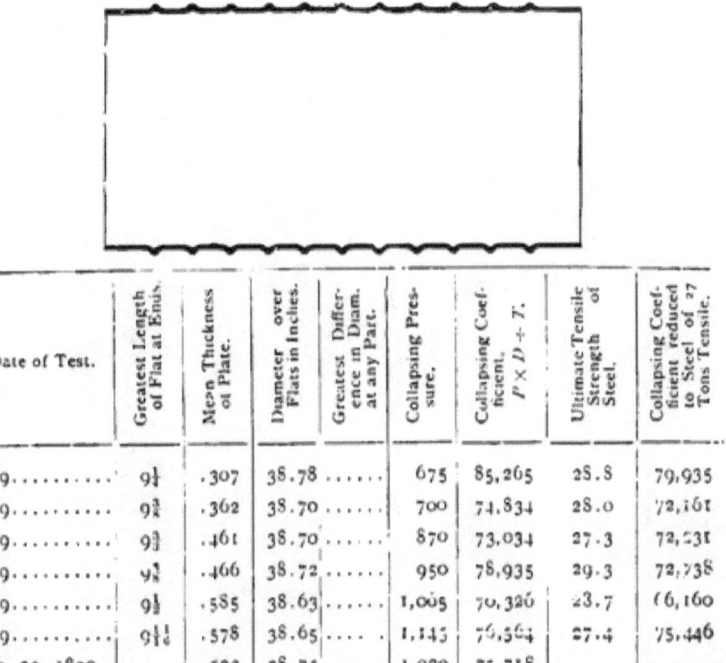

Date of Test.	Greatest Length of Flat at Ends.	Mean Thickness of Plate.	Diameter over Flats in Inches.	Greatest Difference in Diam. at any Part.	Collapsing Pressure.	Collapsing Coefficient, $P \times D \div T$.	Ultimate Tensile Strength of Steel.	Collapsing Coefficient reduced to Steel of 27 Tons Tensile.
1889.........	9¼	.307	38.78	......	675	85,265	28.8	79,935
1889.........	9¾	.362	38.70	......	700	74,834	28.0	72,161
1889.........	9¾	.461	38.70	......	870	73,034	27.3	72,231
1889.........	9⅞	.466	38.72	......	950	78,935	29.3	72,738
1889.........	9½	.585	38.63	......	1,005	70,326	23.7	66,160
1889.........	9¹¹⁄₁₆	.578	38.65	......	1,145	76,564	27.4	75,446
Dec. 23, 1890..	......	.522	38.75	......	1,020	75,718	......	......

Corrugations spaced 9" apart.   Not very full records kept.

Note.—The collapsing coefficients are calculated on diameters of furnaces over flats.

Morison's Suspended Furnace.
Official Tests made at the Leeds Forge, Leeds, in 1891.

Date of Test.	Length of Furnace.	Number of Corrugations.	Greatest Length of Flat at Ends	Thickness of Plate in Inches.				Mean Diameter in Inches.	Greatest Difference in Diameter at any Part	Collapsing Pressure.	Collaps. Coef. $F \times D \div T$.	Ultimate Tensile Strength of Steel.	Collaps.Coef.reduced to Steel of 27 tons Tensile Strength.	Position of Collapse.
				Front End.	Middle.	Back End.	Mean.							
Sept. 25–26	6′ 5½″	9	4¾″	.340	.310	.340	.325	34.156	11/32	795	83.550	27.64	81.615	Flat at the front end
Do.	6′ 7¾″	9	4½	.341	.391	.368	.373	34.265	2/32	900	82.677	27.34	81.648	2d corrugation from front
Do.	6′ 5½″	9	4½	.445	.492	.451	.470	34.444	19/64	1100	80.613	26.92	80.852	Uniformly on weld
Do.	6′ 6¾″	9	4½	.452	.458	.441	.452	34.657	11/32	1050	80.578	26.56	81.913	7th, 8th, and 9th corrugation at weld
Do.	6′ 7¾″	9	4¾	.535	.516	.551	.530	34.719	3/32	1300	85.159	27.52	83.550	1st, 2d, and 3d corrugations
Do.	6′ 7′	9	5	.556	.573	.565	.567	34.078	7/32	1340	80.537	27.09	80.269	Flat at front and first corrugation

NOTE.—The collapsing coefficients are calculated on the mean diameter of furnaces. All these furnaces were annealed in the presence of the Board of Trade officers.

Discussion of Results of Tests on Flues.—The stress in a thin hollow cylinder subjected to external fluid pressure may be calculated by an equation having the same form as that for a cylinder subjected to internal pressure; the equation may be deduced by a similar method. Thus the stress will be

$$s = \frac{pr}{t},$$

in which p is the pressure per square inch, r is the radius and t is the thickness, both in inches. In the table we have a column giving the coefficient of collapse calculated by the expression

$$\frac{PD}{T},$$

in which P is the pressure, D is the diameter, and T is the thickness. The coefficient appears consequently to be twice the compressive stress in the flue at the time of collapsing. This coefficient is fairly regular for each style of furnace, and is somewhere near the tensile strength of the metal from which the flue is made; in some cases it is less and in some more than the tensile strength. Now soft steel in the form of short cylinders will begin to flow when the compressive stress in a testing-machine is about equal to the strength of pieces used for tensile tests. In other words, the tensile and compressive strengths are about equal. The furnaces tested appear, then, to have collapsed when the compressive stress was half the ultimate compressive strength of the metal. Now the limit of elasticity for both tension and compression, for soft steel, is about half the ultimate strength, so that the collapse occurred somewhere about the elastic limit. We should not, however, attribute too much importance to this consideration, but it will be better to follow ordinary practice and consider the equations used for calculating the safe working pressure on flues to be empirical, and to depend directly on experiment.

Rules for Working Pressure on Flues.—There are three sets of rules for working pressure on flues, which we shall consider; namely, those of the *British Board of Trade*, those of *Lloyd's Marine Insurance Underwriters*, and those of the *United States Inspectors of Steam-vessels*. These rules are changed from time to time, and include certain directions to inspectors that need not be given here; if a boiler is built for inspection under these or any other rules the only safe way is to obtain the current edition of the rules and see that the boiler conforms thereto, and also that the boiler is properly proportioned according to the best information that can be obtained by the designer.

Rules for Plain Flues.—Both *Lloyd's* and the *United States Inspectors'* rules use for plain flues an equation in the form

$$P = \frac{89,600 \times T^2}{LD},$$

in which P is the working pressure in pounds per square inch, L is the length in feet, and T and D are the thickness and diameter in inches. This is Fairbairn's equation with 2 instead of 2.19 for the exponent of T, and with a constant

$$89,600 = \frac{806,300}{9},$$

so that the working pressure is made one ninth of the calculated collapsing pressure by Fairbairn's rule. The use of so large a factor as nine shows that the rule is not considered adequate. Flues designed under this rule will probably be strong enough.

The Board of Trade rule differs only in replacing the

factor 89,600 by the approximate figure 99,000. The rules, however, require that the pressure shall not be greater than

$$P = \frac{8800 \times T}{D},$$

which provides that the stress shall not exceed 4400 pounds per square inch. For corrugated, ribbed, or grooved furnaces (such as the several furnaces for which tests are given) both the *Board of Trade* and the *Inspectors'* rules give for the working pressure

$$P = \frac{1400 \times T}{D},$$

in which P is the working pressure in pounds on the square inch, and T and D are the thickness and diameter in inches. This rule makes the working stress 7000 pounds per square inch.

Lloyd's rule for these furnaces is given by the equation

$$P = \frac{C(T-2)}{D},$$

in which T is the thickness in *sixteenths of an inch*, D is the diameter in inches, measured over the corrugations or ribs of corrugated or ribbed furnaces, and over the plain part of Holmes' furnaces. C is an arbitrary constant having the following values:

$C = 1000$ for steel corrugated furnaces when the tensile strength of the material is under 26 tons, and corrugations are 6 inches apart and $1\frac{1}{2}$ inches deep.

$C = 1259$ for steel furnaces corrugated on Fox's or Morison's plans, tensile strength to be between 26 and 30 tons.

$C = 1160$ for ribbed furnaces with ribs 9 inches apart.

$C = 912$ for spirally-corrugated furnaces.

$C = 945$ for Holmes' furnaces, when corrugations are not over 16 inches apart and not less than two inches high.

In this rule the use of $T - 2$ (in sixteenths of an inch) instead of T is practically an allowance for wasting of the plate to the extent of one eighth of an inch. The working stress calculated on the assumed diameter will be found by multiplying by sixteen and dividing by two; in case of the first constant the stress is

$$\frac{1000 \times 16}{2} = 8000$$

pounds per square inch.

Fire-tubes.—The thickness usually given to fire-tubes to insure sound welding and to provide for expanding into the tube-sheets is in excess of that required to prevent collapsing. There appears, however, to be no experiments to show the actual collapsing pressure for such tubes.

The joint made by expanding the tubes into the tube-sheets of locomotive and cylindrical tubular boilers has been found both by experiment and practice to be strong enough to secure the tube-sheet without additional staying. It is, however, the custom to make part of the fire-tubes of marine drum-boilers thick enough to take a shallow nut outside of the tube-plate; without such stay-tubes there is liable to be leakage at the ends of the tubes.

Girders.—When a flat surface cannot conveniently be stayed directly, it is customary to stay the surface to girders properly supported at the ends or elsewhere. The crown-bars of the locomotive-boiler shown on Plate II, and the girders over the combustion-chamber of the marine boiler shown by Fig. 13, page 17, may be taken as examples. Again, the channel-irons which are riveted to the flat heads of the cylindrical boiler shown by Plate I act as girders.

The load which a girder of given material can safely carry depends on the form and dimensions of the girder, and on the manner of supporting and loading the girder. Some girders, like those over the combustion-chamber in Fig. 13, can be calculated by the simple theory of beams; others, like crown-bars for locomotives and the channel-bars on Plate I, can be properly calculated only by the theory of continuous girders.

A proper understanding of the theories of beams and of continuous girders can be obtained from standard works on applied mechanics. An adequate statement of even the theory of beams is out of place in a work on boilers; an incomplete statement is unadvisable, since it is liable to be misleading. One simple example will be worked out as an illustration of the use of the beam theory in boiler-design.

As an example, we will take the girders over the combustion-chamber of the marine boiler shown by Fig. 13, page 17. The girders are spaced 7 inches apart, and each carries three stays spaced $6\frac{1}{4}$ inches apart. The load on each stay-bolt at 160 pounds steam-pressure is

$$7 \times 6\tfrac{1}{4} \times 160 = 7000 \text{ pounds},$$

and the total load on one girder is 21,000 pounds. The supporting force at each end of the girder is 10,500 pounds. The span of the girder is $22\frac{1}{2}$ inches, and the half-span is $11\frac{1}{4}$ inches. The bending-moment at the middle of the girder due to the supporting force acting upward, and to the load on one bolt acting downward, is

$$10,500 \times 11\tfrac{1}{4} - 7000 \times 6\tfrac{1}{4} = 74,375 = M.$$

Each girder is made of two plates each 5/8 of an inch thick, and 7 inches deep. The moment of inertia of the section of the girder at the middle is

$$\tfrac{1}{12} \times 2 \times \tfrac{5}{8} \times 7^3 = I.$$

The distance of the most strained fibre is

$$7 \div 2 = 3\tfrac{1}{2} = y.$$

The working fibre-stress is consequently

$$f = \frac{My}{I} = \frac{74\,375 \times 3\tfrac{1}{2}}{\tfrac{1}{12} \times 2 \times \tfrac{5}{8} \times 7^3} = 7287$$

pounds per square inch.

Stayed Flat Plates. — The method of calculating the stresses in a flat plate supported at regular intervals by stays or stay-bolts, such as the sides of a locomotive fire-box, is treated in the theory of elasticity, under the heading of " indefinite plates which are firmly held at a system of points dividing them into rectangular panels." A complete solution of this problem is possible only when the panels are squares, that is, when the rows of stays are equidistant longitudinally and transversely.

If the steam-pressure is represented by p, the thickness of the plate by t, and the pitch of the stays by a, then the maximum direct stress, which is a tension at certain places and a compression at other, is given by the formula

$$f = \frac{2}{9}\frac{a^2}{t^2}p.$$

The maximum shearing-stress is given by the equation

$$f_1 = \frac{1}{36}\frac{pa^4}{Et^3},$$

in which E is the modulus of elasticity of the material.

If the sheets of a locomotive fire-box, or other stayed plates, have a direct tension or compression, proper allowance must be made for it.

If stays or stay-bolts are in rows that are not equidistant each way, as for example the through-stays in the steam-space in Fig. 13, page 17, then the largest pitch may be used in the above equations. The actual stress will in such case be less than the calculated stress by an unknown amount. If, further, stays are arranged irregularly, the greatest distance in any direction may be used in the equations, but the calculated stress may then be very different from the actual stress; it is, however, always the larger.

As an example, we may calculate the stress in a side sheet of the locomotive fire-box shown on Plate II. Here the rows of rivets are four inches apart each way, the plate is 5/16 of an inch thick, and the steam-pressure is 170 pounds. The maximum stress is

$$f = \frac{2}{9} \frac{4^2}{(\frac{5}{16})^2} 170 = 6190.$$

The shearing-stress in this case is very much smaller.

Now the crown-bars are bedded on and are partly supported by the side sheets of the fire-box. The crown-sheet is 72 inches long and $45\frac{5}{8}$ inches wide, and has an area of

$$72 \times 45\frac{5}{8} = 3285$$

square inches, and is subjected to a pressure of

$$3285 \times 170 = 558{,}450$$

pounds. The distribution of this load between the side sheets and the sling-stays can be determined only by the calculation of the crown-bars as continuous girders, and may be disturbed by the expansion of the fire-box and by other causes. If it be assumed that the side sheets carry half the load on the crown-bars, then one side sheet will carry one fourth of 558,050, or 139,512 pounds. The side sheet is 72 inches long and 5/16 of an inch thick, so that the stress per square inch from the load on the crown-bars is

$$139{,}512 \div 72 \times \tfrac{5}{16} = 6200$$

pounds,—about as much as the stress calculated above. The

total compression on the side sheet is therefore about 12,400 pounds per square inch.

This calculation, which appears sufficiently simple, illustrates the danger of making calculations by formulæ without knowing how they are derived and how they should be applied. The formula for staying given above is often quoted without any reference to tensile or compressive stress on the stayed sheet, from other causes; the use of such a formula by one who is unfamiliar with the theory of elasticity may lead to serious error in design.

Factor of Safety.—The ratio of the working pressure of a boiler to the pressure at which the boiler or any part of a boiler may be expected to fail quickly, is called the factor of safety for the boiler or for that part of the boiler.

It is commonly recommended by writers that a factor of safety of six shall be used for boilers; probably such a factor would be economical for a boiler that is expected to work continuously for many years, as it allows a margin for deterioration. If the stresses coming on the parts of a boiler can be determined, a general factor of five will give sufficient security. If the boiler is carefully watched, a factor of four may be used; many boilers are worked with this factor. The use of an excessively large factor of safety, for example of the factor nine for flues calculated by Fairbairn's equation, shows a lack of confidence in the method. It is proper to make allowance for corrosion of parts like stays: this may be done either by using a larger factor of safety, or by a direct allowance; thus all stays, whatever their diameters, may have an eighth of an inch added to the diameter to allow for corrosion. It is of course proper in any structure to make small but important members, such as stays in boilers, large enough to place them beyond any suspicion of failure.

Hydraulic Tests of Boilers.—It is customary to subject new boilers to a water-pressure considerably in excess of the working pressure, to discover any leaks at riveted joints, at the tube-sheets, or elsewhere; should there be any gross defect of design or workmanship it will be developed by this hydraulic test. Old boilers after repairs are subjected to a hydraulic test for the same purpose, but the pressure is not carried so high as for new boilers.

The pressure applied during a hydraulic test is seldom more than once and a half the working pressure, and as most boilers have an actual factor of safety of not more than five, and frequently of four, it is apparent that the recommendation of some authors, that the test pressure should be twice the working pressure, cannot ordinarily be followed without danger of injuring the boiler. With a factor of safety of six there should be no danger of injuring the boiler by applying a hydraulic pressure equal to twice the working pressure.

It should be borne in mind that some of the worst stresses that come on the different parts of the boilers are due to unequal expansion and contraction, and that such stresses are not set up during a hydraulic test. Finally, the fact that a boiler has successfully withstood a hydraulic test is not a conclusive proof that it is safe; too many unfortunate explosions of boilers, more frequently old boilers, after a hydraulic test, have shown this.

The safety of a boiler is to be insured by careful and correct design, honest and thorough workmanship, and intelligent care in service. Forms and methods of design and construction that do not admit of ready calculation should be avoided; in no case should the ordinary hydraulic test be relied upon to guarantee the strength of parts that cannot be calculated with a fair degree of certainty. If such forms are used in any case, they ought to be tested separately to a pressure of two or three times the working pressure, and some examples of each form and size ought to be tested to destruction.

The boiler undergoing a hydraulic test should be carefully inspected, and any notable change of shape or leakage should be investigated to discover the cause. Frequently small leaks that are developed during a test are stopped at once by calking or otherwise, but it is preferable to mark the place of the leak and calk after the pressure is removed. This of course requires another test to find out if the calking is successful.

The pressure is usually applied by filling the boiler entirely full of water and then pumping in enough water, by hand or by power, to supply the leaks and develop the pressure required. If the pumping is done by hand, it is desirable to carefully remove all air from the boiler to avoid the labor of compressing air up to the test pressure. If the pumping is done by power, the saving of work is of less consequence, and a little air remaining in the boiler will act as a cushion, and lessen the shocks due to the strokes of the pump.

New boilers are tested on the boiler-shop floor; old boilers are commonly tested in their settings, and in such case the inspection during a test is less convenient and efficient.

It is sometimes recommended that hot water shall be used for testing a boiler; but there seems to be no advantage in so doing, as it is unequal expansion, and not merely rise of temperature, that sets up the unknown stresses that are so destructive to the boiler. Of course the use of hot water makes an efficient inspection during the test difficult if not impossible.

When there is no other way of applying the hydraulic test to a boiler in its setting, the boiler may be quite filled with water, and then a light fire may be started in the furnace. The expansion of the water will develop the required pressure at a much less temperature than that of steam at the same pressure, and with less danger should the boiler fail. This method cannot be recommended for general use; and in case it is followed care must be taken not to exceed the desired pressure.

Hydraulic Test to Destruction.—In 1868 a boiler-shell, made to represent a part of the shell of a gunboat boiler, was tested by hydraulic pressure at the Greenock Foundry,* with the intention of bursting it. The shell was 11 feet long and 7 feet $8\frac{3}{16}$ inches mean diameter. It was made of three sections of 19/32 plate, triple-riveted, with butt-joints and double cover-plates at the longitudinal joints, and lapped and double riveted at the ring seams. The rivets were staggered for both longitudinal and ring seams. The end-plates were 20/32 thick, and stayed with through-stays and washers, spaced 14 inches on centres. The stays were $1\frac{7}{8}$ inches in diameter; the screws at the ends of the stays were $2\frac{1}{4}$ inches in diameter. Finally, it may be said that the shell was designed to fulfil the Admiralty specifications for a working pressure of 145 pounds per square inch. The workmanship was of the same degree of excellence usual for boiler-work at that establishment.

First Test.—The shell was first subjected to the working pressure of 145 pounds, and showed a slight alteration of form due to the tendency of internal pressure to give it a true cylindrical form. The pressure was then raised to the Admiralty test pressure of 235 pounds, and then to 300 pounds without developing leaks. There were some minor changes of form due to the increase of pressure. The pressure was then removed and the shell returned to its original dimensions.

Pressure was then raised to 330 pounds, when there was a slight leak at the manhole door. At 450 pounds pressure the leak at the manhole door exceeded the capacity of the pumps. There was also a slight leak at the corners of two butts. The manhole was then strengthened—no other repairs were made.

Second Test.—Pressure was raised to 350 pounds and developed a small leak at the manhole. There were slight

---

* Trans. Inst. Naval Arch., vol. xxx. p. 285.

leaks at the butt-straps, which were calked at the end of the test. The manhole, however, leaked so that the test was stopped.

Third Test.—After additional bolts were put into the manhole cover the pressure was raised to 350 pounds without leakage. At 360 pounds the manhole began to leak, and at 580 pounds the test was stopped on that account. The butt-straps opened visibly at the calking and leaked more than before.

Fourth Test.—The butt-joints were again calked and additional pumps were employed. The shell was again tight at 350 pounds and the pressure was carried to 620 pounds, at which there was a good deal of leakage at the butt-straps. Only one or two rivets showed signs of leakage; there appeared to be no difference between the hand and machine riveting in this respect. At the pressure of 620 pounds the entire capacity of the pumps was required to supply the leakage.

The distortion of the shell was very marked at the higher pressures, and increased with the pressure; thus the ends bulged an inch at 520 pounds, about $1\frac{1}{4}$ inches at 580 pounds, and nearly two inches at 620 pounds. The sides bulged more irregularly, but to the extent of nearly an inch at 620 pounds. The stays drew down uniformly 1/64 of an inch at 520 pounds, 2/64 at 580 pounds, and 4/64 at 620 pounds. They increased in length $2\frac{1}{32}$ inches at 520 pounds, $3\frac{1}{8}$ inches at 580 pounds, and $3\frac{5}{8}$ inches at 620 pounds; this accounts for the bulging of the end-plates.

The mean tensional strength of the plates from which the shell and butt-straps were made may be taken at 61,500 pounds. At 620 pounds the tension on the plates between the rivet-holes was 57,504 pounds, or $93\frac{1}{2}$ per cent of the strength of the solid plate, and there was no serious disturbance of the structure. The ring seams increased in diameter about $\frac{5}{8}$ of an inch, and the shell bulged out between them.

The various portions of the boiler acted in harmony and showed no special weakness at any point. The butt-joints had the rivets spaced 5¾ inches on centres to give a percentage of 83.7 per cent of the plate, and this may have caused the leakage found there. The riveting appeared to be reliable at the extreme pressure reached. This test seems to show that a boiler will give signs of weakness long before it will fail. Such signs of weakness should be carefully investigated: if there is any local weakness or deterioration, repairs or alterations may be made; if there are evidences of general deterioration, the working pressure must be reduced, or better, the boiler may be replaced by a new one.

Boiler-explosions.—The great destruction of life and property that is liable to be caused by a violent boiler-explosion makes it imperative that the causes should be carefully investigated, to the end that explosions may be prevented.

With this in view the boiler and its parts, and any wreck or evidence of destruction caused by the explosion should be left undisturbed until the scene of the explosion can be examined by a competent engineer. Of course if any persons are injured by the explosion they must be rescued and cared for immediately, and also any building or structure that is so injured as to threaten life or safety must be attended to at once; but it should be borne in mind that the examination by the engineer for the purpose of determining the cause of the explosion is also in the interest of humanity, since its aim is to avoid future explosions. All idle or simply curious persons should be excluded from the scene of the explosion, more especially as such persons are apt to disturb or even carry away things that may be of importance in the study of the cause and history of the explosion. If the explosion is accompanied by loss of life or injury to person or property, it will be followed by a legal investigation in which the testimony of the engineer or engineers who have examined the scene of the explosion will be of prime importance, as it will have a large influence in locating responsibility for the disaster.

While various causes may lead to boiler-explosion, it is unfortunately true that by far the greater part of violent explosions are due to the fact that the boiler is too weak to endure service at the regular working pressure. A new boiler may be weak through defective design or workmanship; there can be no excuse for the explosion of a new boiler from weakness, and such explosions in good practice are rare. An old boiler is liable to become weak through local or general corrosion or other deterioration; this amounts to saying that a boiler will eventually wear out.

The length of time that a boiler will endure service depends (1) on the design, (2) on the thickness of plates and the quality of the metal, (3) on the workmanship, (4) on the care given it, and (5) on the quality of the feed-water. Definite figures cannot be given for the life of a boiler, since it depends on so many things. The following table gives the number of years several kinds of boilers can endure regular service if they are properly built and cared for:

Lancashire, low-pressure	15 to 20 years.
Locomotive type, stationary	12 to 15 "
Locomotive-boilers	8 to 12 "
Vertical boilers	10 to 15 "
Vertical boiler with submerged tubes	14 to 18 "
Horizontal cylindrical tubular	15 to 20 "
Scotch marine boiler	12 to 15 "
Water-tube boiler	12 to 16 "
Pipe or coil boiler	5 to 8 "

By water-tube boiler is here meant a boiler with a shell or drum containing a considerable body of water. By pipe or coil boiler is meant a boiler made up of pipe and pipe-fittings, with a separator.

Horizontal boilers will require one, and vertical boilers two extra sets of tubes, before the shell is condemned. A locomotive-boiler will require two extra sets of tubes, and the entire fire-box will be renewed once in the life of the boiler.

If boilers are subjected to careless or ignorant abuse, they may be used up in a fraction of their proper time of service, especially if cheaply built. This will account for the numerous explosions of sawmill boilers and agricultural boilers.

It has been pointed out that leakage is frequently a sign of weakness; a perversion of this idea leads to the assumption that a boiler is safe as long as it can be kept from leaking. Too many boiler-explosions have this history: The boiler, after long and satisfactory service, began to leak; a cheap man was employed to repair the boiler, the repairs consisting mainly of excessive calking to stop the leaks; soon after the repairs, perhaps the first time the boiler was fired up, it exploded violently. A fit conclusion of the history is to ascribe the explosion to some obscure cause or to carelessness of the attendant, if he was killed by the explosion.

Serious injury may be caused by overheating any part of the heating-surface, due to low water, to defective circulation, or to deposits of non-conducting substance on the plates or tubes. The overheated member, or plates, of the boiler may burst or collapse, and such failure may lead to an explosion of the boiler, but frequently the escape of steam and water will check the fire and relieve the pressure on the boiler. Local failures are dangerous to the boiler attendants, especially in a confined fire-room, as on shipboard. Unless there is direct evidence of overheating, either from known circumstances before the explosion or from signs on the boiler after explosion, the cause of the failure should be sought elsewhere.

If a boiler shows signs of low water or of overheating the fire should be checked by any effectual means. The most ready way of checking the fire is to close the ash-pit doors and throw ashes onto the fire. If there are no ashes at hand, then fresh fuel may be used instead, since its first effect is to deaden the fire. There will be time for caring for, or drawing the fire before the fresh fuel is fairly in combustion. An attempt to draw the fire without first deadening it is liable to give a fierce combustion for a short time; moreover, more time is required to draw the fire. If the furnace has a dumping-grate, the fire may be immediately thrown into the ash-pit without waiting to deaden it. The damper should be left open so that if a rupture occurs the steam may escape up the chimney. Meanwhile the steam made by the boiler should be disposed of by allowing the engine to run or by any other means, for example by opening the safety-valve, provided that it is merely a case of overheating, not accompanied by excessive pressure. It will probably be well to start the feed-pumps or to increase the supply of feed-water. Should the introduction of feed-water be badly arranged so that a large volume of cold water will be thrown onto a heated plate, it is possible that starting the feed-pump may cause a contraction which will start a rupture.

It has been found by experiment that boiler-flues that have been purposely allowed to become bare and overheated have been saved by suddenly directing a stream of cold feed-water upon them, though such treatment may make them leak at the joints. The heat stored in such hot plates is insignificant as compared with the heat in the water and steam in the boiler.

Excessive pressure, especially if it is enough to give good reason to fear an explosion, is more difficult to deal with; the chances of success are less and the risks are greater than when the water is low, but the pressure is not excessive. If possible the fire should be checked and the pressure relieved. The first may be done by throwing on ashes or cold fuel, and the second by running the engine at full load. It is at least doubtful whether starting the feed-pump will reduce the pressure fast enough to do much good, and on the other hand there may be cases where such action would start an explosion. It is not best to open the safety-valve, since the sudden opening of a large safety-valve gives a shock which may determine the explosion. Some explosions have been reported that occurred immediately after the safety-valve opened.

A large amount of energy is stored in the steam and water in a boiler in the form of heat. An idea of the amount of energy in any given case may be obtained by a simple calculation. Thus the cylindrical boiler shown on Plate I, at 150 pounds pressure by the gauge, will contain 6600 pounds of water and 22 pounds of steam. Taking 165 pounds absolute to correspond to 150 pounds by the gauge, we find from a table of the properties of steam that 338 thermal units are required to raise one pound of water from freezing-point to 366° F., corresponding to 165 pounds absolute. Now one thermal unit is equivalent to 778 foot-pounds of work. Consequently the energy stored in the hot water in the boiler, calculated from freezing-point, is

$$6600 \times 778 \times 338 = 1,736,000,000 \text{ foot-pounds.}$$

After the water is heated to 366° F. there will be required 855.6 thermal units to vaporize one pound into steam at 165 pounds absolute. But 83.6 thermal units will be expended in changing the volume of the fluid when it passes from water into steam, leaving 772 thermal units for the internal heat of the steam. Consequently the heat stored in a pound of steam is $772 + 338$ thermal units. The equivalent energy stored in 22 pounds of steam is

$$22 \times 778 \times (772 + 338) = 19,000,000 \text{ foot-pounds.}$$

The first point to be noticed is, that there is many times as much energy in the water as in the steam; and the second is, that even a small fraction of the stored energy is sufficient to account for all the destruction caused by a boiler-explosion.

The circumstances of the boiler-explosion will determine how much of the energy stored in the steam and hot water will be developed and how it will be applied. Even in a particular case it is seldom possible to make proper estimates, nor does there appear to be any advantage from doing so. It is, however, curious to know that if the steam and water in the boiler under discussion were placed in a large cylinder with non-conducting walls and allowed to expand behind a piston, down to the pressure of the atmosphere there would be developed 138,000,000 foot-pounds. And further, if this work were expended in raising the boiler and its contents against the attraction of gravity, it could lift them a mile high.

CHAPTER VIII.

BOILER ACCESSORIES.

IN this chapter will be described various fittings, attachments, and accessories for steam-boilers.

Valves are used to control and regulate the flow of fluids in pipes. They are variously named after their forms or uses, such as globe valves, angle-valves, straightway valves, and check-valves.

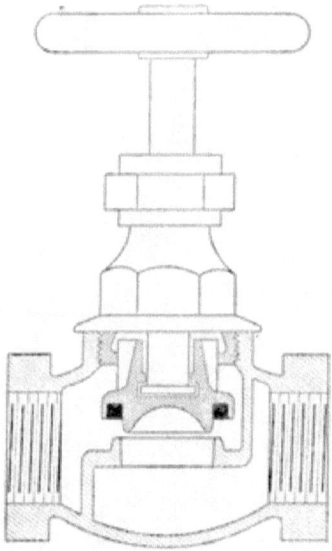

FIG. 83.

Globe Valves are named from the globular form of their cases. The case is separated into two parts by a diaphragm with a passage through its horizontal part, as shown in Fig. 83. The fluid enters at the right, passes under the valve, and out at the left. The valve is shut by screwing down the handle on the valve-spindle. A stuffing-box around the valve-spindle prevents leakage of fluid. In this valve the seat

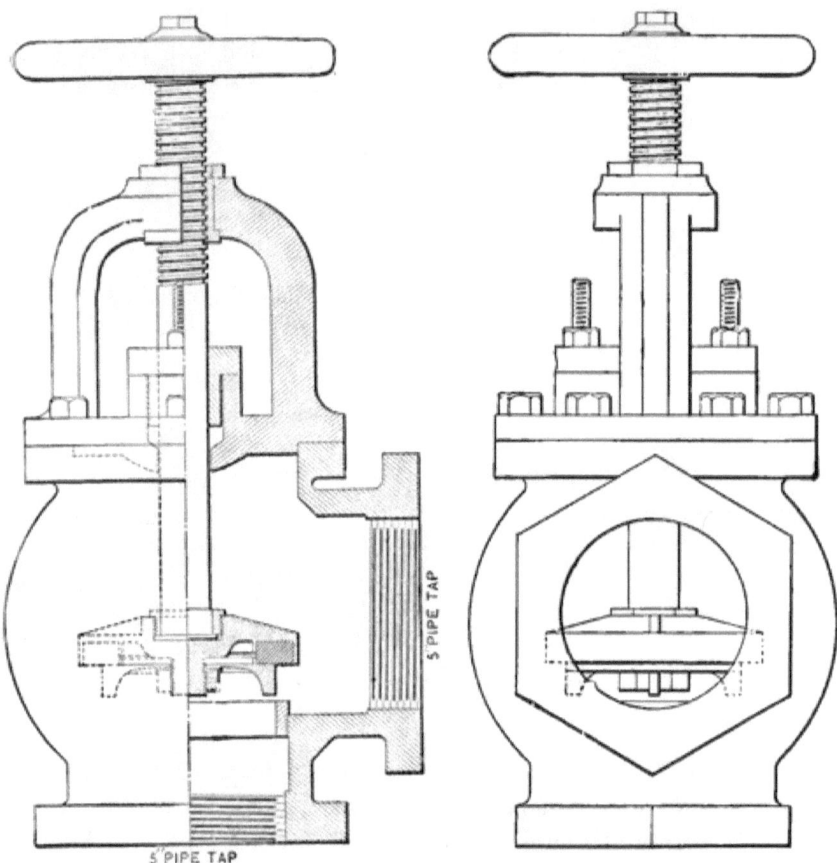

Fig. 84.

is rounded, and the valve face is a ring of a peculiar composition, let into the valve at R. When the valve is shut, this composition is squeezed down onto the seat and makes a tight joint.

If the fluid enters the valve from the right-hand side, the

valve-spindle may readily be packed to prevent leakage while the valve is closed. If the fluid entered the valve at the other end, it would be necessary to shut off the fluid from the entire pipe in order to pack the valve.

Angle-valves.—This form of valve, shown by Fig. 84, has an inlet at the bottom and an outlet at one side, it may take the place of an elbow at a bend in piping. The valve is made in two parts. The upper part carries a ring of soft metal which forms the bearing-surface. The lower part has ribs or wings which enter the opening through the valve-seat and guide the valve to its seat. The valve-spindle has a

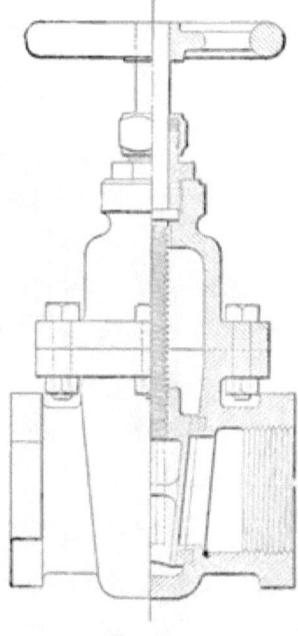

FIG. 85.

screw at the upper end which passes through a yoke entirely outside of the body of the valve.

The body of the valve is made of cast iron. The valve,

valve-seat, valve spindle, and stuffing-box follower are made of brass or composition.

This form of valve is frequently used for the stop-valve between the boiler and the main steam-pipe.

Straightway or Gate Valve.—This form of valve gives a straight passage through the valve, and offers very little resistance to the flow of fluids when it is open. Fig. 85 represents a Chapman valve, in which the valve is wedge-

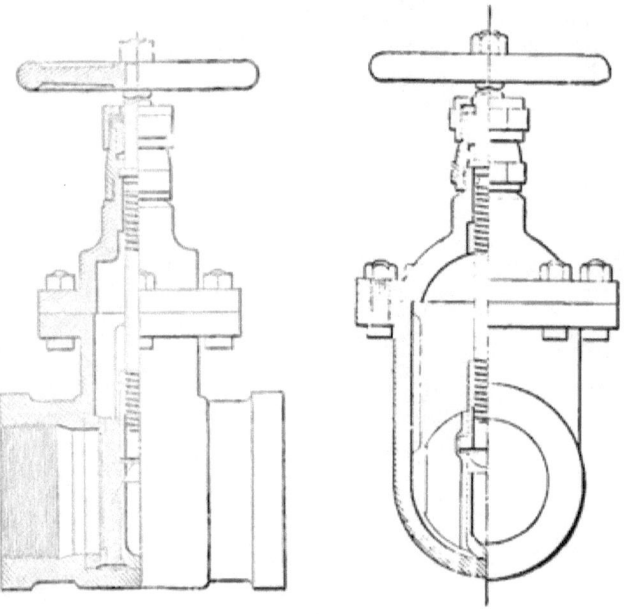

Fig. 86.

shaped and is forced against a wedge-shaped seat. The valve-spindle is held at a fixed height by a collar, and draws up or forces down the valve to open or close it. The body of the valve is of cast iron; the valve, valve-spindle, and stuffing-box are of brass; the valve-seat is a soft composition.

Fig. 86 represents a Peet valve, which has the faces of the valve-seats parallel. The valve itself is made in two pieces, between which is a peculiar casting, U shaped at the bottom and with wedge-shaped lips at the top. When the valve is shut this casting rests on the bottom of the valve body, and the two halves of the valve are thrown against the parallel valve-seats by the wedge-shaped lips of the casting. When the valve is opened this casting hangs between the two halves of the valve by the under side of the wedge-shaped lips.

Check-valves allow fluids to pass in one direction, but not in the other. Fig. 87 represents a lift check-valve; it

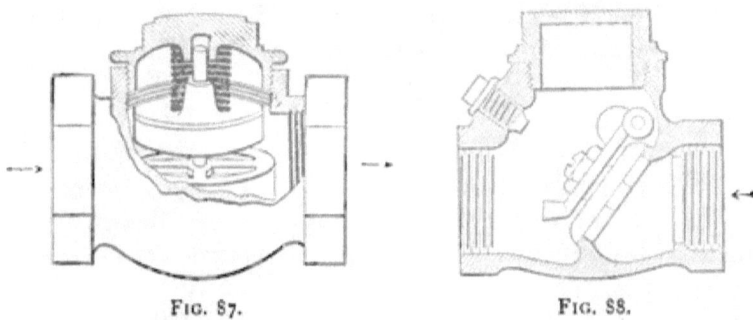

FIG. 87.   FIG. 88.

resembles a globe valve without a valve-spindle. Fluid entering at the left will lift the valve and pass out at the right. Should the current be reversed the valve will be promptly closed.

Fig. 88 represents a swing check-valve. It offers less resistance to the flow of fluid than the valve shown above, and there is less chance that foreign matter will lodge on the valve-seat. The valve has some looseness where it is fastened to the swinging arm, so that it may properly seat itself.

A feed-pipe must always have a check-valve to keep the boiler-pressure from acting on the pump, or injector, when it is not at work. It automatically opens to allow water to pass into the boiler. There should also be a stop-valve (a globe or gate valve) near the boiler which can be shut at will; thus when the check-valve shows signs of leaking the stop-valve may be shut, and then the check-valve may be opened and examined.

Safety-valves are intended to prevent the pressure of steam from rising to a dangerous point. In order to accomplish this, the effective opening of the valve should be sufficient to discharge all the steam that the boiler can make when urged to its full capacity. The effective opening is equal to the circumference of the valve-seat multiplied by the lift of the valve, if the valve-seat is flat; if the valve-seat is conical, the lift should be measured at right angles to the seat. Then if l is the vertical lift and if a is the angle which the seat makes with the vertical, the effective lift is

$$l \sin a.$$

The lift of a safety-valve rarely exceeds 1/10 of an inch. A two-inch pop safety-valve, made by the Crosby Gauge and Valve Co., and tested at the laboratory of the Massachusetts Institute of Technology, was found to lift from 0.07 to 0.08 of an inch. The valve had a conical seat with an angle of 45°. The actual flow was about 93 per cent of the calculated flow for this valve.

The amount of steam that a boiler can make may be estimated from the grate-area, the rate of combustion, and the evaporation per pound of coal. The first item is fixed, and the other two, though somewhat indefinite, may be estimated from the type of boiler and the conditions under which it works.

For example, a factory boiler having a grate 5 feet by 6 feet may be assumed to burn 18 pounds of coal per square foot of grate-surface per hour, and to evaporate 8 pounds of water per pound of coal. It will therefore generate

$$\frac{5 \times 6 \times 18 \times 8}{60 \times 60} = 1.2 \text{ pounds of steam per second.}$$

The amount of steam which will be delivered by a safety-valve may be calculated by an empirical equation proposed by Rankine; it may be written

$$W = A\frac{p}{70},$$

in which W is the weight of steam in pounds delivered per second, A is the effective area of discharge in square inches, and p is the absolute pressure of the steam in pounds per square inch.

If the weight of steam to be discharged per second is known, then this equation may be used to calculate the effective area; and will then read

$$A = \frac{70W}{p}.$$

In the example given above the weight of steam per second is 1.2 pounds. If the steam-pressure is 100 pounds absolute (85.3 by the gauge), then the effective area must be

$$A = \frac{70 \times 1.2}{100} = 0.84$$

of a square inch. If the effective lift be assumed to be 0.075 of an inch, the circumference of the valve-seat should be

$$0.84 \div 0.075 = 11.2 \text{ inches,}$$

and the diameter should be 3.5 inches.

A common rule requires that there shall be an area of 1/3 of a square inch through the valve-seat for each square foot of grate-surface. It so happens that this rule gives almost identically the same result as that just calculated for the above example; thus:

$$\frac{5 \times 6}{3} = 10 \text{ square inches,}$$

$$\sqrt{\frac{4 \times 10}{\pi}} = 3.5 + \text{ inches, diameter.}$$

Should the size of the valve determined by the two methods be different, the larger one must be taken; for the engineer will desire to fulfil the requirements of the first method for the sake of safety, and the requirements of the second method must be fulfilled if the boiler is to pass inspection.

Lever Safety-valve.—The general arrangement and some of the details of a well-made safety-valve are shown by Fig. 89.

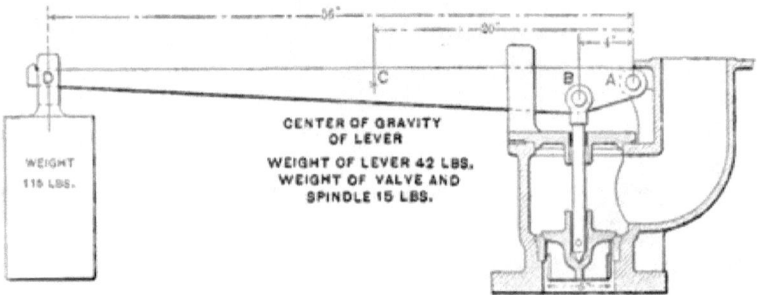

FIG. 89.

The body of the valve is of cast iron, and has an opening at one side from which the escaping steam is led out of the boiler-room through an escape-pipe. The valve and valve-seat are of brass or composition; the bearing-surface is at an angle of 45° with the vertical. The load is applied by a steel spindle, to a point beneath the bearing-surface so that the valve is drawn down to its seat. The spindle passes through a brass ring in the cover to the valve-casing. The load is applied by a lever with a fulcrum at A and a weight at D. It is steadied by guides cast on the cover of the casing; in the figure the valve and body are shown in section but the spindle, lever, guides and weight are shown in elevation.

It is important that the pins at A and B shall be loose in their bearings, and that the spindle shall be free where it passes through the top of the valve-case, so that the valve may not fail to rise even if the working parts are rusted a little.

After a safety-valve has blown off it is liable to leak a little, and such leakage is likely to injure the bearing-surface. In this way safety-valves sometimes get leaky and troublesome. The proper way is to regrind the valve and make it tight, but if the boiler attendant is careless he may try to stop the leak by jamming the valve on its seat. This may be done by hanging on extra weight, or wedging a piece of wood or metal against the lever. To remove temptation, it is well to have the guides for the lever open at the top, and also to cut off the lever to just the proper length so that the weight cannot be slid farther out. A short lever and a heavy weight are better, for this reason, than a lighter weight and a longer lever.

In order to make a calculation of the pressure at which a safety-valve will blow off, we must know the diameter of the valve, the weight of the valve and valve-spindle, the length of the lever and the weight hung at its end, and the weight and centre of gravity of the lever. This last may be found by calculation, or more simply by balancing the lever on a knife-edge.

In the example shown by Fig. 89 the valve has a diameter of 5 inches and an area of

$$\frac{3.1416 \times 5^2}{4} = 19.635$$

square inches, on which the steam presses.

The valve and spindle weigh 15 pounds; this is applied directly at the valve. The weight of 115 pounds at the end of the lever, is 56 inches from the fulcrum at A. It is equivalent to a weight of

$$\frac{115 \times 56}{4} = 1610$$

pounds at the valve. The weight of the lever is 42 pounds, applied at the centre of gravity C, 20 inches from the fulcrum. It is equivalent to a weight at the valve of

$$\frac{42 \times 20}{4} = 210$$

pounds. The total equivalent weight, or the load on the valve, is

$$15 + 1610 + 210 = 1835 \text{ pounds.}$$

Since the area of the valve is 19.635 square inches, the steam-pressure per square inch required to lift the valve will be

$$1835 \div 19.635 = 93.46 \text{ pounds.}$$

Problems concerning the loading of a safety-valve may be conveniently stated and solved by taking moments about the fulcrum; that is, by multiplying each weight or force by its distance from the fulcrum.

Let the weights of the valve, spindle, lever, and weight be represented by V, S, L, and W. Let a be the distance of the weight from the fulcrum and b be the distance from the fulcrum to the valve, while c is the distance of the centre of gravity of the lever from the fulcrum.

The moment of the weight is Wa, and the moment of the lever is Lc. The moment of the valve and spindle is $(V+S)b$. All three moments act downward, and their total effect is equal to their sum,

$$Wa + Lc + (V + S)b.$$

If the diameter of the valve is d, then the area is $\frac{1}{4}\pi d^2$. Representing the steam-pressure above the atmosphere by p, the force acting on the valve is

$$\frac{\pi d^2}{4} p,$$

and the moment of that force is

$$\frac{\pi d^2}{4} pb.$$

This moment acts upward and, when the valve lifts, will be equal to the total downward moment. So that the equation for calculating the load on a lever safety-valve is

$$pb\frac{\pi d^2}{4} = Wa + Lc + (V + S)b.$$

This equation gives for the steam-pressure at which the valve shown by Fig. 89 will lift

$$p = \frac{[Wa + Lc + (V + S)b]}{\pi d^2 b}.$$

$$\therefore p = \frac{4(115 \times 56 + 42 \times 20 + 15 \times 4)}{3.1416 \times 5^2 \times 4}.$$

$$\therefore p = 93.46 \text{ pounds},$$

as found by the previous calculation.

For a second example let us find the distance at which the weight of the valve shown by Fig. 89 must be placed from the fulcrum in order that the valve will blow off at 50 pounds above the atmosphere.

Solving the general equation for a, we have

$$a = \frac{pb\frac{\pi d^2}{4} - Lc - (V + S)b}{W}.$$

$$\therefore a = \frac{50 \times 4 \times \frac{3.1416}{4} \times 5^2 - 42 \times 20 - 15 \times 4}{115}.$$

$$\therefore a = 26.32 \text{ inches}.$$

For a third example find the weight which should be hung at the end of the lever if the valve is to blow off at 30 pounds above the atmosphere.

Here we have

$$W = \frac{pb\frac{\pi d^2}{4} - Lc - (V + S)b}{a}.$$

$$\therefore W = \frac{30 \times 4 \times \frac{3.1416}{4} \times 5^2 - 42 \times 20 - 15 \times 4}{56}.$$

$\therefore W = 26$ pounds.

These last two problems can of course be stated and solved much after the first manner applied to the first problem, but the work, which will amount in the end to the same thing, cannot be so well arranged nor so easily done.

Pop Safety-valve. — A defect of the common lever safety-valve is that it does not close promptly when the steam-pressure is reduced, and it is apt to leak after it has returned to its seat.

The valve shown by Fig. 90 has a groove turned in the flange which projects beyond the bearing-surface, and there is another groove between the outer edge of the valve-seat and a ring which is screwed onto the valve-seat. When the valve lifts the escaping steam is twice deflected, once by the groove in the valve and again by the groove at the valve-seat. The reaction of the steam assists the pressure of the steam on the under surface of the valve, and suddenly opens the valve to its full extent. The valve stays wide open till the steam-pressure in the boiler has fallen a few pounds below the blowing-off pressure, and then the valve shuts as suddenly as it opens.

The ring which is screwed onto the valve-seat has a number

of holes drilled through it to allow steam to escape from the groove at its upper surface. It may also be screwed up or

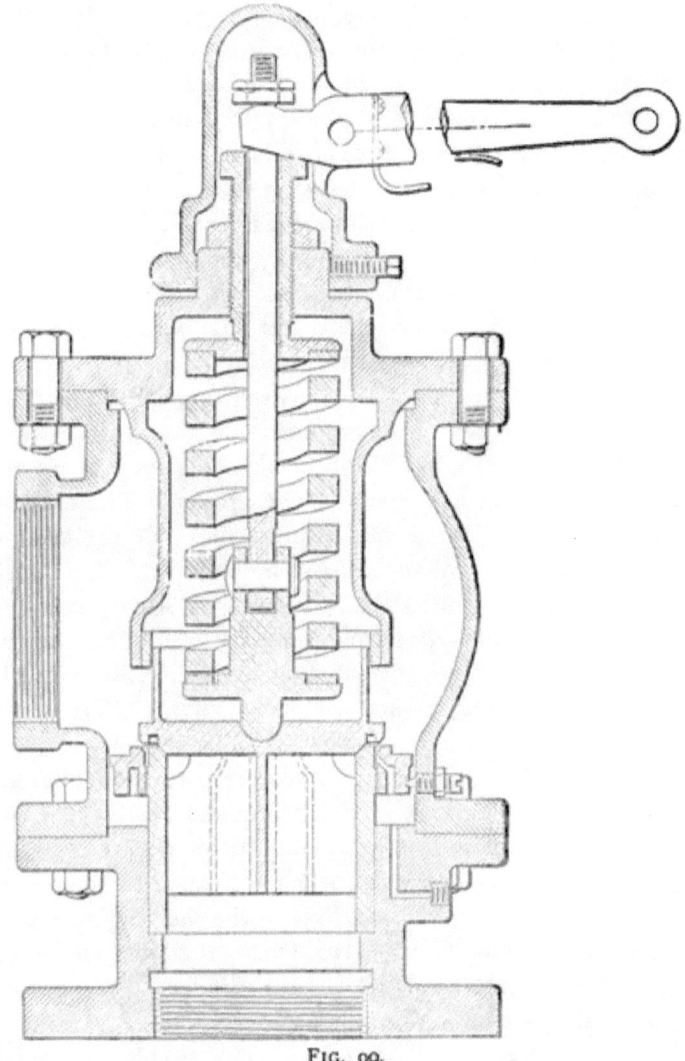

Fig. 90.

down to adjust its position; a screw at the side of the case clamps it when adjusted. The action of the valve is regulated by the number of holes in the ring and by its vertical position.

This valve is loaded by a helical spring. The tension of the spring and the load on the valve is regulated by a sleeve which is screwed down through the top of the valve-case. It is of course possible to load a plain safety-valve in a similar way, or to load a pop-valve with a lever and weight. The valve is extended up in the form of a thin shell to guard the spring from the escaping steam. The valve-spindle is extended through the top of the case, and may be pulled up by a lever when it is desired to ease the valve off from its seat. A drip at the lower right-hand side of the case draws off water which may collect in the case.

The valve and its seat, the adjusting-ring on the seat, the valve-spindle, and the bearing-pieces on the spring are all brass. There is also a brass ring inside the shell that extends down from the cover and incloses the spring. There should be a little clearance between this brass ring and the shell on the valve so that the valve shall not be cramped. The entire valve-casing, which is made in four parts, is of cast iron.

The closeness of regulation by a safety-valve depends mainly on the width of the bearing-surface. Thus a valve with a narrow bearing-surface will close after the pressure in the boiler is reduced a few pounds; a valve with a wide bearing-surface will stay open till the pressure has suffered a serious reduction. By making the bearing-surface very narrow the reduction of pressure may be made as small as two pounds. For example: a certain valve was made to open at 100 pounds and to close at 98 pounds. When the bearing-surface is narrow it must be made of hard, dense metal to endure the pressure concentrated on it. Hard bronzes, compositions and nickel alloys are used for this purpose.

A safety-valve should be set by trial, to blow off at the required pressure as shown by a correct steam-gauge. A safety-valve should occasionally be lifted from its seat to insure that it is in proper condition. An unexpected opening of a safety-valve or continued leakage shows lack of attention to duty on the part of boiler attendants. While the safety-valve for a boiler should be able to deliver all the steam it can make, it may be considered that the proper function of a safety-valve is to give warning of excessive pressure. The safety of the boiler must always depend on the faithfulness and intelligence of the boiler attendants.

Inspection laws commonly require that every boiler shall have two safety-valves, and that one of them shall be locked up in such a manner that it cannot be overloaded by accident or design.

Water-column.—The position of the water-level in a boiler is indicated either by a water-glass or by gauge-cocks or by both. These may be connected directly to the front end of the boiler, or they may be placed on a fitting known as a *water-column* or *combination*. Fig. 91 shows a good form of water-column. It is a cast-iron cylinder connected to the steam-space at the top and to the water-space near the bottom. The normal position of the water-level is near the middle. There is at the bottom a globular receiver into which deposits from the water may settle and be blown out at will.

In one side of the water-column are brass fittings for the water-glass, which is a strong tube of special make. The glass tube passes through a species of stuffing-box in the brass fitting. The joint is made tight by a rubber ring which fits on the tube and is compressed by a follower screwed onto it. Each fitting has a valve by which steam may be shut off when the tube is cleaned or replaced. A cock at the bottom drains water from the tube; for this purpose the lower valve is closed and the cock is opened. Stout wires at the side of the glass tube guard it from injury.

If either valve leading to the water-glass is closed, the level of the water will rise in the tube. If the upper valve is closed, the steam in the upper part of the tube is gradually condensed by radiation, and is replaced by water entering from below. If the lower valve is closed, the condensation of steam from radiation will accumulate and gradually fill the tube.

Gauge-glasses are very brittle and, though carefully annealed, are under considerable stress from unequal cooling.

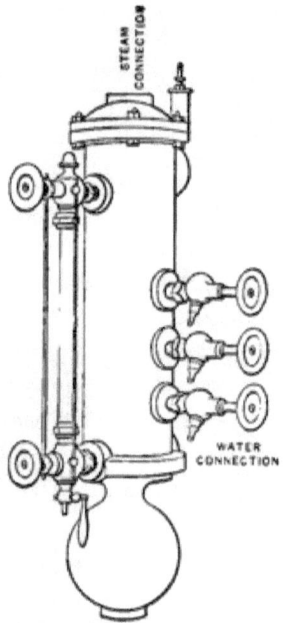

FIG. 91.

Before a tube is put in it may be cleaned by pouring acid through it, or by drawing a bit of waste through on a string. A wire should never be forced through a glass tube, for the slightest scratch may start a break which will end in reducing the tube to small pieces. When a tube is in place it may be cleaned by closing the lower valve and opening the drainage-cock and allowing steam to blow through.

When a boiler is left banked overnight the water-glass

should be shut off, since a breakage may result in drawing the water in the boiler down to the level of the lower end of the tube.

In addition to the water-glass, which shows at all times the level of the water, the water-column carries three gauge-cocks. One is set at the desired water-level, one a little above and one a little below. Steam from the steam-space, through the upper gauge-cock, becomes superheated as it blows into the atmosphere and looks blue. The lower cock discharges hot water from the water-space, which flashes into steam as it escapes, but it has a white color, which is very distinct from that of the jet from the steam-space. A good fireman occasionally tests the position of the water-level by using the gauges to be sure that the indication by the water-glass is not erroneous. Engineers on locomotives, and boiler attendants where very high-pressure steam is used, often prefer to depend entirely on the gauge-cocks, and dispense with the water-glass, which may be annoying or dangerous when it breaks.

The water-column shown by Fig. 91 has an alarm-whistle, which shows above the main casting, at the right. It is controlled by two floats inside the cylinder; one float at the top opens the valve leading to the whistle when the water-level is too high, the other near the bottom blows the whistle when the water-level is too low.

If the fire is stirred up under a boiler which has had the fire banked, the water-level rises in the water-glass; the reason being that the circulation is from the front of the boiler to the rear, and that this circulation is maintained by a difference of level between the front and rear ends. On the contrary, the water-level falls when a boiler which has been steaming freely is checked.

Steam-gauges.—The pressure of the steam in a boiler is shown by a spring-gauge which has the external appearance shown by Fig. 92. The essential part is a flattened brass tube bent into the arc of a circle as shown by Fig. 93. The section of the tube may be an oval, or it may have two longitudinal corrugations as shown by Fig. 94.

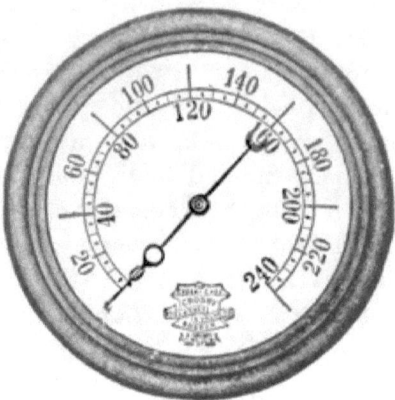

FIG. 92.

Pressure inside of such a tube makes it bulge and tends to straighten it. One end is fixed and is in communication

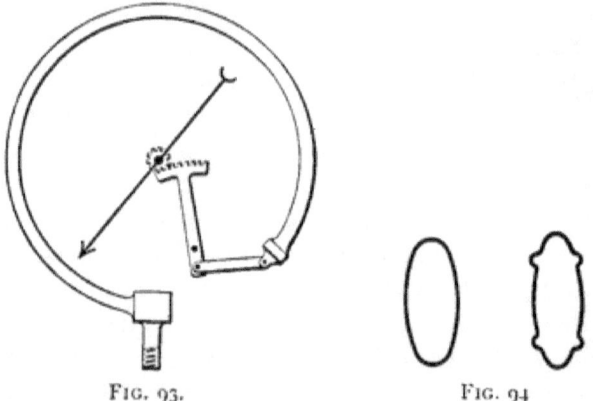

FIG. 93. FIG. 94.

with the space where the pressure is to be measured. The other end is closed and is free to move. It is connected by a link to a lever which bears a circular rack in gear with a pinion. The motion of the free end of the tube is multiplied and is shown by the motion of a needle on the pinion. The scale on the dial is marked by trial to agree with the indications of a mercury column or of a standard gauge. A hairspring on the pinion (not shown in Fig. 93) takes up the backlash of the multiplying-gear.

The long, flexible spring-tube is liable to vibrate to an undue extent when the gauge is exposed to the jarring of a locomotive. To avoid this difficulty, two short stiffer tubes have their ends connected to a more effective multiplying device, shown by Fig. 95. The greater number of joints in this device makes it less sensitive than the other form.

FIG. 95.

Since the spring-tube changes its shape if the temperature changes, hot steam should not be allowed to enter it. An inverted siphon or U tube filled with water is, therefore, interposed between the gauge and the steam from the boiler.

Safety-plugs, or Fusible Plugs, as shown by Fig. 96, are made of brass and provided with a core of fusible metal. If the plate into which they are screwed is in danger of overheating, the fusible metal will melt and run out, and steam and water will blow into the furnace. If the fire is not put out, it will at least be checked and the attention of the fireman will be attracted.

The melting-point of fusible metals is not always certain, and the plugs not infrequently blow out when there is no apparent cause. On the other hand, they sometimes fail to act when the plate is overheated. If the plug is covered with incrustation, the fusible metal may run out without giving warning.

The following are some of the places where a fusible plug is used:

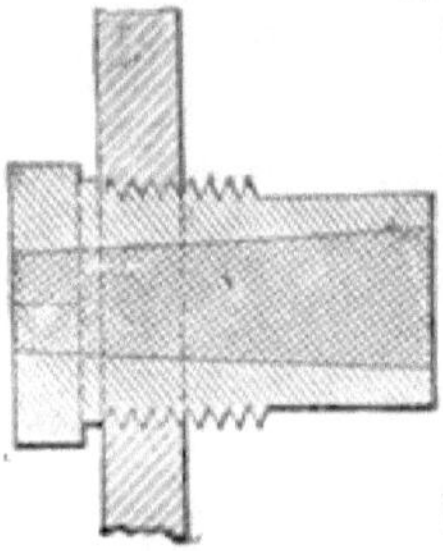

In the back head of a cylindrical tubular boiler, about three inches above the top row of tubes.

In the crown-sheet of a locomotive fire-box.

In the lower tube-sheet of a vertical boiler; or sometimes in one of the tubes a little above that tube-sheet.

FIG. 96.

In the lower side of the upper drum of a water-tube boiler.

The fusible composition has a conical form so that it cannot be blown out by the pressure of the steam.

Foster Reducing-valve.—When steam is desired at a less pressure than that of the boiler, it is passed through a reducing-valve like that shown by Fig. 97. The valve H is held open by the spring at J, acting through the toggle-levers a, until the steam-pressure in the exit-pipe B, pressing on the diaphragm D, is able to overcome the spring and close the valve. The pressure at which this may occur is determined by the tension of the spring, which may be regulated by the screw at K. It is expected that the valve will be drawn up so as to admit just the proper amount of steam to the exit-pipe B to maintain the desired pressure in it. Valves for this purpose are liable to work intermittently, i.e. they close till the pressure falls below the proper point, then they open and raise the steam-pressure above that point. The valve is a species of throttling-valve, and therefore cannot be expected to remain tight. If the machinery supplied by the reducing-valve is liable to

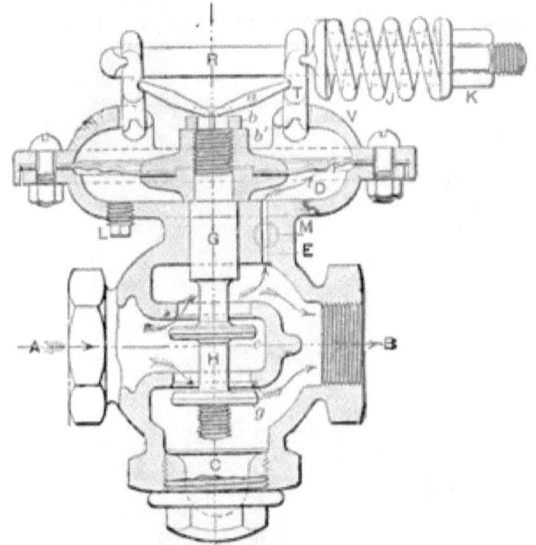

Fig. 97.

be injured by excessive pressure, there must be a stop-valve beyond the reducing-valve. The stop-valve must be closed when no steam is drawn, and must be used to regulate the supply of steam until the amount drawn exceeds the leakage of the reducing-valve.

The Damper-regulator shown by Fig. 98 places the damper in the flue leading to the chimney under the control of the steam-pressure, so that if the pressure of the steam falls, the damper is opened wider to quicken the fire. The pressure of the steam in the boiler is communicated through the pipe a to the lower surface of a diaphragm, and lifts the loaded lever b, which stands half-way between the stops at the middle of its length when the steam-pressure is at the proper point. Should the steam-pressure rise above the proper point it raises the lever and opens a small piston-valve at c, and water from a hydrant flows into d and presses on a piston which lifts the weights at e and so shuts the damper.

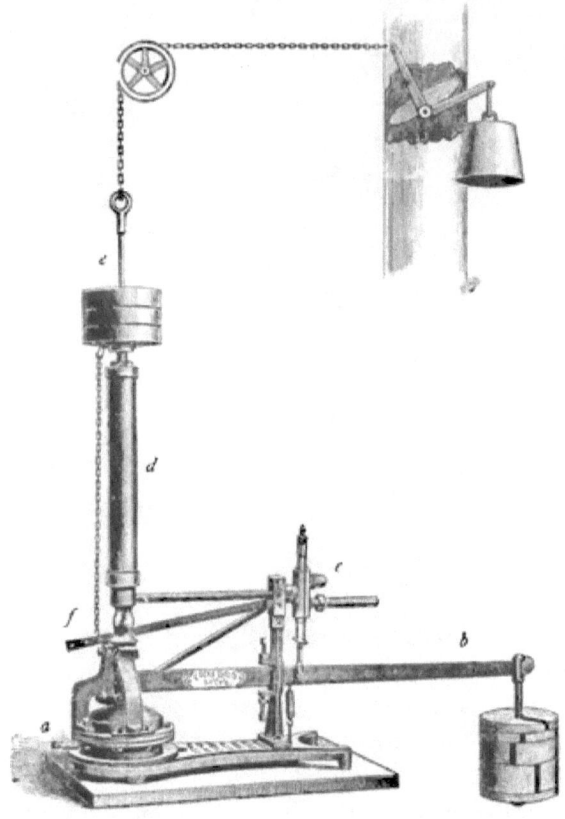

FIG. 98.

The weighted head e of the piston is connected by a chain to the lever f, and closes the valve c as it rises, and so shuts off the water from the hydrant.

A regulator of the same form attached to a throttle-valve acts as a reducing-valve, and regulates the pressure below the valve with a variation of less than one pound. Fig. 99 shows the steam-valve used when the Locke regulator acts as a reducing-valve. The valve is a double valve which is nearly balanced, but with a slight tendency to rise under steam-pressure, as the lower valve is the larger. The cylindrical part of the valve is cut into V notches, so that the supply of steam is regulated to a nicety when the valve is partially open. The cylindrical portion of the valve protects the valve-seat and the valve-face so that the valve may remain tight when closed.

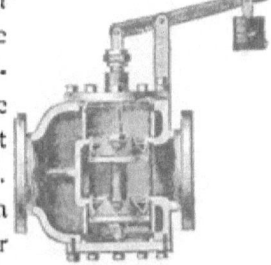

FIG. 99.

Steam-traps. — The object of a steam-trap is to drain condensed water from steam-pipes without allowing steam to escape. As a rule a trap is placed below the pipe to be drained so that the drip from the pipe will run into it. Some traps that return the condensed water to the boiler do not conform to this rule.

Some traps, such as the McDaniels, the Baird, and the Walworth, have a valve under the control of a float, which will allow water to pass but not steam.

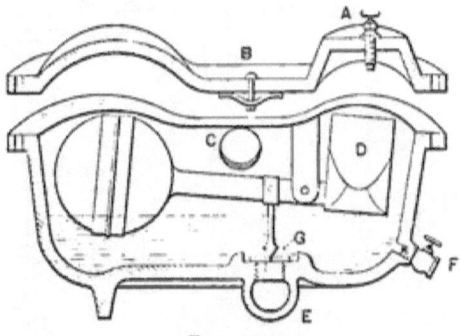

FIG. 100.

The McDaniels trap is shown by Fig. 100. The drip enters at C and escapes through the exit at E when the valve G is open. This valve is raised by the spherical float when the water rises to a sufficient height. When the water is drained from the pipe served by the trap, the water-level in the trap falls and the valve G is closed. D is a counter weight to balance the weight of the spherical float. The valve at G can be opened by screwing down the screw at A

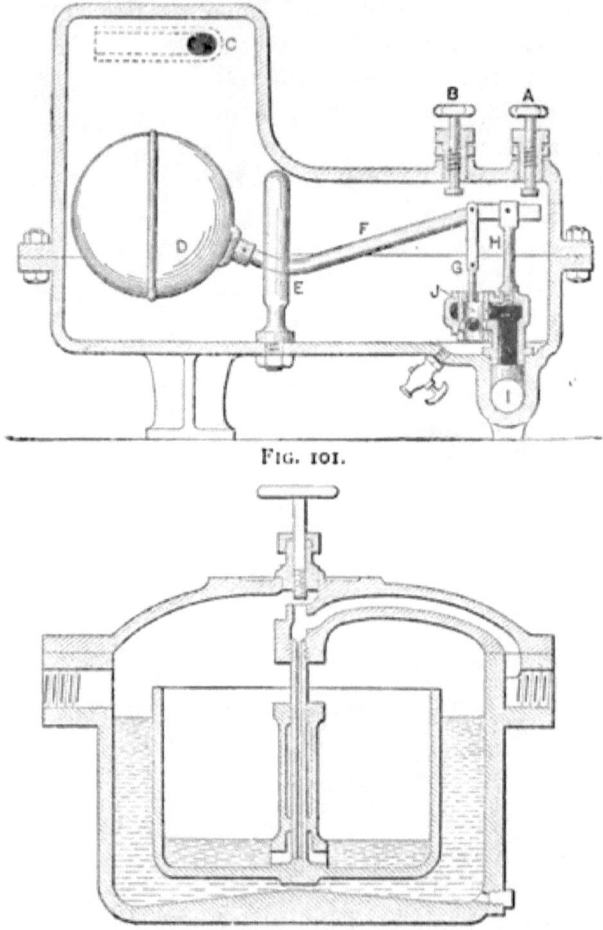

FIG. 101.

FIG. 102.

on to the counterweight. The trap can be emptied through the valve at F.

The Baird trap, Fig. 101, has a spherical float D which controls a piston-valve at J. The inlet is at C, and the outlet at I. The screws A and B allow the valve J to be opened or closed by hand.

The Walworth trap (Fig. 102) has a floating bucket into which the drip overflows after the outer case is partially filled. When the bucket sinks it opens a passage through the central spindle, and the water in the bucket is driven out through this spindle. The hand-wheel and screw at the top control a valve which is closed when the trap is working.

The Flynn trap (Fig. 103) depends for its action on a head of water acting on a flexible diaphragm. Water may enter at the top or the bottom at orifices marked A. It fills the pipe B and the globe C as high as the end of the pipe E, and produces a pressure of about a pound per square inch on the under side of the diaphragm at D. The spring at G produces a pressure of about half a pound per square inch on the upper side of the diaphragm. Consequently the valve leading from the chamber F to the escape-pipe H is closed so long as the pipe E remains empty. But when the water overflows the top of the pipe E and fills the chamber F, the water-pressure on top of the diaphragm will be the same as that on the bottom, and the spring at G will open the valve and allow water to escape. If the supply of water

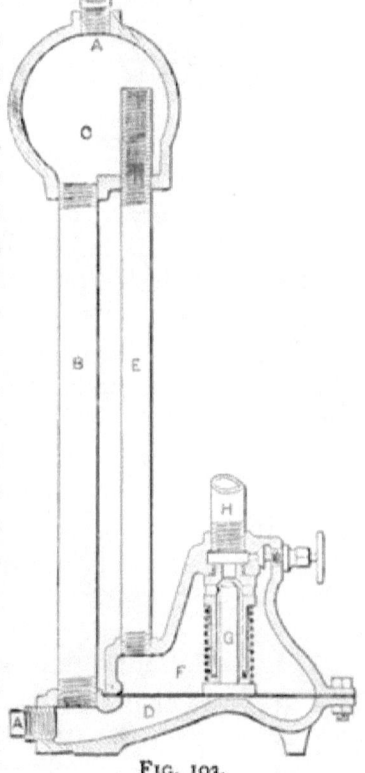

FIG. 103.

at A ceases, the pipe E will be emptied and the valve will be closed under the influence of the pressure on the under side of the diaphragm. In the trap as actually constructed the pipe *H* is about 24 inches long; in the figure it is made shorter in proportion.

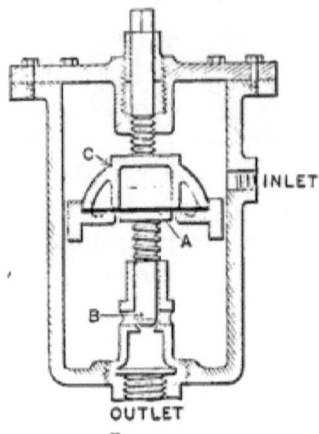

FIG. 104.

The Curtis trap (Fig. 104) has an expansion-chamber at *C* which is closed by a diaphragm *A* at the bottom, and is filled with a very volatile fluid. So long as the expansion-chamber is immersed in water the pressure of the fluid on the diaphragm is balanced by the spring on the valve-spindle *B*. If the water is drained away and the chamber is exposed to the temperature of steam (212° F. or more), the fluid vaporizes and exerts enough pressure on the diaphragm to compress the spring and close the exit-valve.

Return Steam-trap.—The traps thus far considered usually discharge against the pressure of the atmosphere. They may discharge into a closed tank against a pressure that is higher than the atmosphere, but in all cases the pressure in the pipes drained by the trap must be higher than the discharge-pressure. Return steam-traps are arranged to discharge directly into the boiler.

The Bundy return-trap, shown by Figs. 105 and 106, is set three feet or more above the water-line in the boiler. It is so

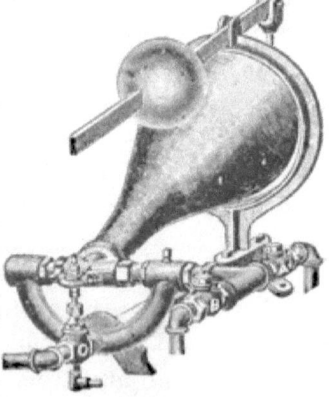

FIG. 105.

made that it is first opened to the pipe to be drained, and fills up under the pressure in that pipe. It is then put in communication with the steam-space and with the water-space of the boiler, and the water previously collected drains into the boiler.

The trap consists of a pear-shaped receptacle or closed bowl, hung on trunnions, through which the bowl is filled and emptied. When empty the bowl is raised by a weight and lever; when filled with water it overbalances the weight and

FIG. 106.

falls. The ring around the bowl limits the motion. The condensed water from the pipe or system of pipes to be drained enters the trap through the check-valve B, which prevents water from flowing back from the trap into the pipe to be drained. The trap is emptied through the check-valve A, which prevents water from the boiler from flowing into the trap. At C is a valve under the control of the trap, which receives steam by a special pipe from the boiler. When the trap is empty and is lifted by the weight and lever, the valve C is thrown down and is shut; water then flows in through the valve B from the pipe to be drained, and air escapes from an air-valve below C, which is open in this position of the trap. A check-valve on the air-pipe prevents air from entering the trap if a vacuum happens to be formed in it. When the bowl is filled it falls and opens the steam-valve C, and steam enters the bowl through a **curved** pipe shown in Fig. 106. The pressure in the bowl is now equal to that in the boiler, and the water collected flows into the boiler by gravity.

Separators.—If steam is carried to a distance in pipes, a considerable amount of water of condensation accumulates. It is undesirable to have this water delivered to a steam-engine in any case, but if the water accumulates in a pocket or a sag in the piping, it may come along with the steam in a body whenever there is a sudden change of steam-pressure, and then the engine will be in danger of injury.

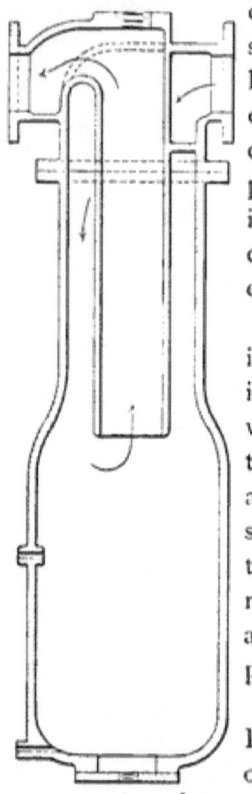

Fig. 107.

A good way of removing such water is to allow the steam to come to rest in a steam-drum of suitable size, from which the water is drained by a steam-trap; the steam meanwhile may flow from a pipe at the top of the drum. A small steam-drum used as separator is likely to fail, from the fact that the steam does not come to rest, or because the entering and leaving currents of steam are not properly separated.

The Stratton separator, shown by Fig. 107, brings in the steam at one side of a cylinder, with a whirling motion that throws the water onto the side of the cylinder; dry steam escapes through a pipe in the middle.

A good steam-separator will remove all but one or two per cent of moisture from steam, even though the entering steam is very wet.

Attention has already been called to the use of separators

with some forms of water-tube boilers which do not have a sufficient free water-surface for the disengagement of steam.

Feed-water Heaters.—The feed-water supplied to a boiler

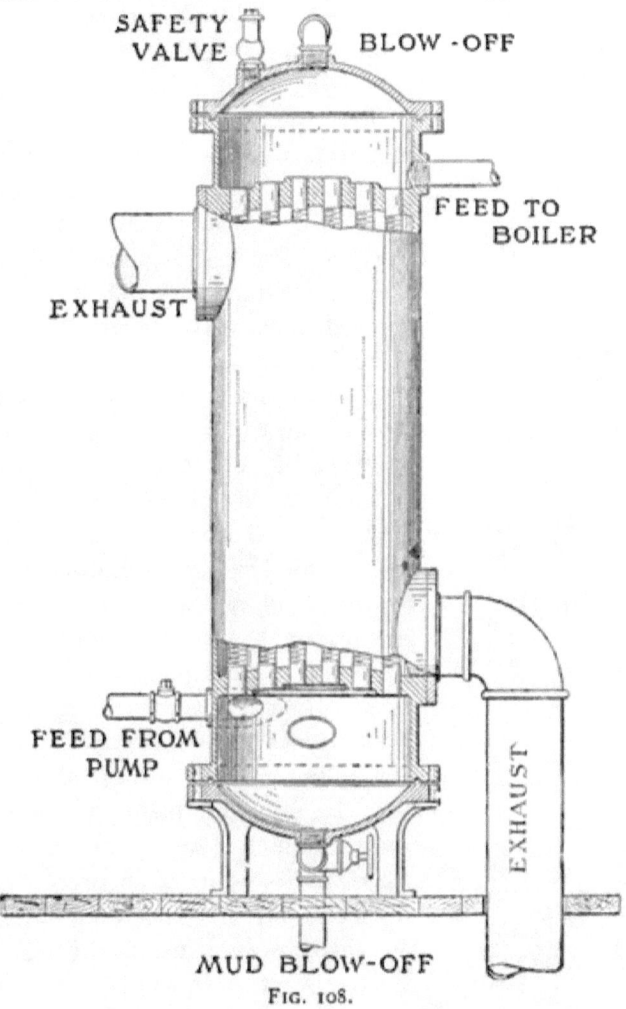

Fig. 108.

may be heated up to the temperature of the exhaust-steam by passing it through a feed-water heater. Feed-water heaters are sometimes made open, i.e., the steam from the engine mingles with and heats the feed-water. Such heaters have the disadvantage that the oil from the engine is carried into the boiler.

A closed feed-water heater resembles a surface condenser, and as the steam and water do not mingle, there is no danger of carrying oil from the engine into the boiler. The Wainwright heater, shown by Fig. 108, has the heating-surface of corrugated copper or brass tubes, of peculiar make, to allow for expansion. The steam from the engine passes around the tubes and the feed-water passes through the tubes.

The Berryman feed-water heater, shown by Fig. 109, is arranged to have the exhaust-steam pass through a series of inverted U tubes, around which the feed-water circulates.

Live-steam feed-water heaters take steam from the boiler to raise the temperature of the feed-water up to, or nearly to, the temperature in the boiler. The principal advantage appears to be that unequal contraction, due to the introduction of cold water, is avoided. It is claimed that with some forms of boilers a better circulation is obtained by aid of such a heater.

The use of a feed-water heater for removing lime-salts from feed-water has been discussed on page 73, and an example of such a feed-water heater was illustrated in connection therewith.

Feed-pipes.—The temperature of the feed-water is usually much below the temperature in the boiler. It thus becomes essential to so locate the inlet, and to so distribute the water, that undue local contractions may not occur; this is of special im-

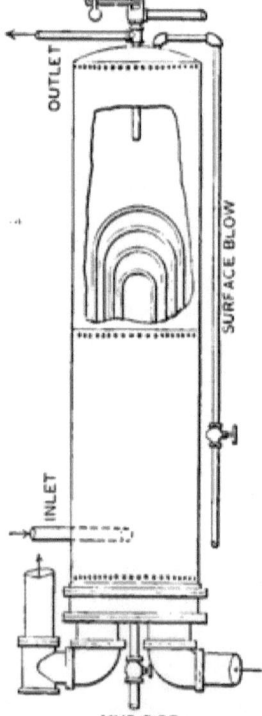

FIG. 109.

portance when the supply is intermittent. The feed-pipe for the cylindrical tubular boiler, shown by Plate I, enters the shell near the water-line, through the front head. It is carried along one side of the boiler for about three fourths of its length, and then is carried across over the tubes and opens downward. A feed-pipe is often perforated to give a better distribution of the feed-water.

The shell is reinforced by a piece of plate riveted on the outside, where the feed-pipe enters the boiler. The end of the pipe has a long thread cut on it, so that it can be secured through the reinforcing-plate and the boiler-shell, and may then receive a pipe-coupling which connects it to the continuation of the feed-pipe inside.

Sometimes the feed-water is delivered to an open trough inside the boiler, from which it overflows in a thin sheet. Or a perforated pipe may deliver the water in form of spray in the steam-space. Either method has the advantage that the water comes in contact with steam and is heated before it mingles with the water in the boiler. There is the disadvantage that the steam-pressure may fall off when the feed-water is turned on or is increased.

It has already been pointed out that the feed-pipe should have a globe valve near the boiler, and a check-valve between the globe valve and the feed-pump.

Feed-pumps.—Boilers are commonly fed by a small direct-acting steam-pump placed in the boiler-room. The steam-consumption per horse-power per hour of such pumps is very large, and yet the total steam used is insignificant. They are cheap and effective, and easily regulated.

Power pumps driven from a large engine are more economical, provided their speed can be regulated; they not infrequently are arranged to pump a larger quantity than required for feeding the boiler, the excess being allowed to flow back to the suction side of the pump through a relief-valve.

When one pump supplies several boilers, a series of difficulties is liable to arise. First, if the boilers are fed singly in rotation, the large intermittent supply of feed-water is likely to give rise to local contraction and the water-level in the boiler fluctuates; there is liability that the water-level will fall too low, endangering the heating-surface, or there may be excessive priming when the water-level is high. It appears advisable that the feed should be delivered to all the boilers simultaneously, the supply to each boiler being regulated by its stop-valve; each branch pipe to a particular boiler should be provided with its own check-valve, and the water-level and rate of feeding of each boiler must be carefully watched by the fireman, or by a water-tender if there are many boilers.

An injector is conveniently used for feeding a boiler if the feed-water is not too hot; it has the incidental advantage that it heats the water as it feeds it into the boiler. An injector should be connected up with unions, so that it may readily be taken down for inspection. At sea an injector is commonly used when the boilers are fed from the sea or from a supply-tank.

Every boiler should have two independent sources of supply of feed-water, so that there may be some resource if the usual supply gives out. There may be two pumps, or a pump and an injector. A locomotive usually has two injectors.

Blow-off Pipe.—The blow-off pipe draws from the lowest part of the boiler, or from some place where sediment may be expected to collect. On the blow-off pipe there is a cock or a valve which is opened to blow out water from the boiler. Sometimes there are both a cock and a valve. A cock has the disadvantage that it may give trouble by sticking; a valve may leak and the leak may not be detected.

The pipe should be carried beyond the cock, so that the attendant is not liable to be splashed with hot water, but the pipe should end in the boiler-room or where discharge through the pipe on account of a leaky cock or valve may be sure to attract attention. Each individual boiler should have its own blow-off pipe.

The blow-off pipe where it passes through the back connection is covered with magnesia, asbestos, or fire-brick. In spite of this protection the blow-off pipe may burn off. The device shown by Fig. 110 is used to overcome this difficulty.

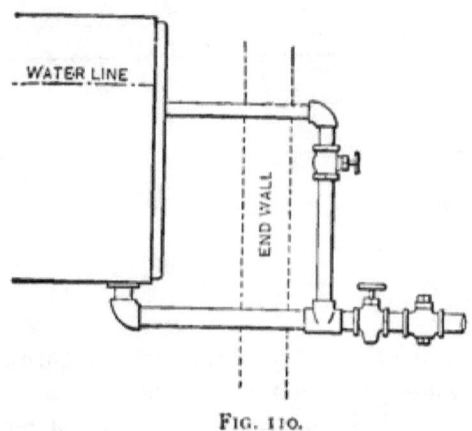

Fig. 110.

When the blow-off cock is shut and the valve on the vertical branch is open, there is a continuous circulation of water which keeps the pipe from burning. The valve on the vertical branch is closed before the blow-off cock is opened.

If a blow-off pipe burns off and water begins to escape, the feed-pump should be run at full capacity to keep water in the boiler and guard the plates from burning, if that is possible. The fire should then be checked by throwing on wet ashes or by other means, unless escape of steam from the break in the blow-off pipe prevents.

Piping to carry steam from a boiler to an engine, for heating buildings, and for other purposes is too important to be considered as accessory to the boiler. A few remarks, however, may not be out of place.

The expansion of the pipe due to changes of temperature should be provided for, or else cracks in the pipe or fittings, or leakage at the joints may be expected. A common way of allowing for expansion is illustrated by Fig. 111, which shows

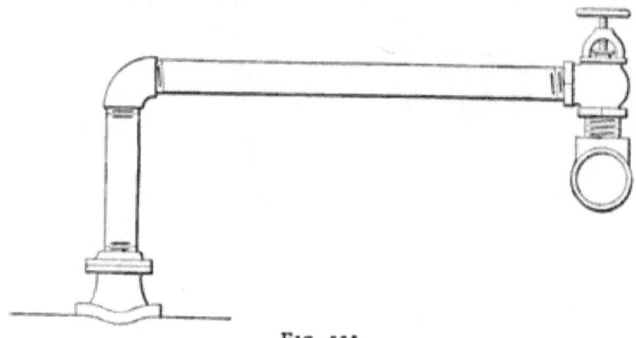

Fig. 111.

the connection from a boiler to the main steam-pipe. When the main steam-pipe expands or contracts, the short nipple between it and the angle-valve turns a little at one or at both ends; in like manner the vertical pipe turns a little at the nozzle or at the elbow. The motion is so small and so distributed as not to give any trouble unless the expansion to be provided for is very large. A large and long straight steam-pipe may require an expansion-joint. A slip-joint may be made of a brass pipe inside a shell with packing-box and follower, arranged something like the piston-rod of an engine. It is essential that the slip-joint shall be in line or it will be cramped and give trouble. For this purpose the joint may be carried and guided by a cast-iron bed-plate.

Fig. 111 is so arranged that there is no space where water can collect when the boiler is shut off from the main steam-pipe. If the stop-valve were in the vertical pipe, as is sometimes the case, then the pipe over the valve would fill up with water when the boiler is shut off, and that water would be suddenly blown into the steam-main when the stop-valve is next opened. A pipe so situated should always have a drip-pipe to draw off condensed water before the valve is opened. As a special example we may mention the pipe leading to an engine, which always has a drip-pipe above the throttle-valve. Pipes that are likely to be troubled by condensation should be continuously drained by a steam-trap.

Horizontal pipes are sometimes arranged so that water may collect in them, due to a sag in the pipe or to the fact that they do not properly drain through a side branch. Though the water may lie quiet in such a pocket while the draught of steam is steady, a sudden increase in the velocity of the steam, or a rapid opening of the valve supplying steam to the pipe, will sweep the water up and carry it along with the steam. The danger from the inrush of water to an engine is readily seen, but it is not so well known that the water thus violently thrown against elbows and other fittings give rise to leaks, if it does not burst the fittings. It is to be remembered that steam offers little or no resistance to the movement of water in a pipe, as it is readily condensed either from a slight increase of pressure or by mingling with colder water. Again, water at the temperature corresponding with the pressure easily separates, forming bubbles of steam, which as easily collapse, and the shock of impact of the water gives rise to pressures that search out all weak places in the pipe, even at some distance.

Drawings for piping commonly represent the work as though it were all in one plane. There is little liability of confusion since the actual piping could usually be swung into one plane, turning in tees and elbows and other fittings. Lengths are given from centre to centre of pipes represented, because the fittings may differ in length.

Piping up to two inches in diameter can be cut by hand. Larger sizes are cut by machine. Sizes of pipe are named by the inside diameter; but the actual diameter, especially of small sizes, may be larger than the nominal diameter.

Pipe sizes are $\frac{1}{8}$, $\frac{1}{4}$, $\frac{3}{8}$, $\frac{1}{2}$, $\frac{3}{4}$, 1, $1\frac{1}{4}$, $1\frac{1}{2}$, 2, $2\frac{1}{2}$, 3, $3\frac{1}{2}$, 4, 5, 6, 7, 8, 10, 12, etc. Brass piping is nearer the nominal size than iron piping. Boiler-tubes are named from the outside diameter.

Pipe-hangers.—When a pipe needs support it is commonly hung from an overhead beam by a wrought-iron ring, a little

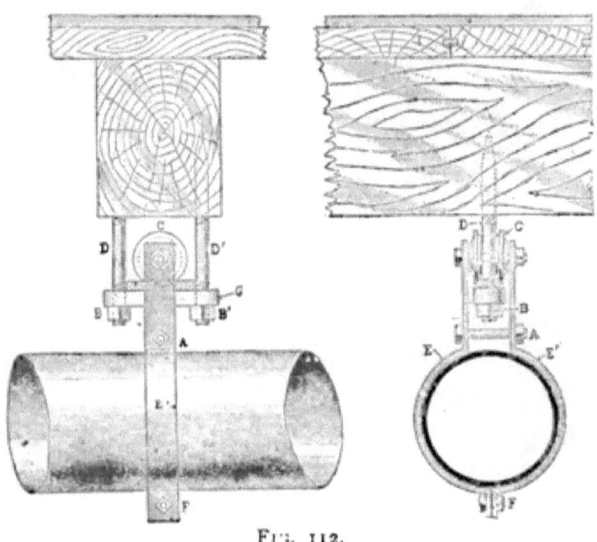

FIG. 112.

larger than the pipe, which is held up by a lag-screw in the beam. If the pipe is long, the expansion is likely to cramp the ring on the pipe and then bring an awkward side strain on the lag-hook; or if the hook is open in the direction of the expansion, the ring may be lifted out of the hook and so the support at that point may be lost. The hanger shown by Fig. 112 has the supporting ring carried by a roller. The track for the roller is carried by lag-screws. In some cases the lag-screws can be advantageously replaced by bolts which pass clear through the beam. Various modifications of this device may be used. For example, the pipe may rest on a roller with a hollow face; the roller is on a horizontal bolt which is supported by straps to an overhead beam.

Area of Steam-pipe.—In order that the loss of pressure in a steam-pipe due to friction may not be excessive, it is customary to limit the velocity to 5000 or 6000 feet per minute. If there are many bends or elbows in the pipe, the velocity may be 4800 feet per minute, or less.

Example.—Required the diameter of the main steam-pipe leading from a battery of boilers having an aggregate of 3000 boiler horse-power. Assume the pressure to be 100 pounds by the gauge, or about 115 pounds absolute. Assume also that a boiler horse-power is equivalent to 30 pounds of steam per hour. Then the steam drawn from the boiler in one hour is

$$30 \times 3000 = 90,000$$

pounds. The steam per minute is consequently 1500 pounds.

Now one pound of steam at 115 pounds absolute has a volume of 3.862 cubic feet. Consequently

$$1500 \times 3.862 = 5793$$

cubic feet of steam per minute must pass through the steam-main. With a velocity of 5000 feet per minute the area of the pipe must be

$$5793 \div 5000 = 1.157$$

square feet, or 166.6 square inches. The corresponding diameter is $14\frac{1}{2}$ inches. The next larger size of pipe is 16 inches, which will be used.

CHAPTER IX.

SHOP-PRACTICE.

THE method of work in a boiler-shop depends on the size and arrangement of the shop and on the class of work. There are, however, certain general principles which can be recognized in all modern shops.

The materials, especially the plates, are received at one end of the shop, near which is a storeroom, and a bench for laying out work. The plates, after they are laid out, pass in succession to the several machines, where they are sheared, punched or drilled, planed, rolled, and riveted. The machines for performing these operations are arranged in order with proper spaces for handling and working. Space is provided where boilers may be assembled and receive their tubes and furnaces. Machines which, like the punch, have much work to do, compared with other machines, may be duplicated.

There should be an efficient system for handling the material at the machines and for passing it on from one machine to the next. A good arrangement is to have a swing-crane near each machine; the spaces served by the several cranes overlap, so that one crane takes material from the next, and so on. It is advantageous, especially in large shops, to have a travelling crane that can handle the largest boiler made, and which can serve any part of the shop.

Flanging and smithing are usually done in a separate shop or room. A few machine-tools are needed for doing work on steam-nozzles, manhole rings and covers, etc.

A boiler-shop will have an office, a drawing-room, and a

pattern-room, also a storeroom for patterns. These may be conveniently located in the second story.

A Boiler-shop.—The application of the general principles just stated and the explanation of details can be best given by aid of an example. A medium-size shop for making cylindrical boilers has been chosen for this purpose; the shop is capable of making any shell boiler of moderate size. This shop will employ sixty or seventy men and can turn out two 100-horse-power boilers per day. It will take about three days to finish one boiler, so that there may be six or more boilers in process of construction at one time.

The shop which is represented by Fig. 113 has one end on the street and has a driveway or yard at one side. Plates are received at the street-door by a travelling crane and stored near at hand. The same crane takes plates to the laying-out bench and from there to the crane which serves the shearing-machine. Along one side of the shop are arranged in succession a shearing-machine, two punches, a plate-planer, a set of plate-rolls, and a riveting-machine. Between the punches and nearer the wall is a flange-punch; near the planer is a forge for scarfing. This series of machines is served by four swing-cranes, and there are also two hydraulic cranes near the riveting-machine. These cranes, which are at the top of a tower thirty feet high, are operated from the working platform of the riveter. There are two shipping-doors where the finished boilers are delivered to teams, and at each door there is a jib-crane for handling the boilers. These jib-cranes and the hydraulic cranes at the riveter have a capacity of eight or ten tons; the swing-cranes may be much lighter. A shop where large marine boilers are made will have more powerful cranes.

The machine-shop is near the receiving-door. Here are the lathes, planers, and drills for doing work on manholes, nozzles, and other fittings; also a bench for fitting up boiler-fronts. Two drills for boring tube-holes in tube-plates, and

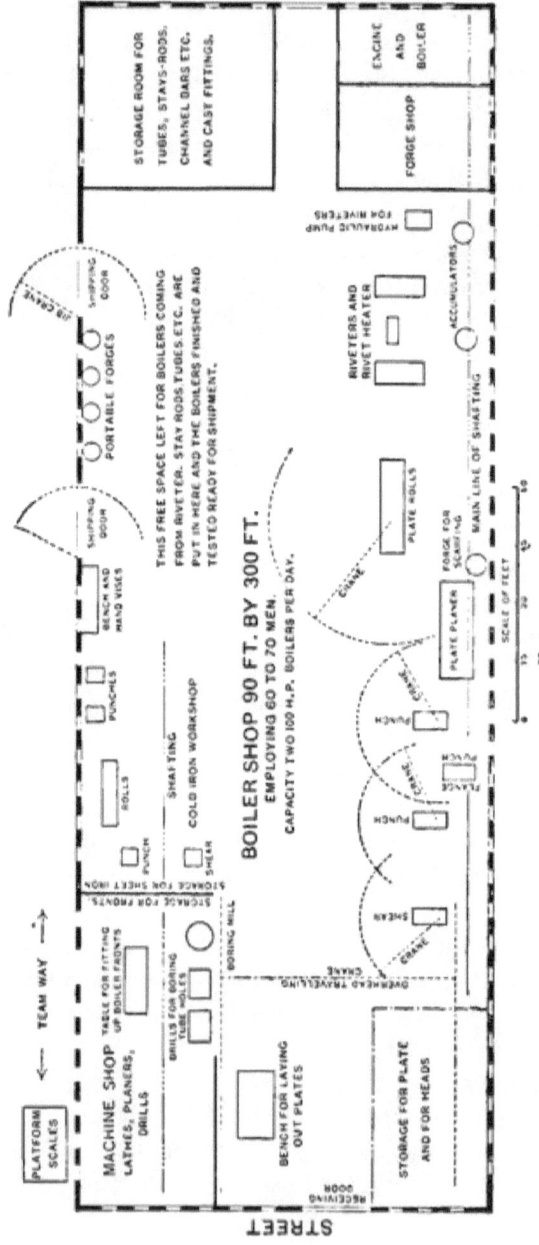

Fig. 113.

a boring-mill for facing off the flanges of boiler-heads, are placed in the entrance to the machine-shop, where work can be conveniently brought to them from the boiler-shop. At the end partition of the machine-shop are places for storing boiler-front castings and sheet-iron. The corner of the boiler-shop near the machine-shop is known as the cold-iron shop; here the uptakes, flues, and dampers are made. This shop has a shearing-machine, three punches, and a set of rolls suitable for sheet-iron work; also a bench with hand-vises.

At the rear of the boiler-shop there is in one corner a store-room for tubes, stay-rods, channel-bars, and finished fittings. In the opposite corner are the forge-shop and the engine-room. These are separated from each other and from the boiler-shop by glass partitions which do not cut off the light, and yet keep the smoke and dust from the forge out of the other rooms.

The main line of shafting is near the wall over the shearing-machine, punches, and rolls. The shafting for the machine-shop and cold-iron shop is driven by a belt from the main shaft, near the front end of the building. A space is left near the riveter where the plates from the rolls can be assembled and bolted together before going to the riveter. In front of the riveter there is a space about 60 feet wide and 120 feet long where boilers are deposited after leaving the riveter. Here the boilers receive their stays and tubes, here they are calked and receive all fixtures that are permanently attached to the shell. At this place the boilers are tested by hydraulic pressure, usually to one and a half times the working pressure. When complete the boilers are painted and oiled, ready for shipment.

To illustrate the method of building a boiler more in detail, the different steps in making a horizontal boiler will be followed in order.

Flanging Heads.—Regular sizes of boiler-heads flanged at one operation by machinery can now be bought on the market, and all except the largest shops are in the habit of buying them. The flanging-machine has a former and a die between which the plate is formed under hydraulic pressure while at the proper flanging temperature. No strains due to unequal heating or cooling are set up in this process, and the plate, which is allowed to cool gradually, does not need to be annealed.

Irregular sizes and shapes are made in the shop on a special cast-iron anvil, which is about six inches deep, flat on top, and curved at one side to about the radius of the head to be flanged. The corner of the anvil or former is rounded so as not to cut the plate. It is placed near a special low forge where the plate is heated.

In flanging, the plate is first marked at short distances on the inner circle of the bend with a prick-punch. A portion of the plate is then heated to a good heat, and the plate is taken to the anvil or former. After adjusting so that the depth of flange overhangs the right distance from the edge of the former, the heated portion of the plate is beaten down

FIG. 114.—Lifting-dogs.

against the side of the former by wooden mauls and then smoothed with a flatter and sledge. The plate is then heated in a new place and another portion bent. To straighten the head and also to remove the strains set up by this way of flanging, it should be heated to a dull red and allowed to cool gradually.

The lifting-dogs represented by Fig. 114 are used in lifting and placing the head during the flanging, and in handling plates during other operations.

Fig. 115 represents crane-lifts which are used when plates are lifted and carried by cranes.

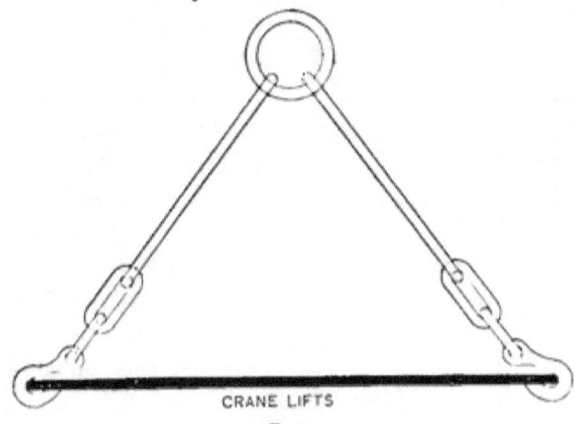

FIG. 115.

After the head is flanged, holes for rivets, stay-rivets, and tubes are marked, and all the rivet-holes are punched.

Flange-punch.—The holes in the flange are punched by a special machine shown by Fig. 116. The punch is carried

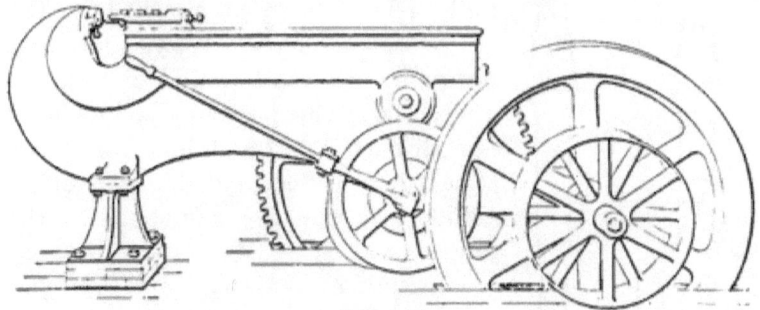

FIG. 116.

by a horizontal wrought-iron plunger which is operated by a cam. The die is carried by a hooked extension of the frame. The head is held horizontal with the flange down; the flange is dropped between the punch and the die and the lever is pulled to throw the cam into play; the plunger then makes a stroke and punches a hole. The machine is driven by a belt, with a fast-and-loose pulley. On the shaft with these pulleys is a heavy fly-wheel. A pinion and spur-gear give a slow powerful stroke to the gear which moves the cam.

Punch and Holder.—The punch (Fig. 117) is made of a solid piece of tool-steel. It has a flat head and a conical shoulder by which it is held onto the plunger, a short straight body, and a slightly coned point. The point is larger at the cutting edge than back toward the straight body, to avoid friction in the hole. A tit in the middle of the face of the punch catches in the centre-punch mark and centres the hole punched.

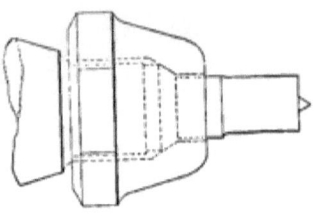

Fig. 117.

The holder is made of wrought iron. It screws onto the end of the plunger, grips the punch by the conical shoulder on its head, and draws it down firmly against the plunger.

Tube-holes.—There are two ways of cutting the holes for the tubes in boiler-heads. Sometimes a small hole is punched at the centre of the hole. A tool like that shown by Fig. 118 is then put in the drill-press. The post in the middle is run through the small hole previously punched or drilled, and the two cutters rapidly cut out the tube-hole to the proper size.

The other way is to punch the tube-holes at once to the proper size by a helical punch

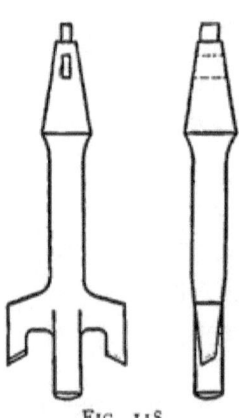

Fig. 118.

Fig. 119.

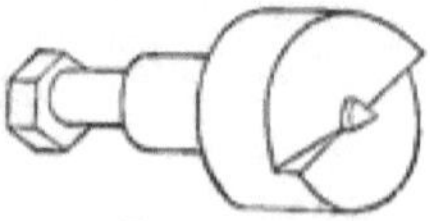

shown by Fig. 119. The die is made in the form of a ring with a flat face, so that the punch begins to cut at the corners, and the metal is removed by a shearing cut. Though not always done, the holes ought to be punched a little under size and then reamed out to give a fair surface against which the tubes may be expanded.

Finishing the Flange.—The boiler-heads are placed on the platen of a boring-mill like that shown by Fig. 120, and the edge of the flange is turned off. The heads of marine boilers are often turned to a true cylinder at the flange to insure that they shall exactly fit the cylindrical shell into which they are riveted. This also gives a good surface to calk against.

Boring-mill.—A simpler machine than the boring-mill shown by Fig. 120 would answer to turn off the flanges of the boiler-heads. But the machine is useful in other ways and may do the work which is commonly done on a large lathe.

The platen is driven much in the same manner as the head of a lathe, through gearing and cone pulleys, to provide for various speeds. This gearing is not well shown in the figure, as it is hidden by the frame. The cutting-tool is adjusted and controlled much like the tool of a planer. The tool-carriage is on a horizontal cross-head which is supported at the side frame and on a round vertical bar at the middle. The tool can be traversed in and out on the cross-head, and the cross-head may be raised or lowered.

For doing some classes of work the cross-head may be set vertically on the guides that are shown on the horizontal bars of the frame near the right-hand end. Or, again, a tool may be carried by the central rod, which can be fed down by the screw at the top.

Laying on the Plates.—The first and one of the most important steps in the work on the shell is the marking out of the plates. Generally one man in each shop does all the laying out. After squaring the sheet, he marks off the length and locates the rivet-holes by means of gauges. These gauges have to be made by trial, a suitable allowance being made in them on account of the thickness of the plate for the

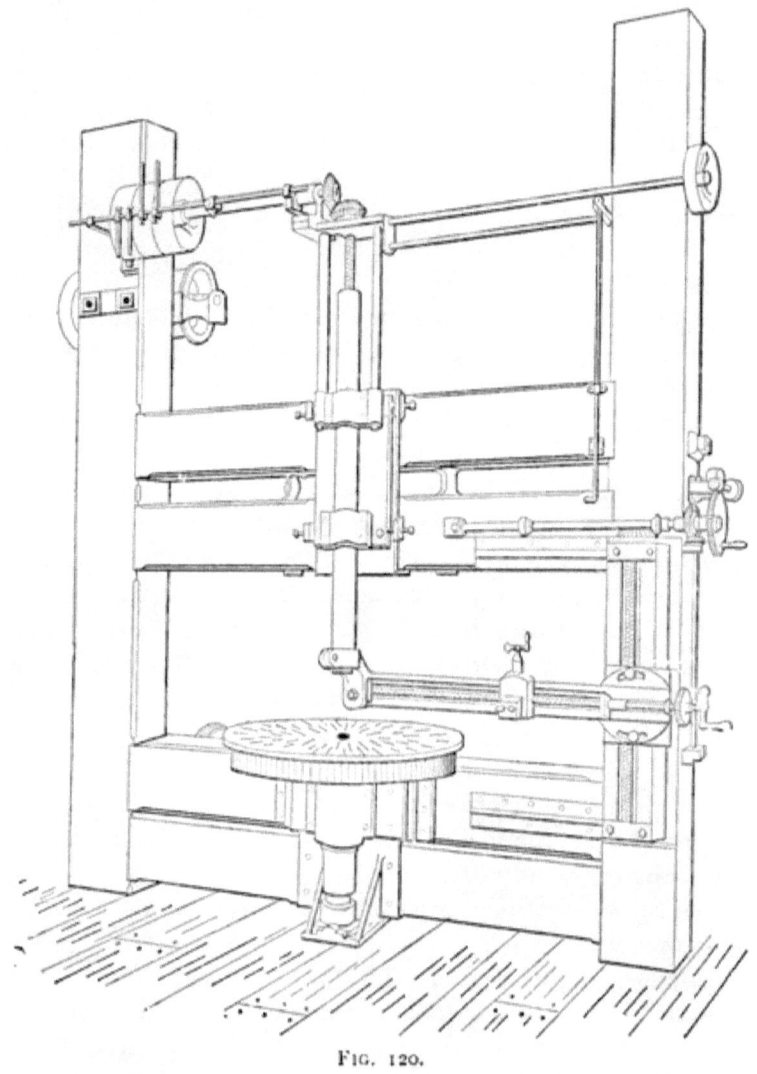

Fig. 120.

change in length due to rolling. There is a gauge for each course, or a set of gauges for each size boiler, and also sets for the same size, but with different thickness of shell. The plates are marked either with a piece of soapstone or with a slate-pencil. Rivet-holes are prick-punched at the centre.

Shearing.—When the plate is laid out it is taken from

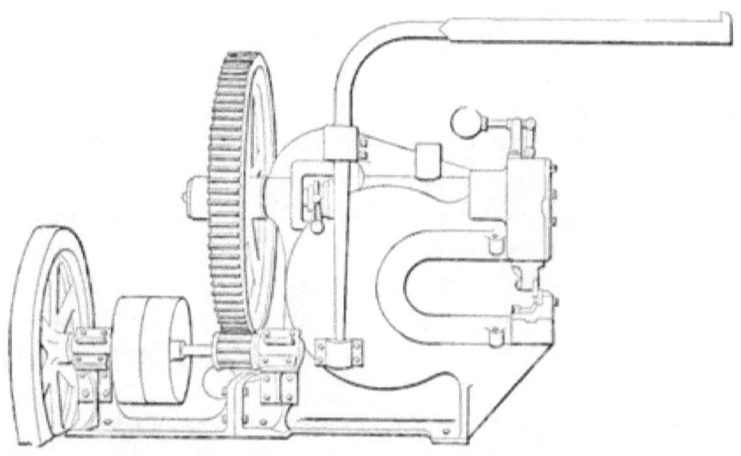

FIG. 121.

the bench to the shears and any superfluous stock is cut off. A shearing-machine is shown by Fig. 121. The lower knife is fixed and the upper knife is moved by an eccentric inside the head. The eccentric-shaft is coupled to the gear-shaft by a clutch that is controlled by a treadle. The weight of the sliding-head is counterbalanced by a weight and lever at the top. Lugs are shown on the casting near the knives; when the machine is required to do extra-heavy work, wrought-iron bolts are put through the lugs and screwed up to strengthen the frame.

The machine is driven by a belt with a fast-and-loose pulley; the shaft carrying these pulleys has a pinion gearing into a large gear to give the necessary power for shearing. A flywheel steadies the motion of the machine; it must be able to supply the power for shearing-plates without a large reduction in speed.

Punch.—After the plate is sheared to size it is taken to one of the punches and all the rivet-holes are punched. Larger openings for man-holes and other fittings are cut out by punching overlapping holes, thus leaving a ragged edge which is afterwards chipped smooth. The plate is not entirely cut away at such large openings, but the piece to be removed is left hanging at three or four places until after the plates are rolled into cylindrical form. If the pieces were removed, there would be less resistance to the rolls at such places and the plates would have a conical form instead of a true cylindrical form.

The punches resemble the shears shown by Fig. 120, with a punch and die instead of the knives. Machines are often so made that they either punch or shear.

Planing.—After the plate is sheared and punched the edges are planed to a slight angle to give a good calking edge.

The planer shown by Fig. 122 has a long narrow bed on which the edge of the plate is laid and to which it is clamped by a follower; the follower is forced down by screws which pass through a beam as shown. The tool-carriage is drawn back and forth by a leading-screw; the tool is made to cut on both strokes, and is fed by hand between the cuts.

Scarfing.—When the plates are joined by a lap-joint the proper corners of each plate are heated in a portable forge near the planer, and are drawn down or scarfed so that the overlapping plates may come close together and not leave a space.

Plate-rolls.—The plates for forming the cylindrical shell are bent to shape cold by running them through bending-rolls The horizontal roll represented by Fig. 123 has two parallel rolls below that are driven in the same direction by gearing. The upper roll is adjusted at each end separately, and some care is required or the shell will receive a conical shape instead of a true cylindrical shape. The bearing at one end of the

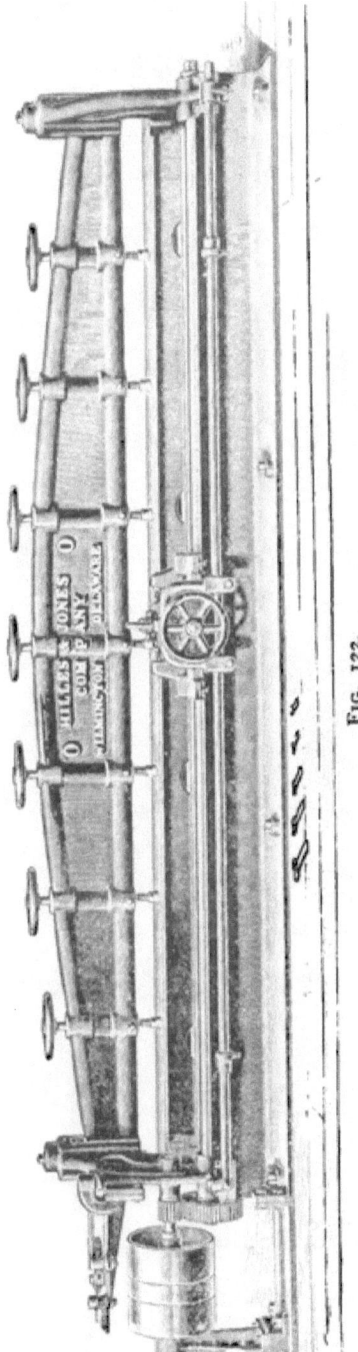

Fig. 122.

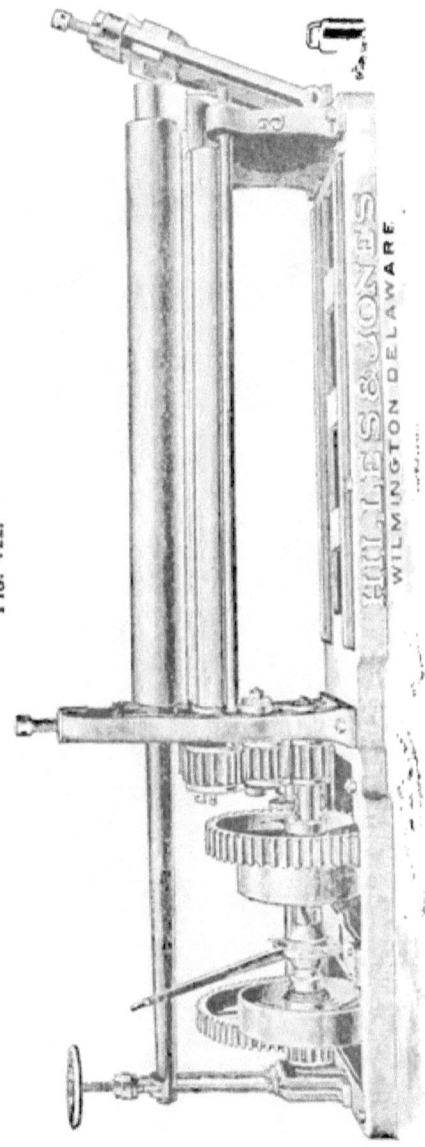

Fig. 123.

roll can be swung out, as shown by the figure, to remove the plate after it is rolled.

The rolls may be driven in either direction by crossed and open belts. The plate to be rolled has one edge introduced

FIG. 124.

between the upper and lower rolls, the upper roll is brought down and the rolls are started up. The plate is run through nearly to the other edge then the top roll is screwed down farther and the rolls are reversed. Thus the plate is run back and forth and the todp roll is gradually rawn down till the plate acquires the proper form.

The extreme edges of the plate are not bent in this process; they are commonly bent afterwards by hammering them with sledges. Some rolls have a special device for bending the edges; it consists of two short overhanging rolls about fifteen inches long, one concave and the other convex. The ends of the plate are fed through these rolls sideways, and are bent before they are introduced into the long rolls.

Vertical rolls, shown by Fig. 124, are coming into use in boiler-shops. They take up less floor-space, and the plate after it is rolled up into cylindrical form is easily hoisted off from the front roll. For this purpose the front roll is counterbalanced and the top end can be swung out clear from the housing. The figure shows the rolls as erected by the builders; in the boiler-shop the plate at the lower end of the rolls is flush with the floor of the boiler-shop.

The width of plate that can be rolled by either horizontal or vertical rolls depends on the length of the rolls. The length of the rolls and the reach of the riveter (to be mentioned later) determine the width of plate that can be handled in the shop.

Assembling and Riveting.—When the plates for a boiler have been punched, planed, and rolled they are assembled in courses, and bolted together ready for riveting. Formerly boilers were commonly punched and riveted; now it is customary to punch the rivet-holes one eighth of an inch smaller than the finished size and then drill to the right size after the boiler is assembled. This is more expeditious than drilling directly, and as all the metal affected by punching is removed it gives as good results. It is the custom in most shops to drill the holes out at the riveting-machine immediately before the rivets are driven and thus each rivet-hole is sure to be true.

The shells of heavy marine boilers are drilled after the plates are assembled without previous punching. A few holes are drilled before the plates are rolled and serve for bolting the plates in place when the boiler is assembled. There are two forms of machines for drilling marine-boiler shells. In one the boiler is placed horizontal on rollers so that it may be readily turned. There are two or three upright frames each carrying a drill. The frames may be adjusted lengthwise of the boiler, and the drills may be set at any height or turned at an angle. When a longitudinal seam is drilled the boiler is rotated to bring a row of rivets to a drill, and the frame is traversed from hole to hole. When a ring-seam is drilled the drill is brought to the proper place, and the boiler is rotated so as to bring the rivet-holes in succession to the drill. The other machine has the boiler placed on one end and the vertical frames carrying the drills can be rotated into place, and the boiler can be turned on a vertical axis.

If plates are punched and riveted *without* drilling, the holes should be punched from the side of the plate which comes in contact with the other plate. The reason for this is that the die is always a little larger than the punch and the hole is slightly conical, larger at the side where the die holds up the plate. If the smaller ends of the holes in two plates are brought together, then the rivet fills the hole better and draws the plates up more perfectly as the rivet cools. It is clear that three or more overlapping plates should always be drilled, as punched holes cannot always be brought together in a proper manner. This is aside from the desirability of drilling all rivet-holes.

Returning now to the assembling of a cylindrical boiler, the process is as follows: The back head is put in the rear course or ring of the shell, and is bolted with six or eight bolts through the punched holes. The head and ring are hoisted up to the drill near the riveter, and six or eight holes are drilled at about equal distances around the seam holding the head into the ring or course, and rivets are driven by the machine in these holes. The bolts are now taken from the punched holes, and all the remaining holes are drilled and riveted, completing the ring-seam through the flange of the back head. The reason for driving a few rivets first, at equal intervals, is that the errors of spacing, when any exist, are distributed, and are removed during the subsequent drilling; while such errors might accumulate and give trouble if the seam were riveted in succession beginning at one point, without first driving a few rivets at intervals.

After the ring-seam through the flange of the head is completed, the longitudinal seam or seams are drilled and riveted. Here again a few rivets are driven at intervals before the seam is riveted up. A few holes at the ends of the seams are left for convenience in joining onto the next course.

The head and first course are now lowered onto the next course, which has been assembled in readiness. A few bolts are put through the punched holes, and the two courses are hoisted up, drilled and riveted in the way already described for the rear course.

When all the courses are riveted together the front head is put in with the flange out so that the rivets in that flange can be driven on the machine. The closing seams on a boiler which, like the Scotch boiler, has both heads set with the flange in, must be riveted by hand.

Rivets are heated in a small forge near the riveter and are passed to a man inside the boiler, who picks them up in tongs, thrusts them through the holes from within and guides the head of a rivet up to the die which is inside the boiler. Sometimes the rivets are thrust through from without, in which case the man inside the boiler guides the point to the die. On the platform of the machine stand the riveter and two or three helpers. They adjust the boiler so that the rivet is brought between the dies, and the riveter pulls the lever which controls the ram, and the outer die is driven against the rivet, forming the head and closing up the rivet in the joint.

The holes are drilled about one sixteenth of an inch larger than the rivets. The pressure of the dies varies from 20 to 70 tons, depending on the thickness of the plate; enough to compress the rivet and fill the hole completely. The rivets, as they cool, shrink and draw the plates firmly together.

Riveting-machines.—There are four types of riveting-machines used for boiler-work, depending on the method of moving the ram or plunger which carries the movable die. The motion may be derived from—

1. A cam and toggle.
2. A hydraulic cylinder,
3. A combination of a hydraulic cylinder with a cam and toggle.
4. A steam-cylinder.

The *cam and toggle* riveter is now seldom used. In it the ram carrying the movable die is driven by a toggle-joint that is closed by a cam, which in turn is driven by a belt and gearing. The adjustment for different thicknesses of plate is made by a wedge behind the ram, which can be set by aid of a screw. The pressure on the rivet is controlled by the elasticity of the frame of the machine and the setting of the wedge; it cannot be regulated satisfactorily.

The *hydraulic* riveter, in one form or another, is most commonly used at the present time. With it a definite pressure can be applied to each rivet whatever the thickness of plate. Fig. 125 represents a hydraulic riveter with a reach of 96 inches which can apply a pressure of 150 tons. It consists essentially of two heavy cast-iron levers or beams, bolted together near the middle and at the lower end. One beam carries the fixed die at its upper end; the other carries the ram and hydraulic cylinder. The stroke of the ram can be adjusted and is controlled by a single lever. The ram moves in straight girders, and may apply an eccentric pressure without rotating or springing.

Some hydraulic riveters have a hydraulic closing device

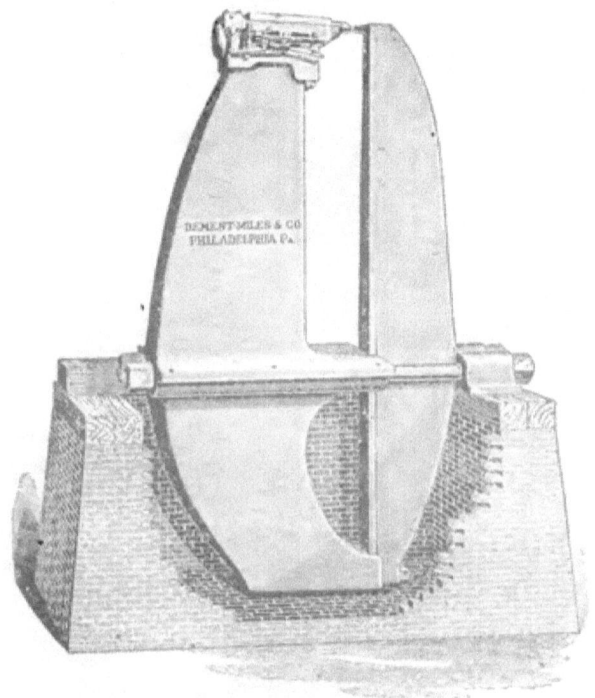

FIG. 125

for holding the plates together while the rivets are driven. Even when furnished it is commonly not used.

The *reach* of a riveting-machine is the distance from the dies to the bed-plate at the middle of the machine. It limits the width of plate that can be riveted by the machine.

A portable hydraulic riveter is shown by Fig. 126, which has a reach of 12 inches and can apply a pressure of 75 tons. It can be swung into position by a crane and can be turned to any angle by the gear at the trunnion. This type of machine is used largely for bridgework; it is sometimes used for riveting nozzles, manhole-rings, brackets, and reinforcing-plates onto boilers.

The power for working a hydraulic riveter is derived from either a steam-pump or a power-pump. A heavy geared

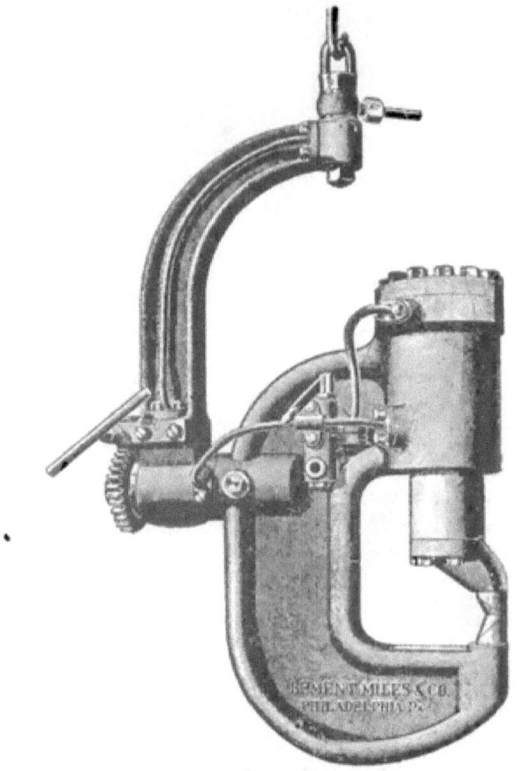

FIG. 126.

power-pump is shown by Fig. 127; it is run continuously and delivers water to an accumulator from which water is supplied to the hydraulic cylinder which moves the ram. The accumulator consists essentially of a loaded piston or plunger. Water is pumped into the cylinder of the accumulator, and is drawn out by the hydraulic cylinder as needed. When the accumulator reaches the end of its stroke it closes a valve on the pipe from the pump so that it receives no more water; at the same time it opens a by-pass from the delivery to the suction of the pump which continues to run, but has at that time very little resistance to overcome. When

FIG. 127.

some water has been withdrawn from the accumulator the by-pass is closed and the valve on the delivery-pipe is opened. When a steam-pump is used there is a device for shutting off steam from the pump when the accumulator is near the end of its stroke, and letting it on again when more water is required.

An accumulator, shown by Fig. 128, is loaded by scrap-iron in a plate-iron cylinder. Inside the plate-iron cylinder is a cast-iron cylinder which is closed at the top and which moves on a fixed plunger. This plunger passes through a stuffing-box and is carried by a cast-iron bed-plate. When water is

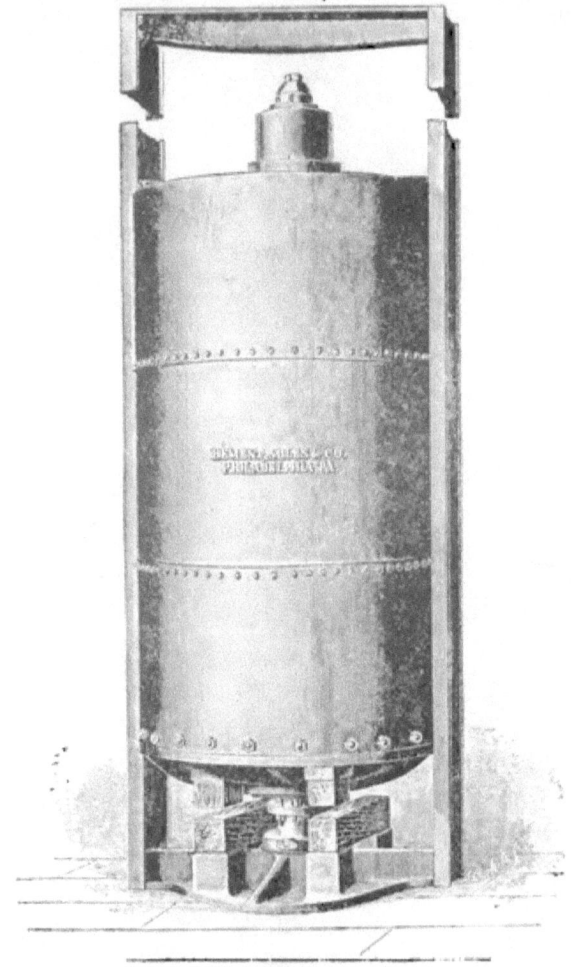

FIG. 128.

pumped into the cylinder through a passage in the fixed plunger, the whole weight of the cylinder, plate-iron casing, and scrap-iron load are lifted. The pressure required to do this depends on the load; it is the pressure which is exerted on the plunger of the hydraulic cylinder moving the ram. The frame of I beams at the sides forms a guide for the accumulator-cylinder and its load.

Another form of accumulator, loaded with heavy cast-iron blocks and without any exterior guides, is shown by Fig. 129.

FIG. 129.

The *hydraulic riveter with toggle and cam* combines the simplicity of the cam-and-toggle machine with the advantage of a definite and determinable pressure on the rivet, which is the best feature of the hydraulic machine. The toggle bears against the ram at the front end, and against the plunger of a hydraulic cylinder at the back end. The cylinder is connected with an accumulator which is loaded to give the desired pressure on the rivet. Suppose that pressure to be 30 tons; then when the cam closes the toggle, the rear end, resting against the hydraulic plunger, remains at rest, and the front end drives the ram and compresses the rivet till a pressure of 30 tons is reached. When that pressure is reached the hydraulic plunger yields, forces water into the accumulator and raises the load on it. When the cam releases the toggle, the hydraulic plunger moves forward and the load on the accumulator falls and drives water into the cylinder. The stroke of the hydraulic plunger may be very short, as the principal part of the stroke of the ram is made before the plunger yields.

FIG. 130.

There is no loss of water except by leakage, which may be made up from time to time by a hand-pump. This machine gives a definite pressure on the rivet whatever the thickness of the plate, like the plain hydraulic riveter. It has no pump and the accumulator is smaller. If the plunger has a large area, the load on the accumulator need not be very great.

A *steam-riveter*, shown by Fig. 130, has the same external appearance as a hydraulic riveter, except that the power is applied by the direct pressure of steam on a piston, which must have a large area to give sufficient pressure to drive the rivets properly. The steam-valve is balanced so that it can be easily moved by the working-lever. If the valve is opened slowly, the ram is first moved forward against the rivet and then full pressure is applied to close the joint; but if the valve is opened promptly, the ram strikes a blow like that of a hammer. There is no reason why this cannot be guarded against if the valve is small and the machine is operated carefully. The fact that the machine is commonly so used that it strikes a blow, and the fact that it is wasteful of steam, have brought the steam-riveter into disrepute except for small or for portable machines. The ram is moved back by the steam before escaping, after a rivet is driven.

Hand-riveting.—In a modern boiler-shop almost all the riveting is done by machine because it is cheaper and, especially on heavy work, is more likely to be well done. There are, however, a good many rivets on any boiler that must be driven by hand. In such case the rivet, which may be heated entirely or at the point only, is thrust through the hole from within and is held up by a man inside, who has for this purpose a hammer or weight which weighs about 20 pounds on a long handle. He has also an iron hook which he hooks into a rivet-hole, and against which he gets a purchase to hold the rivet up while it is driven. Two men with hammers that weigh about 5 pounds drive the rivet, striking in turn. A few heavy blows are struck to close the joint and partially form the head, then the head is finished in the shape of a straight-sided cone with lighter hammers. If the rivet is long enough to form a good head, and if it is driven with care and skill, hand-riveting may be equal to machine-riveting. If the heads are ill-formed, or if they are too low, the work may be very inferior.

Snap-riveting.—This method of riveting, which is especially convenient for driving rivets in contracted spaces, has some resemblance to machine-riveting. The rivet is thrust through the hole and held up from within the boiler. The joint is closed and the head is roughly formed by a few blows of a heavy hammer, then a *snap* or die is held on the rivet and driven with sledge-hammers. For large rivets the section of the snap should be a parabola, and the head should be relatively small in diameter and high, because this form causes the rivet to fill the hole better and makes sounder work.

Tube-expanders.—The tubes are expanded into the tube-sheets to make a steam-tight joint, beginning at the least accessible end. They are commonly a little too long and are cut off at the projecting end by a tube-cutter. The tubes extend through the tubes a slight amount, and are beaded over, after they are expanded, by a special tool. The expanders most commonly used are known as the Prosser and the Dudgeon expanders.

The *Prosser* expander, represented by Fig. 131, is made up

FIG. 131.

of a number of steel segments held in place by a spring on a cylindrical extension of the segments. The acting part of the segments have the form to be given to the tube after it is expanded. The inside of the segments forms a straight hollow cone into which a steel taper pin fits. The expander is forced into the tube and is expanded by driving in the pin with a hammer. This should be done gradually so as not to distress the metal of the tube too much, and the expander should be frequently slacked back and shifted part way round on account of the spaces between the segments.

The *Dudgeon* expander, Fig. 132, has a set of rolls, three or more, in a frame. The rolls are forced out against the sides

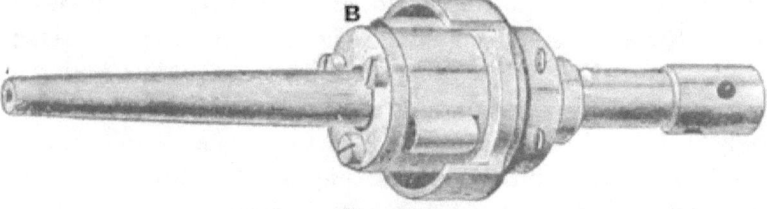

Fig. 132.

of the tube by driving in a taper pin. The pin and frame are rotated as the pin is driven, and the rolls gradually force the tube against the tube-plate.

Although the two expanders accomplish much the same result, the action is different. The Prosser causes an abrupt

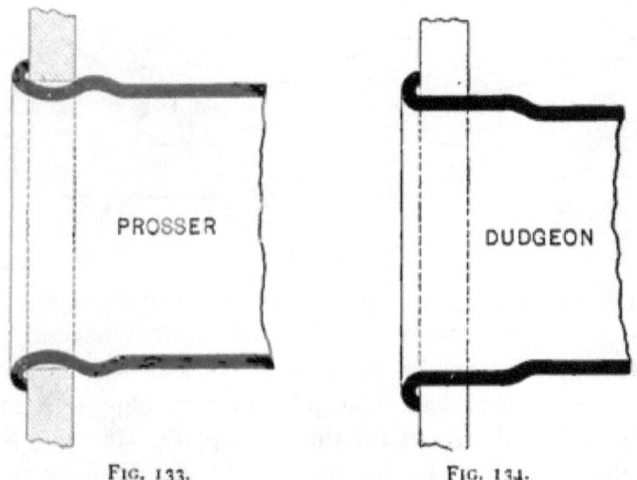

Fig. 133. Fig. 134.

stretching of the tube and leaves the tube as shown by Fig. 133, bearing at the corners of the plate only. The Dudgeon enlarges the end of the tube and makes it bear against the entire thickness of the tube-sheet.

After the tubes are expanded the ends are beaded over by

a special tool, as represented in both figures, which adds to their grip on the plate when they act as stays.

A vacuum may possibly be found in a boiler, if it is allowed to cool without admitting air. The Prosser method has an advantage in such case, when the tubes act as struts between the heads. The Dudgeon method will then act by friction only. The rollers might be shaped to give an expansion just inside the plate, instead of making them straight; there is, however, no evidence of trouble from this source in practice.

Calking.—The riveted seams of a boiler are made steam-tight by calking, which consists in driving the lower part of the planed edge forcibly against the plate beneath. Fig. 135 shows the form of calking-tool used in hand-calking, the posi-

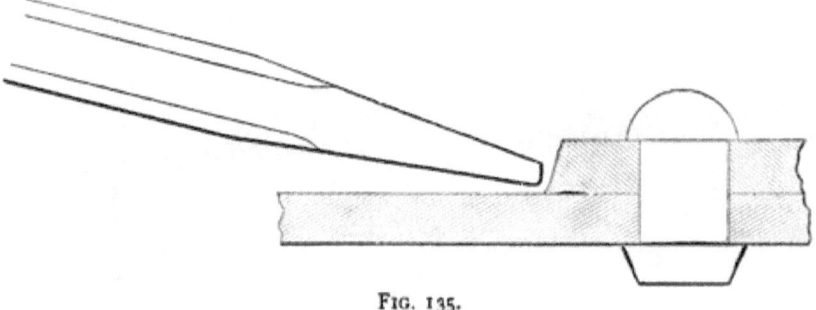

FIG. 135.

tion in which it is held, and the way the extreme edge of the plate is compressed against the plate beneath. The acting surface of the tool, which is about an inch wide, is ground at an angle of somewhat less than 90°, and the edge is rounded slightly so that it will not cut the lower plate. The tool is slid along the under plate against the edge of the upper plate and struck with a hammer. If the tool is ground to a sharp edge and used carelessly, a groove may be cut in the under plate and serious injury may be done.

A pneumatic calking-machine or tool is now used for doing most of the calking in boiler-shops. In general principle it resembles a rock-drill, and consists of a cylinder in which works a piston and rod on the end of which is the calking-tool. Air is supplied for working the piston, at a pressure of 60 or 80 pounds, through a flexible tube. It makes about 1500 working-strokes a minute, 3/16 of an inch long. The calker, which is about $2\frac{1}{2}$ inches in diameter outside and 15 inches long over all, is held by a workman who presses it slowly along the seam to be calked. The edge of the tool is well rounded so as not to injure the lower plate. Work can be done four times as rapidly with the pneumatic calker as by hand.

Cold-water Test.—After the boiler is calked it is tested to about once and a half the working pressure, with cold water. During the test the boiler is carefully watched to detect any notable change of shape or other sign of faulty design or construction, and important leaks are marked; small leaks are of no consequence, as they will fill up with rust. Important leaks must be calked after the pressure is relieved; if necessary, pressure may be applied again to see if they are stopped. The method of making this test and the precautions to be observed are given on page 224.

If the boiler is examined by a boiler-inspector, he makes his inspection before the boiler is painted, and stamps certain letters on the head or over the fire-door to show that the boiler has passed inspection.

Finally the boiler is painted and oiled ready for shipping.

CHAPTER X.

BOILER-TESTING.

THE main object of a boiler-test is to determine the amount of water evaporated per pound of coal, or, more exactly, the amount of heat transferred to the boiler per pound of coal burned. For this purpose it is necessary to determine:

1. The number of pounds of water pumped into the boiler during the test.
2. The number of pounds of coal burned, and the weight of ashes left.
3. The temperature of the feed-water when it enters the boiler.
4. The pressure of the steam in the boiler.
5. The per cent of moisture in the steam discharged from the boiler.

It is desirable to determine the conditions of combustion, such as the draught, the weight of air supplied per pound of coal, the composition of the products of combustion, and the temperature of the escaping flue-gases. It is also desirable to have determinations made of the composition of the coal and its total heat of combustion, but, as was explained in Chapter II, these determinations should usually be intrusted to a chemist and to a physicist.

Water.—The best and most satisfactory way is to weigh the feed-water directly, in proper tanks or barrels on scales. There should be two barrels or tanks large enough so that the filling, weighing, and emptying may proceed without haste.

The scales should be adjusted and tested with a standard weight and should be known to be correct and sensitive. Good commercial platform scales are sufficient for this purpose.

The weighing-barrels should be placed high enough to discharge into a tank or reservoir from which the feed-water is drawn by a pump or injector. This tank should hold more than both weighing-barrels, so that when it is about half empty an entire barrelful of water may be discharged into it without danger of overfilling it and wasting water. The barrels are emptied through large quick-opening lever-valves; this point should receive attention, as any delay caused by small valves is very annoying.

The weighing-barrels are filled either from a water system or by a special pump from a well or reservoir. When a direct-acting steam-pump is used, a quarter-inch by-pass should be carried from the delivery-pipe to the suction-pipe; the pump will then run slowly when the valves on the pipes leading to the weighing-barrels are shut; when one of these valves is opened the pump starts away promptly, and it slows down again when the valve is shut. If a power-pump is used, it may be convenient to arrange so that it shall run all the time at full power, discharging into the well or reservoir when neither barrel is filling.

Weighing water, though simple enough, requires care and intelligence, as any blunder will spoil the test. The observer should proceed systematically. He will naturally start with both barrels filled, weighed and recorded before the test begins. When the level in the feed-tank has fallen so that it can receive a barrelful of water he will open the discharge-valve from one barrel, which should be marked and designated as Barrel No. 1. When that barrel is emptied, he will close the valve and weigh the barrel; the weight empty is set down and subtracted from the weight full to get the weight discharged. The record of weights is kept in a table containing columns for the name of the barrel, weights full, weights empty, weights discharged, and time at which discharged. The weight of the barrel empty must be taken each time, as the barrel will not drain completely in the time that can be allowed.

Water may now be turned on to fill Barrel No. 1, and Barrel No. 2 may be emptied, as occasion demands. Then one barrel may be filling when the other is emptying, and the work may proceed rapidly but without confusion. The errors that a novice is liable to are either to forget to record the weight of a barrelful of water, or to empty a barrel that has not been weighed.

It is convenient and almost necessary to have some sort of an index or telltale to show the water-weigher where the water-level is in the feed-tank. For this purpose we may use a float, with a string that runs up over a pulley and is kept taut by a small weight moving over a scale, which is placed in front of the weighing-barrels. This float is not used to determine the level of the water in the feed-tank at the beginning and end of the test.

At the beginning of the test the level of the water in the feed-tank is marked, and at the end of the test the level is brought to the same mark, so that all the water delivered by the weighing-barrels is drawn out of the feed-tank by the feed-pump. A good way of marking the water-level is to fasten to the side of the tank a piece of wire bent into a hook, with its point projecting slightly above the water-level. This hook will commonly be placed in position before the test begins, and the tank will be filled up to the level so marked before water is drawn from the feed-tank.

If water cannot be weighed directly, it may be measured in tanks of known capacity which are alternately filled and emptied. Or the water may be measured by a good watermeter, which must be tested under the conditions of the test to determine its error. Care must be taken to keep the meter free from air or it will record more than the amount of water which actually passes. Boiler-tests on steamships can scarcely be made without using meters.

At the time when the test begins, the water-level is noted at the water-glass, and at the end of the test the water-level is brought to the same place. The best way is to fix a wooden scale near the water-glass and record the height of the water above an arbitrary point on the scale. Sometimes a string is tied around the glass at the water-level when the test is started; in such case the distance of the string from some fixed point on the fittings of the water-glass must be recorded, so that the string can be replaced if it happens to be moved or if the glass tube breaks. If the water is not brought exactly to the same level at the end as at the beginning of the test, the difference is noted and allowance is made. It has already been pointed out that the apparent height of the water depends to a certain extent on the rate of vaporization and on the rapidity of circulation in the boiler; consequently the boiler must be making steam at the same rate at the times when the water-level is observed for beginning and ending the test.

All pipes leading water to or from the boiler, except the feed-pipe, must be disconnected. Steam may be taken for any purpose and through any pipe, so far as the boiler-test is concerned.

Frequently the steam used by an engine is determined by weighing the feed-water for a boiler which is used exclusively for that engine. If the boiler is fed by an injector, the steam for running the injector should be taken from the boiler, for it will be condensed by the feed-water and returned to the boiler. A very small amount of the heat (less than two per cent) in the steam supplied to an injector is used in pumping the feed-water; the remainder is used in heating the feed-water and is returned to the boiler. The temperature of the feed-water must be taken before it goes to the injector. If the boiler is fed by a direct-acting steam-pump, that pump should be run with steam taken from some other source. If that cannot be done, then the steam used by the pump must be determined and allowed for, unless the exhaust from the pump can be turned into and condensed by the water in the feed-tank, in which case the pump is in the same condition as an injector. The best way of determining the amount of steam used by a steam-pump is to condense it in a small surface condenser, and to collect and weigh the condensed water. Or the steam may be run into a barrel filled with cold water, which is weighed before and after steam is run in. This method requires that the barrel shall be emptied when the water begins to vaporize, and filled afresh with cold water. Steam used by a calorimeter for determining the amount of water in steam must be ascertained also; the methods will be given in connection with a description of the instruments.

Coal and Ash.—The coal required during a boiler-test should be brought in as required in barrows; it may be fired from the barrow or dumped and fired from the floor. The barrow should be weighed full and empty, and the difference should be recorded together with the time; the latter to serve as a check on the record and make sure that a barrow-load is not neglected. The weight of the barrow is usually the same throughout the test. Any coal left unburned is weighed back

It is essential that the condition of the fire shall be the same at the beginning and at the end of the test. There are two methods in vogue for trying to attain this result; if the test is 24 hours long or more, the condition of the fire is estimated by its appearance; if the test is 10 or 12 hours long, the test is started and stopped with the grate empty. These are for tests of factory boilers with a combustion of 15 to 20 pounds of coal per square foot of grate per hour. For tests on marine or locomotive boilers, where the rate of combustion may be twice or five times as rapid, the duration of a test may be correspondingly reduced.

Coal in solid mass will weigh 70 or 80 pounds to the cubic foot; when lying on a grate it will weigh 50 or 60 pounds. It is difficult to estimate the thickness of the bed of coal on a grate nearer than two inches. But a layer of coal two inches thick will weigh 8 or 10 pounds, which is about half the rate of combustion for a factory boiler. If a test is only ten hours long, the error resulting from a wrong estimate of the thickness of the fire may readily be five per cent. If the test lasts twenty-four hours, the error will probably not be more than two per cent, provided a proper method is used.

If the condition of the fire is estimated at the beginning and end of the test, the fire should be cleaned and freed from ashes and clinker shortly before the test begins, and should then be spread in rather a thin even layer of clean glowing coal. Its height above the grate should be estimated with reference to some mark in the furnace that can be recognized readily. Just as long before the end of the test the fire should be cleaned and levelled in the same manner, and the thickness should be estimated with reference to the mark chosen at the beginning. The fireman is sure to have a clean bright fire at the beginning of the test, but he is apt to have a fire with much the same appearance that is half clinker at the end. The error from estimation may be very serious in such case, even though the test is 24 hours long.

If the test is started and stopped with the grate empty, the boiler must be brought into good working condition about an hour before the test is to start, with all the brickwork thoroughly heated. The fire is allowed to burn low, and the steam-pressure is maintained by reducing the draught of steam from the boiler. Twenty or thirty minutes before the test starts, the fire is drawn or dumped and the grate and ash-pit are cleaned out. A new fire is started with wood, and coal is thrown on as soon as the wood is well alight. The time when coal is thrown on is counted as the beginning of the test. If the steam-pressure falls while the fire is drawn, the stop-valve may be nearly or quite closed to keep it from falling much below the working-pressure. Toward the end of the test the fire is allowed to burn low, and at the end of the test it is drawn out on the boiler-room floor and quenched with as little water as may be, not enough to leave it wet. The unburned coal is picked out by hand and weighed back, the clinker and ashes are separated and weighed together with the clinker withdrawn during the test and the ashes in the ash-pit.

If any appreciable amount of coal falls through the grate, a sample from the ash-pit may be picked over by hand to estimate the proportions of unburned coal in the ash. The coal in the ash is allowed for in calculating the per cent of ash in the coal, but is not added to the coal weighed back, for there is no way of burning coal thus lost through the grate. When a test is started with a wood fire, more or less coal is apt to fall through the grate in starting. This is drawn from the pit and fired over again.

It is customary to allow the fire to burn low before drawing the fire at the end of the boiler-test, both because it brings the fire more nearly to the condition at the beginning, and because it is a hard and unpleasant job to draw a thick fire. But the fire should be maintained at its normal condition until the end of the test approaches, and should be a good fire when drawn. Extraordinary results may be obtained by allowing the fire to burn nearly out at the end of the test, a very considerable amount of steam being formed by heat given out by the boiler-setting. It is unnecessary to say that such results are entirely misleading.

The wood used for starting the fire is weighed and allowed for on the assumption that a pound of wood is equivalent to 0.4 of a pound of coal. The total weight of wood used is not large.

Temperature of Feed-water.—The temperature of the feed-water is taken by a thermometer in a cup filled with oil, screwed into the feed-pipe close to the check-valve. If the temperature varies, it may be read every five minutes; if it is found to be steady, less frequent intervals will do.

Pressure of Steam.—The steam-pressure must be very nearly the same at the beginning and end of a test, and should remain nearly constant throughout the test. Readings are commonly taken every fifteen minutes, but the fireman should be required to keep the pressure nearly constant at all times.

The steam-pressure is taken by a spring-gauge like that shown by Fig. 92 on page 252. The gauge should be compared with a mercury column or a standard gauge both before and after the test, and a correction should be applied if necessary. If the pipe carrying pressure to the gauge fills up with water, allowance for the pressure of that column of water must be made. Each foot of water will give a pressure of about 0.43 of a pound per square inch.

The reading of the barometer should be taken two or three times during a test. The reading in inches of mercury can be reduced to pounds per square inch by multiplying by the weight of a cubic inch of mercury, which is about 0.491 of a pound.

Very commonly the pressure of the steam is obtained indirectly by aid of a thermometer set in the steam-pipe. The absolute pressure corresponding to the temperature is then obtained from a table of the properties of saturated steam. The thermometer is readily standardized, and is not so likely to become unreliable as a steam-gauge.

Most vertical boilers and some water-tube boilers give superheated steam; in such case there should be both a thermometer and a gauge on the steam-pipe, to indicate temperature and pressure. The excess of the temperature by the thermometer above that corresponding to the absolute pressure of the steam, as found in a table of properties of steam, is the degree of superheating.

Specific Heat of Superheated Steam.—The mean value

given by Regnault for the specific heat of superheated steam is 0.4808, or approximately 0.48. This property of steam can be used in calculating the amount of heat in steam due to superheating.

For example, let the pressure by the gauge be 65.3 pounds, and let the temperature be 350° F. by the thermometer. The absolute pressure corresponding to 65.3 pounds is 80 pounds, at which saturated steam has the temperature of 311°.8 F. The superheating is consequently

$$350° F. \quad 311°.8 F. = 38°.2 F.$$

The heat due to the superheating is

$$0.48 \times 38.2 = 18.3 \text{ B. T. U.}$$

When the steam is superheated, the formula for equivalent evaporation is changed from the form given on page 135 to

$$w\frac{0.48(t_s - t) + r + q - q_0}{965.8},$$

in which t_s represents the actual temperature of the superheated steam, and t is the temperature corresponding to the absolute pressure of the steam determined from the reading of the gauge.

Priming.—A boiler which has sufficient steam-space and free water-area will deliver steam which contains less than two per cent of moisture.

Professor Denton[*] has pointed out that a jet of steam blowing into the air from a petcock will give a characteristic blue color if there is less than two per cent of water in the steam. If there is more than two per cent of moisture, the jet will be white. Since steam seldom contains less than one per cent of moisture under the usual conditions of ordinary practice, it is possible by this method to estimate the condition of steam with a probable error of one per cent.

---

[*] Trans. Am. Soc. Mech. Engs., vol. x. p. 349.

The most ready way of determining the condition of steam is by the aid of a *throttling-calorimeter*, devised by Professor Peabody,* which depends on the fact that the total heat of steam increases with the pressure, so that dry steam becomes superheated when the pressure is reduced by throttling. If the steam is only slightly primed, superheating will still take place, and the amount of priming can be determined from the temperature and pressure of the steam after it is throttled. If there is much moisture in the steam, it fails to superheat.

A good form of this apparatus is shown by Fig. 136, consisting of a reservoir A to which the steam to be tested is admitted through a half-inch pipe b with a throttling-valve near the reservoir. The steam flows away through an inch pipe d. At f is a gauge for measuring the pressure, and at c there is a deep cup for a thermometer to measure the temperature. The boiler-pressure may be taken from a gauge on the main steam-pipe near the calorimeter. It should not be taken from a pipe in which there is a rapid flow of steam as in the pipe b, since the velocity of the steam will affect the gauge-reading, making it less than the real pressure. The reservoir is wrapped with hair-felt and lagged with wood to reduce radiation of heat

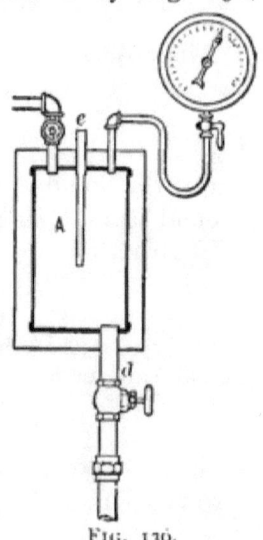

FIG. 136.

When a test is made the valve on the pipe d is opened wide (this valve is frequently omitted), and the valve at b is opened wide enough to give a pressure of five to fifteen pounds in the reservoir. Readings are then taken of the

---

* Trans. Am. Soc. Mech. Engs., vol. x. p. 327.

boiler-gauge, of the gauge at f, and of the thermometer at e. It is well to wait about ten minutes after the instrument is started before taking readings, so that it may be well heated.

The method of calculation can be readily understood from the following

Example.—The following are the data of a test made with a throttling calorimeter:

> Pressure of the atmosphere.......... 14.8 pounds.
> Pressure by the boiler-gauge......... 69.8 "
> Pressure by the calorimeter-gauge.... 12.0 "
> Temperature in the calorimeter...... 268°.2 F.

The absolute pressure in the boiler was

$$69.8 + 14.8 = 84.6 \text{ pounds},$$

at which the heat of vaporization is 892.7 B. T. U. and the heat of the liquid is 285.3 B. T. U. So that with x part of a pound steam (and $1 - x$ priming) the heat in one pound of moist steam was

$$892.8x + 285.3,$$

in which x was to be determined. The absolute pressure in the calorimeter was

$$12 + 14.8 = 26.8 \text{ pounds},$$

at which the temperature was 243°.9 F, and the total heat was 1156.4 B. T. U. The heat due to superheating was

$$0.48(268°.2 - 243°.9) = 11.7 \text{ B. T. U.},$$

and the heat in one pound of steam in the calorimeter was

$$1156.4 + 11.7 = 1168.1 \text{ B.T.U.}$$

But the process of throttling neither adds nor subtracts heat, consequently

$$892.8x + 285.3 = 1168.1,$$
$$\text{or} \quad x = 0.988,$$

and the priming was

$$100(1 - 0.988) = 1, 2 \text{ per cent.}$$

The calculation can be conveniently expressed by an equation in which r and q are the heat of vaporization at the absolute boiler-pressure, and λ_i and t_i are the total heat and the temperature at the absolute pressure in the calorimeter, all taken from a table of proportions of steam; while t_s is the temperature of the superheated steam in the calorimeter. Then

$$xr + q = \lambda_i + 0.48(t_s - t_i);$$

$$x = \frac{\lambda_i + 0.48(t_s - t_i) - q}{r}.$$

It has been found by experiment that no allowance need be made for radiation from the calorimeter if made as described, provided that 200 pounds of steam are run through it per hour. Now this quantity will flow through an orifice one fourth of an inch in diameter under the pressure of 70 pounds by the gauge, so that if the throttle-valve be replaced by such an orifice the question of radiation need not be considered. In such case a stop-valve will be placed on the pipe to shut off the calorimeter when not in use; it is opened wide when a test is made. If an orifice is not provided, the throttle-valve may be opened at first a very small amount and the temperature in the calorimeter noted after a few minutes; the valve may be opened a trifle more, whereupon the temperature will usually rise, showing too little steam used. If the valve is opened little by little till the temperature stops rising, it will then be certain that enough steam is used to reduce the error from radiation to a very small amount.

Various modifications of the throttling-calorimeter have been proposed, mainly with a view of reducing its size and weight. Almost any of them will prove satisfactory in practice, but some will be found to be liable to error from radiation or from the fact that there is not sufficient opportunity for the steam to come to rest and properly develop the superheating due to throttling. One great advantage of this instrument is that ordinary care with ordinary gauges and thermometers gives sufficient accuracy. For example, with 100 pounds absolute boiler-pressure and with atmospheric pressure in the calorimeter, an error of half a degree by the thermometer, or half a pound by the boiler-gauge, or a third of a pound by the calorimeter-gauge will each give an error of one-tenth of a per cent in the priming.

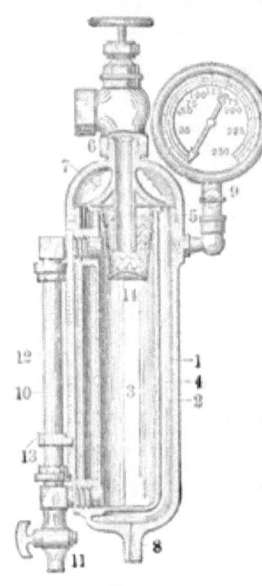

FIG. 137.

If steam contains more than three per cent of priming, the amount of moisture can be determined by a good separator, which will remove nearly all the moisture. It remains then to measure the steam and water separately. The water may be best measured in a calibrated vessel or receiver, while the steam may be condensed and weighed, or may be gauged by allowing it to flow through an orifice of known size. A form of this instrument devised by Professor Carpenter * is shown by Fig. 137.

Steam enters a space at the top which has sides of wire gauze and a convex cup at the bottom. The water is thrown against the cup and finds its way through the gauze into an inside chamber or receiver, and rises in a water-glass outside. The receiver is calibrated by trial so that the amount of water may be read directly from a graduated scale.

---

* Trans. Am. Soc. Mech. Engs., vol. XVII. p. 608.

The steam meanwhile passes into the outer chamber which surrounds the inner receiver, and escapes from an orifice at the bottom. The amount of steam may either be calculated, by a method to be explained, from the diameter of the orifice and the pressure of the steam, or it may be condensed and weighed or measured. The latter is the more accurate way, and it has the advantage that then there is no error from radiation, for the inner receptacle is well protected by the outer chamber, and condensation in the outer chamber is collected and weighed with the steam. If the instrument is well wrapped and lagged, and if a sufficient quantity of steam is used, then the error from radiation can be neglected, just as was found to be the case with the throttling-calorimeter. This instrument, for want of a better name, is called a *separator calorimeter;* it is a question whether either it or the throttling-calorimeter are properly calorimeters at all, and whether it would not be better to call both *priming-gauges.*

It is customary to take a sample of steam for the calorimeter or priming-gauge through a small pipe leading from the main steam-pipe. The best method of securing a sample is an open question; indeed it is a question whether we ever get a fair sample. There is no question but that the composition of the sample is correctly shown by either of the priming-gauges described. It is probable that the best way is to take steam through a pipe which reaches at least half-way across the main steam-pipe, and which is closed at the end and drilled full of small holes. It is better to have the samping-pipe enter the steam-pipe at the side or at the top of the main, so that any water that may trickle along the bottom of the main shall not enter the calorimeter. Again, it is better to take a sample from a pipe through which steam flows upward. The sampling-pipe should be short and well wrapped to avoid radiation.

If the steam from the boiler can be wasted during the test, then the entire steam delivered by the boiler may be passed through a large priming-gauge, and the difficulty of getting a sample may be avoided.

Flow of Steam.—It has been shown by Rankine * that the flow of steam through an orifice into the atmosphere may be represented by an empirical equation,

$$W = A\frac{p}{70},$$

in which W is the number of pounds of steam per second, A is the area of the orifice in square inches, and p is the absolute pressure of the steam. This equation, which has already been mentioned in connection with safety-valves, can be applied only when the absolute steam-pressure is more than double the pressure of the atmosphere; that is, the pressure of the steam must be 15 pounds by the gauge, or more. Experiments made in the laboratory of the Massachusetts Institute of Technology † show that this equation is liable to an error of about two per cent, but this error may be determined by direct experiment for a given orifice under various pressures, and then a correction can be applied which will reduce the error to a fraction of one per cent.

It appears then that the use of an orifice to determine the amount of steam in Professor Carpenter's separator priming-gauge is at least questionable unless direct experiments are made to determine the correction to be applied. On the other hand, the amount of steam used by a throttling priming-gauge may be very properly determined by allowing it to flow through an orifice, since the total amount of steam used by the calorimeter is small.

The same equation may be used for calculating flow of steam from one reservoir to another provided that the pressure in the second reservoir is less than half that in the first

---

* *The Engineer*, vol. XXVII. p. 359, 1869.
† Trans. Soc. Am. Engs., vol. XI. p. 187.

reservoir. This allows us to gauge small quantities of steam used for any purpose, at a pressure that is less than half the boiler-pressure; for example, for running a steam-pump. A convenient arrangement for gauging the flow of steam in an inch pipe consists of a reservoir three feet long, made up of three-inch piping, and fittings divided at the middle by a brass plate through which there is an orifice of proper size. If the pipe carries steam at 100 pounds absolute, at a velocity of 100 feet a second it will deliver

$$\frac{\pi d^2}{4} \times 100 = \frac{3.1416 \times (\frac{1}{12})^2}{4} \times 100 = 0.5455$$

cubic feet per second. The density or weight of one cubic foot of steam at 100 pounds absolute is 0.2271 pounds. So that the pipe will carry

$$0.5455 \times 0.2271 = 0.124$$

of a pound of steam per second. If this weight is put for W in Rankine's equation, and if A is replaced by $\frac{1}{4}\pi d^2$, we shall have

$$0.124 = \frac{\pi d^2 \times 100}{4 \times 70},$$

or

$$d = \sqrt{\frac{0.124 \times 4 \times 70}{3.1416 \times 100}} = \frac{1}{3}$$

of an inch, nearly, for the diameter of the orifice for gauging the flow of steam. With an orifice of approximately the right size, the flow of steam may be regulated by a valve below the gauging device; for example, by the throttle-valve of the pump.

Flue-gases.—At frequent intervals samples of flue-gases should be taken from various places, such as back of the bridge, from the uptake, and from the chimney. These samples are analyzed as soon as may be by Orsat's apparatus, as described on page 56.

Though not commonly done, it would be well if a continuous sample could be taken in a reservoir from which samples for analysis could be taken at intervals.

Draught-gauge.—The draught given by a chimney is seldom more than an inch or an inch and a half of water. It can be measured roughly by a simple U tube filled with water. An instrument for accurate determinations of draught should be at once simple and certain in its action.

The draught-gauge shown by Fig. 138, devised by Prof.

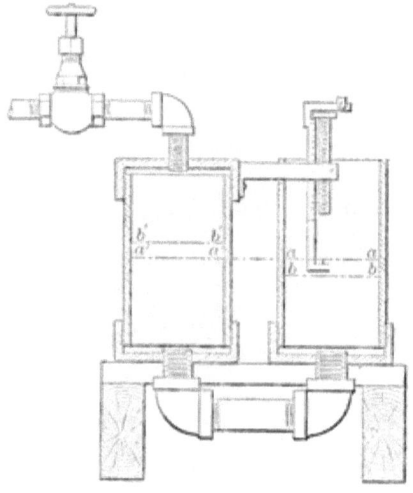

FIG. 138.

Miller, has been used with satisfaction for this purpose. It consists of two pieces of three-inch brass pipe connected by a half-inch pipe at the bottom. One of the pipes is closed at he top and can be connected to the chimney by a small pipe with a valve as shown. The other piece of brass pipe is open and has a hook-gauge, reading to 1/1000 of an inch, suspended in it. In preparing for a reading, the closed tube or leg is shut off from the chimney and opened to the atmosphere; the water then stands at the same height aa, $a'a'$, in both legs. The closed leg is now shut off from the air and connection is made with the chimney, whereupon the level falls to bb in the open leg and rises to $b'b'$ in the closed leg. As the two legs have exactly the same internal diameter, the fall ab is half the draught, measured in inches of water. The hook-gauge is set to the level aa when the closed leg is open to the air, and to the level bb when it is connected to the chimney. The difference of the readings multiplied by 2 is the draught in inches of water. The reading by the hook-gauge can readily give an acuracy of 1/1000 of an inch, which is sufficient for this purpose.

Pyrometers.—The determination of high temperatures, as in flues and chimneys, is difficult and uncertain. Most commercial pyrometers, depending on the unequal expansion of metals, are unreliable if not misleading; not only is the scale of such a pyrometer likely to be incorrect, but the zero of the scale is liable to change during use.

The Chatelier pyrometer has been used with satisfaction at the Massachusetts Institute of Technology for measuring temperatures in flues and chimneys. It consists essentially of a thermoelectric couple made by joining the ends of two wires, one of platinum and the other of platinum alloyed with ten per cent of rodium. All but about four inches of the wire at the junction is incased in fire-clay inside an iron pipe about four feet long. From the wires of the pyrometer connection is made to a sensitive galvanometer in a separate observing-room. The deflection of the galvanometer is indicated by a ray of light reflected from a mirror on the needle and moving over a graduated scale. The scale is set to read zero when the junction of the wires is at the temperature of the atmosphere. The junction is then immersed successively in baths of substances which melt at various high temperatures, such as sulphur and naphthaline. The readings of the ray of light when the juncture is in such baths fix known points on the arbitrary scale from which intermediate temperatures may be estimated directly. It is convenient to use a curve for this purpose with scale-readings for abscissæ and with corresponding temperatures for ordinates. After the scale is determined the pyrometer may be introduced into the place or places where temperatures are to be measured, and readings are taken from which the temperatures are determined by interpolation on the curve just described.

Air-supply.—The air for a furnace may be made to enter through a temporary mouthpiece fitted to the ash-pit doors. This mouthpiece may be of galvanized iron, circular in section and about three feet long. Its cross-section should have an area equal to that of the door or doors leading to the ash-pit. The velocity of the air passing through the mouthpiece can be measured by an anemometer. The area of the mouthpiece multiplied by the velocity in feet per second gives the volume of air supplied to the ash-pit in cubic feet per second. From this may be calculated the volume and weight of air supplied to the ash-pit per hour or for the entire test; which weight divided by the total coal consumption gives the air per pound of coal burned.

It should be noted that the anemometer is liable to an error of from two to five per cent, and further that air entering through the fire-doors and elsewhere than through the ash-pit is not measured.

Sample Test.—The test given on page 319, made at the Massachusetts Institute of Technology, may serve as an example of a convenient arrangement for reporting the data and results of a boiler-test.

The average pressure of the air and of the steam in the boiler are liable to vary slightly during the test; the average pressures were obtained from readings taken at regular intervals during the test. The same may be said of the temperature of the feed-water.

BOILER-TEST.

DATE, *Dec. 28, '01 – Jan. 2, '02.*

Duration of Test...................................... _138_ hours.
Average pressure of air.............................. _14.85_ lbs. per sq. in.
   "   gauge-pressure................................ _100.0_ " " " "
   "   temperature of feed-water................. _133°.92_ F.
Kind of coal used....................................... *Lackawanna.*
Per cent of moisture in coal........................ _____

		Boiler No. _1_	Boiler No. __
1.	Description of Boilers: *Babcock & Wilcox, No. 1.*		
	108 tubes 4" dia., 17' 8" long; outside area, 1007.3		
	12 " 4" dia., 14' 6" " " " 50.5		
	2 drums 3' dia., × 17'; one-half of shell, 100.2		
2.	Grate-surface, { No. _1_, _81_ in. by _88_ in. / No. _1_ in. by __ in. } Area, feet........	51.3	
3.	Water-heating surface, feet......................	2211	
4.	Ratio of water-heating surface to grate-surface.....	43.10	
5.	Lbs. coal fired, including coal equivalent of wood.....	61,630	
6.	Unburned fuel...	*	
7.	Coal burned, including coal equivalent of wood.....	61,630	
8.	Average coal burned for _15_ minutes............	130.2	
9.	Total refuse from coal...............................	8660	
10.	Total combustible....................................	53,070	
11.	Average combustible for _15_ minutes...........	109.3	
12.	Average lbs. of air for _15_ minutes.............	1081	
13.	Air per lb. of coal....................................	12.92	
14.	Air per lb. of combustible...........................	15.38	
15.	Quality of steam, saturated steam taken as unity....	0.987	
16.	Total water pumped into boiler and apparently evaporated..	548,791	
17.	Water apparently evaporated per lb. of coal burned....	8.90	
18.	Water actually evaporated, corrected for quality of steam....	530,105	
19.	Equivalent water evaporated into dry steam from and at 212° F..	614,700	
20.	Equivalent water evaporated into dry steam, from and at 212° F., per lb. of coal burned................	9.5	
21.	Equivalent water evaporated into dry steam, from and at 212° F., per pound of combustible...............	11.0	
22.	Coal burned per sq. foot of grate surface per hour.....	9.8	
23.	Water evaporated, from and at 212° F., per sq. foot of heating-surface per hour................................	2.17	

       * Fires not drawn.

The description of the boiler under item 1 is brief and yet sufficient to identify it, and gives the data for calculating heating-surface. The grate-surface, heating-surface, and their ratio are calculated from the dimensions of the boiler and furnace, and given in the 2d, 3d, and 4th items. The 5th item gives the total weight of coal fired; as the fires were not drawn, no wood was used and no coal was withdrawn at the end of the test. Consequently the 7th item, coal burned, is the same as the 5th.

The 9th item gives the weight of all the clinker and ashes produced during the test. The coal burned, minus the refuse, gives the total combustible for the test, set down at item 10.

The air-supply is calculated at intervals of 15 minutes during the test, from the anemometer readings and the condition of the atmosphere as it enters the galvanized-iron temporary mouthpiece of the furnace. This is likely to vary considerably, being greatest immediately after fresh coal is fired. Item 12 gives the average from the several calculations during the test. The coal and combustible for 15 minutes given by items 8 and 11 are calculated for comparison with the air for the same time. Thus the air per pound of coal is calculated by dividing item 12 by item 8; and in like manner the 14th item is calculated from the 11th and 12th.

The quality of the steam was obtained from time to time during the test by a throttling-calorimeter, like the one for which a description and calculation are given on page 309. The average from the several determinations is given by item 15. The priming was

$$100(1.000 - 0.983) = 1.7 \text{ per cent.}$$

The equivalent evaporation for the total coal (given by item 18) was calculated, by a method like that given on page 133, from the temperature of the feed-water, the pressure in the boiler, and the quality of the steam; using the total water apparently evaporated given by item 16.

The absolute boiler-pressure was

$$109.9 + 14.85 = 124.8 \text{ pounds.}$$

The corresponding heat of vaporization and heat of the liquid are 871.8 and 315.1; the heat of the liquid at 122°.9 (the temperature of the feed-water) is 91.0. Consequently the total equivalent evaporation from and at 212° F. was

$$\frac{548{,}794(0.983 \times 871.8 + 315.1 - 91)}{965.8} = 614{,}300 \text{ pounds.}$$

The equivalent evaporation per pound of fuel (item 20) is obtained by dividing the quantity just found by the total coal burned (item 7). In like manner the equivalent evaporation per pound of combustible is obtained from item 10.

The coal burned per square foot of grate-surface per hour is obtained by dividing the total coal burned by the area of the grate and by the duration of the test. Thus

$$\frac{64{,}639}{51.3 \times 128} = 9.8 \text{ pounds.}$$

The equivalent evaporation per square foot of heating-surface per hour (item 23) is obtained by dividing the total equivalent evaporation (item 19) by the heating-surface and by the duration of the test. Thus

$$\frac{614{,}300}{2214 \times 128} = 2.17 \text{ pounds.}$$

Remark.—In this chapter are given the observations that are required and the precautions to be taken in making an ordinary boiler-test. It is, however, intended rather as a description for the student than as a guide for the engineer, who must learn how to make tests by experience. Many of the processes and observations are so simple that they may be intrusted to any careful and intelligent person; the conduct of the test must receive the attention of a competent engineer, for there is no expert work that an engineer may be called upon to do in which there is more chance for error and deception than in making a boiler-test.

CHAPTER XI.

BOILER DESIGN.

IN order to bring together the principles and methods which have been given in the preceding chapters, they will be applied to the design of a boiler. Designing of any sort is an art that is guided and controlled by practical considerations and theoretical principles, and which can be acquired by practice only. The design of a boiler, like many other designs, is further modified to meet the requirements of government boards of inspection, or to conform to the inspection-rules of insurance companies. These rules and requirements vary from place to place and from time to time; they must be known to the designer, but they have no place in a text-book. A simple and common type of boiler has been chosen for design; the methods, with proper modification, can be applied to other types, and the general principles illustrated are much the same for all types.

Type of Boiler.—The kind of boiler used in a given locality depends on custom, on the kind of water used, and on the cost and quality of fuel. Deviation from common practice should be made only for sufficient reason. Where water is bad or where fuel is cheap, the plain cylindrical boiler or a flue-boiler will be chosen. With clean, soft water the cylindrical tubular boiler, like that shown by Plate I, has been found to be convenient, economical, and cheap. All these boilers have external furnaces, so that the shell is in part exposed to the fire. Now plates exposed directly to the fire should not be more than half an inch thick; 3/8 of an inch is preferable. Though thicker plates are sometimes used, this consideration limits the size of boilers of this type when high pressures are used. The importance of high efficiency for the longitudinal riveted joint becomes apparent in this connection.

Internally-fired boilers, like the Lancashire or the Scotch marine boiler, are not limited in diameter by this reason. The marine boiler sometimes has plates an inch and a quarter thick; the fact that so great a thickness is undesirable sometimes serves as a check on the size of such boilers.

General Proportions.—Whatever may be the type of boiler chosen, there must be provided—

1. Sufficient grate-area to burn the fuel required under the available draught.

2. Suitable combustion-space to properly burn the fuel.

3. Sufficient area of flues or tubes to carry off the products of combustion.

4. Sufficient heating-surface to absorb the heat generated.

5. Proper water-space to prevent too great a fluctuation of the water-level when there is an irregular demand for steam.

6. Suitable steam-space to prevent too great a fluctuation of pressure when steam is taken at intervals, as for the cylinder of a steam-engine.

7. Sufficient free-water area for disengagement of steam.

The last three conditions are not fulfilled by most water-tube boilers; some such boilers depend on a separator for disengaging steam from water.

Problem for Design.—Let it be required to determine the main dimensions and some of the details of a horizontal cylindrical tubular boiler to develop 80-horse power A. S. M. E. standard (page 135). Let the working-pressure be 150 pounds per square inch by the gauge, and the test-pressure 225 pounds, or once and a half the working-pressure.

Assume that anthracite coal will be used, and that it will give an equivalent evaporation of 9 pounds of water per pound of coal from and at 212° F. Assume further that 12 pounds of coal will be burned per square foot of grate-surface per hour.

The heating-surface may be about thirty-seven times the grate-surface. Tubes 16 feet long will be used, which length should not much exceed sixty times the diameter.

The area through the tubes will be made about 1/7.5 of the grate-area.

Grate-area.—The A. S. M. E. standard requires that 34.5 pounds of water per hour shall be evaporated from and at 212° F. for each horse-power. The total equivalent evaporation will consequently be

$$80 \times 34.5 = 2760 \text{ pounds per hour.}$$

With an equivalent evaporation of 9 pounds of water per pound of coal the coal burned will be

$$2760 \div 9 = 307 \text{ pounds per hour.}$$

With a rate of combustion of 12 pounds of coal per square foot of grate surface per hour, the grate-area must be

$$307 \div 12 = 25.6 \text{ square feet.}$$

Tubes.—A common rule for finding the diameter of tubes is to allow one inch for each four feet of length when soft coal is used, and five feet when hard coal is used. A tube three inches in diameter will very nearly fulfil this condition.

The table of proportions of flue-tubes in the Appendix, gives the area of the internal transverse section of such a tube as 6.08 square inches; the external area is 7.07 square inches. The internal circumference is 8.74 inches, and the external circumference is 9.42 inches.

The area through the tubes has been chosen as 1/7.5 of the grate-area, equal to

$$\frac{25.6 \times 144}{7.5} = 492 \text{ square inches.}$$

Since the area through one tube is 6.08 square inches, there will be required

$$492 \div 6.08 = 80.8,$$

or, more properly, 81 tubes. It may be found convenient in laying out the tube-sheet to use more than this number of tubes; a less number is of course improper.

Steam-space.—A good rule for this type of boiler is to allow from 0.8 to 1 cubic foot of steam-space per horse-power, which gives from 64 to 80 cubic feet for this boiler. We will assume 80 cubic feet.

For sake of comparison, calculations will be made also by rules given on page 132. Thus for certain boilers working at moderate pressures it is found that the steam-space may be made equal to the volume of steam used by the engine in 20 seconds. Suppose that this boiler, though designed for 150 pounds pressure, may run at 70 pounds pressure, and may supply an 80 horse-power engine which uses 30 pounds of steam per horse-power per hour.

Now the volume of one pound of steam at 70 pounds by the gauge, or 85 pounds absolute, is 5.125 cubic feet. So that the engine will use

$$80 \times 30 \times 5.125 = 12,300$$

cubic feet of steam in an hour, or

$$\frac{20}{3600} \times 12300 = 68$$

cubic feet in 20 seconds. This is about the lower limit by the rule used above. It is clear that the steam-space would be very small if determined by this rule for an engine using steam at 150 pounds pressure.

Another rule makes the steam-space from 50 to 140 times the volume of the high-pressure cylinder of the engine; 50 for very high pressure and high speed, 140 for slow speed and low pressure. For medium speeds and pressures 60 to 90 may be used.

The boiler under consideration may supply steam to a triple-expansion engine which has a high-pressure cylinder 9 inches in diameter by 30 inches stroke, so that the volume is 1.105 cubic feet. According to this the steam-space needed is 66 to 99 cubic feet.

Diameter of Boiler.—For this type of boiler the steam-space is commonly made one third and the water-space two thirds of the contents of the boiler. To the contents of the boiler there must be added the space occupied by the tubes to find the volume of the cylindrical shell. Now we have decided to use 81 tubes 3 inches in diameter and 16 feet long. The area of the external transverse section has been found to be 7.07 square inches. The space occupied by the tubes is consequently

$$\frac{81 \times 7.07 \times 16}{144} = 64 \text{ cubic feet.}$$

To this add steam-space,	80 " "	
and water-space,	160 " "	
Making in all,	304 " "	

The cylinder is 16 feet long, so that its transverse area is

$$304 \div 16 = 19 \text{ square feet;}$$

which corresponds to a diameter of 59.02 inches, or nearly 60 inches. This will be taken as the trial diameter; it may require change in proportioning other parts of the boiler.

The method of determining the main dimensions of a

boiler from the steam-space will require modification if it is applied to any other type of boiler. Even when applied to a given type it leaves much to the judgment of the designer, who may find difficulty in using it unless he is accustomed to working on that particular type. If the designer has at hand the dimension of several boilers of a given type, he may prefer to select the main dimensions for a new design directly, with the reservation that such dimensions may be modified as the design proceeds. This is commonly done by the designers of marine and locomotive boilers.

Heating-surface.—The heating-surface of a cylindrical tubular boiler consists of all the shell below the supports at the side wall, all the inside of the tubes, and part of the rear tube-plate. Usually half of the cylindrical part of the shell is heating-surface. In the case in hand the heating-surface, exclusive of the tube-plate, will amount to

$$\text{Shell} \ldots \quad \frac{1}{2} \times \frac{3.1416 \times 60 \times 16}{12} = 125.7 \text{ sq. ft.}$$

$$\text{Tubes} \ldots \quad 81 \times \frac{8.74 \times 16}{12} = 943.9 \text{ " "}$$

$$\text{Total} \ldots \ldots \ldots \ldots \ldots \ldots \ldots 1069.6 \text{ " "}$$

The grate-surface is to be 25.6 square feet, so that the ratio of grate-surface to heating-surface will be at least as good as

$$25.6 : 1069.6 :: 1 : 41\tfrac{1}{2}.$$

The actual ratio will be more favorable as it will appear advisable to use more than 81 tubes, and the back tube-sheet remains to be allowed for.

Water-level.—It is now necessary to determine the position of the water-level to see if there will be sufficient free-water surface and sufficient distance from the water-level to the shell above it.

Since the whole boiler is cylindrical, the area of the head of the boiler exposed to steam and to water will have the same ratio as that of the steam-space to the water-space. Consequently the area of the head above the water-level must be one third of the total area of the head less the combined areas of the tubes.

The area of a circle having a diameter of 60 inches is 2827.4 square inches. The area of 81 tubes each having an external cross-section of 7.07 square inches will be

$$81 \times 7.07 = 572.7$$

square inches. The area of the head exposed to steam is consequently

$$\frac{2827.4 - 572.7}{3} = 751.6$$

square inches. We need now to know the height of a segment of a 60-inch circle, which has the area of 751.6 square inches. The second problem in the explanation of the use of a table of segments (see Appendix) gives for the tabular number corresponding to the area

$$\frac{751.6}{60 \times 60} = 0.2088;$$

for which the ratio of the height to the diameter is 0.312. The height of the segment is therefore

$$0.312 \times 60 = 18.7 \text{ inches.}$$

This gives sufficient height above the water, and sufficient free-water surface. The water-level will be

$$30 - 18.7 = 11.3$$

inches above the centre of the boiler.

Factor of Safety.—It has been pointed out that the actual factor of safety of boiler-shells is usually four or five when the boiler is built. The apparent factor of safety for some parts like stay-bolts may be greater, but such factors are illusory because the stays may be subjected to considerable irregular stress from unequal expansion. The apparent stress on stay-rods and bolts, from steam-pressure only, is frequently limited by inspection-rules or by law.

The factor of safety of a boiler which has been at work for some years is much affected by corrosion, which acts upon different parts of the boiler very differently, even when the corrosion is uniform. Thus a plate half an inch thick will have 7/8 of its original strength after it has lost 1/16 of an inch by corrosion. The weakest part of the plate, that is, the riveted joint, seldom suffers as much from corrosion as the whole plate at a distance from the joint, because the plate is protected to some extent by the rivet-heads. Some forms of joint have an internal cover-plate, which protects the plate at the joint and the joint may be nearly as strong after corrosion as before. Very often old weak boilers fail by tearing the corroded plate outside the riveted joint.

Stay-rods and bolts suffer much more from corrosion than plates. Thus a rod one inch in diameter has an area of 0.7854 of a square inch. After corrosion to the extent of 1/16 of an inch has taken place the diameter is 7/8 of an inch and the area is 0.6013, which is

$$0.6013 \div 0.7854 = 0.766$$

of the original area. Compare this with the plate which retains 7/8 or 0.875 of its thickness after the same amount of corrosion. Of course a smaller stay will suffer more, and a larger one less, in proportion.

After the sizes of the parts of a boiler are decided upon it is well to make calculation to see that a factor of safety of four will remain after a reasonable amount of corrosion. Or, as in the case of stay-rods, the size may be calculated with a proper factor, and then the diameter may be increased to allow for corrosion.

Thickness of Shell.—The final decision of the proper thickness of the shell for the boiler under consideration cannot be made until the efficiency of the joint is known; but the efficiency of any of the complex joints now in vogue can be found only when the thickness of the plate is known. It is therefore convenient to assume a factor of safety of about six and make a preliminary calculation.

Thus for the boiler in hand we will get for the thickness (page 183)

$$t = \frac{150 \times 30}{55,000 \div 6} = 0.49$$

of an inch. A similar calculation with a factor of five gives

$$t = \frac{150 \times 30}{55,000 \div 5} = 0.41$$

of an inch. The shell will be either 7/16 or 1/2 an inch thick. Seven sixteenths will give an apparent factor of safety of

$$\frac{55,000 \times 7/16}{150 \times 30} = 5.35.$$

After the allowance for the efficiency of the joint has been made this factor will be found to be about $4\frac{3}{8}$.

Longitudinal Joint.—The shell-plate is made as thin as possible because it will be exposed to the fire. Consequently the efficiency of the longitudinal riveted joint must be high if the real factor of safety is to be satisfactory. The strength of triple-riveted joints like that shown on page 201 ranges from 85 to 90 per cent. The joint with two cover-plates shown by Fig. 139, will be chosen. Following the method given on page 201, it appears that this joint may fail in one of five ways, for which the resistances are as follows:

A. Tearing at outer row of rivets:

Resistance $= (P - d)tf_t$.

B. Shearing four rivets in double shear and one in single shear:

$$\text{Resistance} = \frac{9\pi d^2}{4} f_s.$$

C. Tearing at the middle row of rivets and shearing one rivet:

$$\text{Resistance} = (P - 2d)tf_t + \frac{\pi d^2}{4} f_s.$$

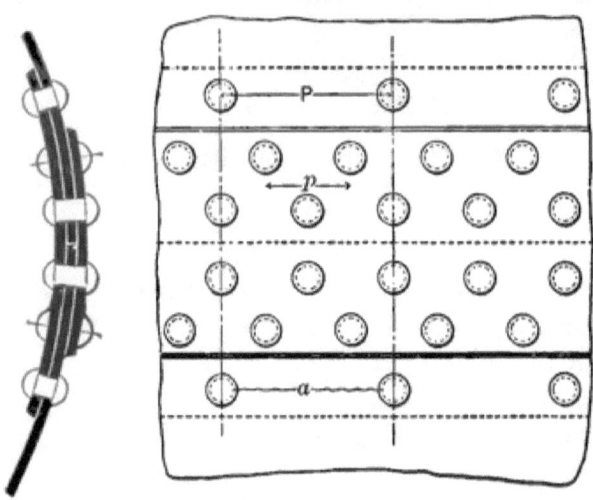

Fig. 139.

D. Crushing four rivets and shearing one:

$$\text{Resistance} = 4dtf_c + \frac{\pi d^2}{4} f_s.$$

E. Crushing five rivets:

$$\text{Resistance} = 4dtf_c + dt f_c.$$

The diameter of rivet will be found by equating the resistances A and C.

$$\therefore (P - d)tf_t = (P - 2d)tf_t + \frac{\pi d^2}{4} f_s.$$

$$\therefore d = \frac{4tf_t}{\pi f_s} = \frac{4 \times \tfrac{7}{16} \times 55,000}{\pi 95,000} = 0.68.$$

The rivet which was used was 13/16 of an inch when driven.

There are several methods in which we may find the way in which the joint will fail, and then find therefrom the efficiency. One is that shown on page 202 by assuming a pitch and calculating the resistance of the joint to failure in each of the five several ways. Another method is to equate the five several resistances two and two and calculate the pitch; the least pitch thus found must not be exceeded. Thus

Equating B and C,

$$\frac{9\pi d^2}{4} f_s = (P - 2d) t f_t + \frac{\pi d^2}{4} f_s.$$

$$\therefore P = \frac{8\pi d^2}{4t} \frac{f_s}{f_t} + 2d$$

$$= \frac{8 \times 3.1416 \times \left(\frac{13}{16}\right)^2}{4 \times \frac{7}{16}} \times \frac{45,000}{55,000} + 2 \times \frac{13}{16} = 9.4.$$

Equating A and B,

$$(P - d) t f_t = \frac{9\pi d^2}{4} f_s.$$

$$\therefore P = \frac{9\pi d^2}{4t} \frac{f_s}{f_t} + d$$

$$= \frac{9 \times 3.1416 \left(\frac{13}{16}\right)^2}{4 \times \frac{7}{16}} \times \frac{45,000}{55,000} + \frac{13}{16} = 9.5.$$

Equating A and D,

$$(P - d) t f_t = 4 d t f_c + \frac{\pi d^2}{4} f_s.$$

$$\therefore P = \frac{4df_c}{f_t} + \frac{\pi d^2}{4t} \times \frac{f_s}{f_t} + d$$

$$= 4 \times \frac{13}{16} \times \frac{95{,}000}{55{,}000} + \frac{3.1416 \times \left(\frac{13}{16}\right)^2}{4 \times \frac{7}{16}} \times \frac{45{,}000}{55{,}000} + \frac{13}{16} = 7.4.$$

Equating A and E,

$$(P - d)tf_t = 4dtf_c + dt_c f_{c\cdot}$$

$$\therefore P = 4d\frac{f_c}{f_t} + \frac{dt_c f_c}{t\ f_t} + d$$

$$= 4 \times \frac{13}{16} \times \frac{95{,}000}{55{,}000} + \frac{13/16 \times 3/8}{7/16} \times \frac{95{,}000}{55{,}000} + \frac{13}{16} = 7.6.$$

Here t_c, the thickness of the cover-plate, is taken to be 3/8 of an inch.

The greatest allowable pitch at the outer row of rivets is evidently 7.4 inches.

Instead of going to the labor of solving all four of the above equations, we may find by some other method how the joint is likely to fail, and make up an equation involving those resistances only. Thus a rivet in the outer row may fail by shearing or by crushing at the cover-plate, which is here made thinner than the shell-plate. Equating the resistances of the two methods, we have

$$\frac{\pi d^2}{4} f_s = t_c d f_{c},$$

or for a cover-plate 3/8 of an inch thick

$$d = \frac{4 \times \frac{3}{8}}{\pi} \times \frac{95{,}000}{45{,}000} = 1.01.$$

A rivet 1.01 inch in diameter will consequently be just as

likely to fail by crushing as by shearing. But the resistance to shearing increases as the square of the diameter, while the resistance to crushing increases as the diameter. It is therefore evident that a rivet larger than 1.01 of an inch will fail by crushing, while a smaller rivet will fail by shearing.

A similar calculation at the inner row, when the rivet bears against a cover-plate both inside and outside, and will consequently crush against the shell-plate, gives

$$\frac{2\pi d^2}{4} f_s = t d f_c;$$

$$d = \frac{2 \times \frac{7}{16}}{\pi} \times \frac{95{,}000}{45{,}000} = 0.6.$$

Here a rivet larger than 0.6 will crush, and one smaller will shear. It is now evident that a 13/16 rivet will shear at the outer row and will crush at the inner row. That is, for this joint the failure will occur by the method D, but not by the methods B or E. Then equating the resistances A and D, and solving for P, we get for the pitch at the outer row 7.4 inches as before. The corresponding pitch at the calking edge of the outer cover-plate is 3.7 inches; we will choose for that pitch $3\frac{5}{8}$ inches, making the pitch at the outer row $7\frac{1}{4}$ inches.

The efficiency of the joint is

$$100 \frac{P - d}{P} = 100 \times \frac{7\frac{1}{4} - 1\frac{3}{16}}{7\frac{1}{4}} = 88.8 \text{ per cent.}$$

In the preceding article the apparent factor of safety based on the whole strength of the shell-plate is 5.35. Allowing for the efficiency of the longitudinal joint, the real factor of safety when the boiler is new is

$$0.888 \times 5.35 = 4.75.$$

With this style of joint the shell-plate is protected from corrosion by the inner cover-plate, and the joint will lose little if any efficiency from corrosion. If it be assumed that the plate loses 1/16 of an inch by corrosion during the life of the boiler, then the strength of the plate will be one seventh less after corrosion, and the corresponding factor of safety will be

$$5.35 \times \tfrac{6}{7} = 4.6,$$

which may be considered to be sufficient.

Ring-seam.—The stress on a transverse section of a homogeneous hollow cylinder from internal fluid pressure is one half the stress on a longitudinal section. It will in general be found that a single- or a double-riveted ring-seam is sufficient for any cylindrical boiler-shell. Marine boilers commonly have double-riveted ring-seams; externally-fired horizontal boilers seldom have the shell more than half an inch thick, and for that thickness, or less, single-riveted ring-seams are used.

It is found in practice that ring-seams of horizontal externally-fired boilers may have a pitch of about $2\tfrac{3}{16}$ inches for all thicknesses of plate from 1/4 to 1/2 of an inch. The diameters of rivets for such seams may be made about the size given in the following table:

Thickness of plate	$\tfrac{1}{4}$	$\tfrac{5}{16}$	$\tfrac{3}{8}$	$\tfrac{7}{16}$	$\tfrac{1}{2}$
Diameter of rivet	$\tfrac{5}{8}$	$\tfrac{11}{16}$	$\tfrac{3}{4}$	$\tfrac{7}{8}$	$\tfrac{7}{8}$

The ring-seam in question has a circumference of about

$$3.1416 \times 60 = 188.2$$

inches, which will allow us to use 84 rivets with a pitch of about 2.24 inches. This joint will fail by shearing the rivets. The efficiency of the joint is consequently the ratio of the resistance of a single rivet to shearing, to the resistance of a strip of plate as wide as the pitch. Consequently the efficiency is

$$\frac{\frac{\pi d^2}{4} f_s}{p t f_t} = \frac{\frac{1}{4} \times 3.1416 \times (\frac{13}{16})^2 \times 45{,}000}{2.24 \times \frac{7}{16} \times 55{,}000} = .433,$$

which is more than half of the efficiency of the longitudinal seam, and will consequently be sufficient.

Lap.—The lap, or distance from the centre of the rivet to the edge of the plate, is usually taken as 1.5 times the diameter of the rivet used, which makes the distance of the edge of the hole from the edge of the plate equal to the diameter of the rivet. For the single-riveted ring-seam this makes the lap equal to

$$1.5 \times \tfrac{13}{16} = 1.22.$$

It is customary to calculate the width of lap required on the assumption that the metal between the rivet and the edge of the plate may be treated as a beam of uniform depth, fixed at the ends and loaded uniformly by the force which would be required to shear or crush the rivet, taking, of course, the larger. The width of the beam is the thickness of the plate, the depth is the distance from the edge of the hole to the edge of the plate, and the length is the diameter of the rivet.

Rivets in single-riveted seams fail by shearing. The load is consequently the shearing resistance

$$\frac{\pi d^2}{4} f_s.$$

The maximum bending moment for a beam of uniform section fixed at the ends and uniformly loaded is equal to the load multiplied by one eighth of the span. The moment of resistance is equal to

$$f \frac{I}{y},$$

in which f is the cross-breaking strength (about 55,000), I is the moment of inertia of the section, and y is the distance of the most strained fibre from the neutral axis. Here we have

$$I = \frac{th^3}{12}, \quad y = \frac{h}{2},$$

representing the distance from the edge of the hole to the edge of the plate by h.

Equating the bending moment to the moment of resistance,

$$\tfrac{2}{3}d \times \frac{\pi d^2}{4} \times f_s = \frac{fth^2}{6}.$$

$$\therefore h = \sqrt{\frac{3\pi d^3}{16t} \times \frac{f_s}{f}}$$

$$= \sqrt{\frac{3 \times 3.1416 \times 13^3}{16 \times \frac{7}{16} \times 16^2} \times \frac{45,000}{55,000}} = 0.77$$

for the case in hand. The lap is consequently

$$0.77 + \tfrac{1}{2} \times \tfrac{13}{16} = 1.18$$

inches for the ring-seam, which is somewhat less than that by the arbitrary rule that it should be once and a half the diameter.

A similar calculation for the cover-plates with the same diameter of rivet, but with a plate 3/8 of an inch thick, gives for the lap 1.24 or $1\tfrac{1}{4}$ of an inch, while the arbitrary rule gives 1.03 of an inch. It is probable that the lap may be considerably smaller than is given by the calculation by the beam theory, but for lack of direct experimental knowledge on this question it is not wise to make the lap much less than the calculation gives; we will consequently use $1\tfrac{1}{4}$ of an inch for the lap of the cover-plates.

The rivets of the inner rows pass through both cover-plates and are in double shear, and consequently fail by crushing as is shown on page 335. The load to be used for calculating the lap is therefore the resistance to crushing in front of the rivet; that is, we here have for the load tdf_c. The equation of bending moment and moment of resistance gives

$$\frac{1}{8}d \times tdf_c = f\frac{th^2}{6}.$$

$$h = d\sqrt{\frac{3f_c}{4f}} = \frac{13}{16}\sqrt{\frac{3 \times 95,000}{4 \times 45,000}} = 0.926.$$

The lap is consequently

$$0.926 + \frac{1}{2} \times \frac{13}{16} = 1.27,$$

or a little more than $1\frac{1}{4}$. The lap used is $1\frac{3}{8}$ of an inch.

Tube-sheet.—The next step in the design is to lay out the tube-sheet on the drawing-board. If possible, the tubes should be arranged in horizontal and vertical rows as shown on Plate I. The distance between the tubes should not be less than three fourths of one inch; one inch is better. On Plate I the horizontal rows are spaced one inch apart, while the vertical rows are only three fourths of an inch apart; wider spacing for horizontal rows is more favorable for the free circulation of water and the disengagement of steam. The circulation is improved by having a space in the middle as shown on Plate I

If a very large number of tubes are required for a given boiler, they may be arranged in vertical rows and in rows at 30° with the horizon, as on Plate II. This arrangement is commonly used for locomotive boilers, but is not favored for stationary boilers.

The common range of fluctuation allowed for the water-line with this type of boilers is six inches, three above and three below the mean water-level. The tops of the tubes are set about three inches below low water-level.

The tubes should nowhere be nearer than three inches from the shell, and the bottom row should be from four to six inches from the bottom of the boiler.

The hand-hole near the bottom of the head should be placed as low as possible; the flat surface for the gasket should be at least 3/4 of an inch wide. No tube should be nearer than an inch from its edge.

The tube plate is usually from 1/16 to 1/8 of an inch thicker than the shell-plating. The internal radius of the flange should not be less than half an inch. For plates half an inch thick or less the outside radius is commonly made one inch.

In applying these principles to the tube-sheet for a boiler 60 inches in diameter, as shown on Plate I, it appears that 84 tubes may be used, spaced four inches horizontally and $3\frac{3}{4}$ vertically and with a space at the middle for circulation, provided that the top of the upper row of tubes is $6\frac{1}{2}$ inches above the centre-line of the boiler. This brings the water-level

$$6\tfrac{1}{8} + 6 = 12\tfrac{1}{2}$$

inches above the middle of the boiler, instead of 11.3 as calculated on page 329; that is, the water-level is raised 1.2 of an inch or 1/10 of a foot. At 12 inches above the middle, the boiler is about $4\frac{1}{2}$ feet wide; the layer of water added has consequently a volume of

$$1/10 \times 4.5 \times 16 = 7.2$$

cubic feet. The effect is to reduce the steam-space from 80 cubic feet (see page 326) to 72.8 cubic feet. But the rule used gave from 64 to 80 cubic feet, so that 72.8 cubic feet is a fair allowance. If the tubes were spaced nearer together in the horizontal rows and the space for circulation were omitted, the required number of tubes could be easily provided for without raising the water-level. If in any case a satisfactory arrangement of tubes cannot be made with the diameter assumed from preliminary calculations of steam- and water-space, or from some other method, then a larger diameter must be used.

Area of Uptake.—The area of the uptake, like the total area through the tubes, is made from 1/7 to 1/8 of the grate area. On page 326 the area through the tubes was found to be 492 square inches. The uptake may be made 12 inches deep, measured from front to rear. It will then be

$$492 \div 12 = 41$$

inches wide, measured transversely. The opening through the top of the projecting shell at the front end will be made 12 inches deep, as shown on Plate I, and must be cut down till it is 41 inches wide. The projecting end of the shell is made long enough so that a space of about one inch is left between the uptake and the calking edge of the front tube-sheet.

Length of Sections.—The length of the rings or sections of the cylindrical shell is limited by the reach of the riveting-machine and by the width of plate obtainable. The sections are often made the same length, though there is no other reason for this than the convenience in ordering material. The two rear sections on Plate I are each made 68 inches from centre to centre of riveted joints, or, allowing $1\frac{1}{4}$ of an inch for lap at each end, the plates when finished are $70\frac{1}{2}$ inches wide. The front section is

$$14 + 54\frac{3}{8} + 1\frac{1}{2} = 69\frac{7}{8}$$

inches wide. In this case the plates could all be ordered about 72 inches wide.

The front course which comes over the fire is an outside course, so that the flames may not strike directly against the edge of the plate at the ring-seam. The length of the grate is commonly about one third of the length of the boiler, which brings the first ring-seam over the bridge, where the fire is the hottest. It is well to avoid this by making the front section shorter, and the other sections longer.

Manholes, Hand-holes, and Nozzles. — These fittings should be strong enough and stiff enough to carry the stresses which come from the direct steam-pressure and from the tension in the pieces to which they are fastened; for example, the manhole-ring must be able to take the place of the piece of plate cut away at the hole.

All these fittings can now be bought in the form of steel forgings, made by a hydraulic flanging or forging machine. Gun-iron and cast steel are, however, much used.

The determination of stresses in a manhole-ring, even if approximate methods are used, is both difficult and uncertain, and will not be considered here. Forms and dimensions that have been used in good practice may be taken for a guide in designing. A rule used by boiler-makers for forged rings, which, like that shown on Plate I, lie close to the shell-plate, is to make the section of the ring, exclusive of the lip, equal at least to the section of the plate cut away. The aid given by the lip against which the cover bears is considered to offset eccentric loading, etc. The ring of a steam-nozzle may be treated in the same way, though it is more efficiently aided by the cylindrical portion. Gun-iron manhole-rings should be $1\frac{1}{2}$ of an inch thick, and nozzles may be $1\frac{1}{4}$ of an inch thick.

An approximate calculation of the stress in the manhole-cover may be made by treating it as a beam supported at the ends and loaded by the steam-pressure and by the pull of the bolt at the middle; this last must be assumed, as it cannot be known. The calculated stress will be in excess of the actual stress, since the plate is supported all around. The handhole-plate may be treated in a similar way. Handhole-covers are frequently drawn up by a taper key instead of a bolt and nut, because the nut is exposed to the fire, and often cannot be removed with a wrench, after it has been in place some time.

The bearing-surfaces of the manhole-cover and the lip against which it bears should be machined to make them true and smooth, though this is not always done. The hand-hole-cover may be finished, but it bears directly against the plate, which of course is not finished. In any case the joint is made tight by a gasket which may be 3/4 of an inch wide for the hand-hole and from that width to an inch for the manhole.

Staying.—As is pointed out on page 222, the calculation of stresses in a flat plate supported at intervals can be determined only by the application of the theory of elasticity; and the only determinate case is that in which the supported points are in equidistant rectangular rows, dividing the surface into squares. This case applies directly to the staying of the fire-box of a locomotive by stay-bolts. Whatever system of arranging the supported points is finally chosen, it is convenient to make a calculation for the determinate case, with the points in equidistant rows, in order to get a standard with which the chosen system may be compared.

The equation for finding the stress in a flat plate supported at points in equidistant rectangular rows is

$$f = \frac{2}{9} \frac{a^2}{t^2} p,$$

in which a is the distance of points in a row, t is the thickness of the plate, and p is the steam-pressure in pounds per square inch. In the design in hand $t = 9/16$ of an inch and $p = 150$ pounds. Assuming

$$f = \tfrac{1}{10} \times 55{,}000 = 5500,$$

and solving for a, we have

$$a = \sqrt{\frac{9}{2} \frac{t^2 f}{p}} = \sqrt{\frac{9 \times 5500 \times 9 \times 9}{2 \times 150 \times 16 \times 16}} = 7+ \text{ inches.}$$

If the distance between supported points is made less than 7 inches, whatever the system of arrangement may be, we may be confident that the stresses will not exceed 5500 pounds; in this case stresses in the plate are due only to the pressure on the plate, since the shell of the boiler is self-supporting.

In the several ways of staying the flat ends of boilers shown on pages 150 to 154 the plate is riveted to channel-bars, angle-irons, or crowfeet, which in turn are supported by stay-rods. The rivets are in direct tension, and are subject to initial stresses due to the contraction when they cool; it is customary to limit the apparent working stress to 6000 pounds. Rivets less than 3/4 of an inch are seldom used, since in practice they are found to be too much affected by initial stress due to cooling. Large rivets are also considered to be undesirable. We will choose here 13/16 for the rivets.

If each rivet sustains the pressure on a square a inches wide, then the stress per square inch on the rivet will be

$$\frac{\pi d^2}{4} f_t = 150 \times a^2,$$

in which d is the diameter and f_t is the tensional stress. Assuming $f_t = 6000$ and $d = 13/16$, and solving for a, we have

$$a_1 = \sqrt{\frac{\pi \times 13 \times 13 \times 6000}{150 \times 16 \times 16}} = 4.55 \text{ inches.}$$

This gives for the limiting distance of rivets 4.55 inches. Of course a less distance may be used if convenient.

In some cases the pitch of the rivets may be controlled by the system of staying. For example, the rods used with crowfeet are seldom more than $1\frac{1}{4}$ of an inch in diameter, because larger rods may bring too large a local stress where they are riveted to the cylindrical shell. Rods one inch or an inch and an eighth are frequently used. A double crowfoot has four rivets, each of which will carry one fourth of the load on the stay-rod. A stay-rod $1\frac{1}{4}$ of an inch in diameter, and limited to a stress of 7500 pounds, may carry a pull in the direction of its length of

$$7500 \times (1\tfrac{1}{4})^2 = 11,720 \text{ pounds.}$$

If the rod makes an angle of 20° with the shell-plate, the pull which it will exert perpendicular to the head will be

$$11,720 \cos 20° = 11,720 \times 0.93969 = 11,013$$

pounds, so that each rivet will carry about 2750 pounds. If each rivet supports a square having the side a_2 exposed to the pressure of steam at 150 pounds, then

$$11,013 = 150 \times a_2^2,$$

or

$$a_2 = \sqrt{\frac{11,013}{150}} = 3.8 \text{ inches.}$$

Laying out Stays.—Having selected the form of staying to be used, the plan must be laid out on the drawing-board, giving proper attention to practical considerations, such as the way in which the stays are to be inserted, and taking care that accessibility is not too much interfered with. Fig. 140 repeats the upper part of the head of the boiler shown by Plate I, with certain additional dotted lines, which will be referred to in the explanation of calculations. The area to be stayed is considered to be limited by the upper row of tubes, and by a dotted line drawn $1\frac{1}{4}$ of an inch from the inside of the shell. This line is drawn at the right only; it is very nearly the place where the rounded corner of the flange joins the flat surface of the head. The distance of the lowest row of rivets from the top row of tubes, and of the outer row of rivets from the dotted line, may be as great as their maximum distance from each other. Rivets should not be placed nearer than 3 inches from the tubes, lest the expansion of the tubes should start leaks. Rivets may be placed near the dotted line, if that is convenient. For example, the outermost row of rivets in crowfoot staying (Fig. 45, page 152) may be at a distance a_2 from the dotted line; for $1\frac{1}{4}$ inch stay-rods $a_2 = 3.8$ inches.

The method of staying selected consists of channel-bars riveted to the head and supported by through-stays; the upper channel-bar is assisted by an angle-iron. The channel-bars selected are six inches wide, and the horizontal rows of rivets in each bar are $3\frac{1}{4}$ inches apart, which brings them as near the flanges of the bar as they can be driven. The middle of the lower channel-bar is $5\frac{5}{8}$ inches above the top of the tubes, so that the lowest row of rivets is

$$5\tfrac{5}{8} - \tfrac{1}{2} \times 3\tfrac{1}{4} = 4$$

inches above the top row of tubes. But the plate cannot be properly considered to be rigidly supported at a line drawn through the tops of the tubes; we will assume the line of support to be a fourth of the diameter lower down. This makes a space of $4\frac{3}{4}$ inches, instead of the 4.55 inches calculated for 13/16 rivets. The excess may be considered to be offset by the fact that the other row of rivets in the channel-bar is only $3\frac{1}{4}$ inches distant.

The upper channel-bar is placed 8 inches above the lower one, so that the stay-rods are

$$30 - (6\tfrac{1}{2} + 5\tfrac{5}{8} + 8) = 9\tfrac{7}{8}$$

inches below the shell. If these upper rods are much less than 10 inches from the shell access to the boiler will be difficult. The space immediately above the upper channel-bar is stayed by aid of an angle-iron which is riveted to the channel-bar.

The distance of the lower row of rivets in the upper channel-bar, above the upper row in the lower bar, is

$$8 - 3\tfrac{1}{4} = 4\tfrac{3}{4}$$

inches—the same as the distance assigned to the lowest row of rivets above the assumed line of support at the top row of tubes. The top row of rivets in the angle-iron is only a little more than four inches below the dotted boundary-line.

Lower Stay-rods.—In order to determine the load carried by the lower stay-rods, we will assume that half the load on the plate between the lowest row of rivets and the top row of tubes is carried by the rivets, and that the load on the plate between the channel-bars is divided equally between them. Now we have assumed that the line of support at the tubes is a quarter of their diameter below their tops, and have found this line to be $4\frac{3}{4}$ inches below the lowest row of rivets. Half of $4\frac{3}{4}$ is $2\frac{3}{8}$. Again, the distance between the top row of rivets in the lower channel-bar and the bottom row in the upper bar is $4\frac{3}{4}$ inches, of which half is $2\frac{3}{8}$. The distance apart of the two rows of rivets in the channel-bar is $3\frac{1}{4}$ inches. The total width of plate supported by the channel-bar may therefore be considered to be

$$2\tfrac{3}{8} + 3\tfrac{1}{4} + 2\tfrac{3}{8} = 8 \text{ inches.}$$

The length of the lower channel-bar at the middle is 52 inches, as measured on Fig. 142; but it is convenient to space the rods $13\frac{1}{2}$ inches apart, and to consider the bar to have four equal spaces, which leads to an assumed length of 54 inches.

The load on the lower channel-bar is considered to be

$$150 \times 8 \times 54 = 64{,}800 \text{ pounds.}$$

We will treat the channel-bar as a continuous girder with four equal spaces and five points of support, of which three are at the stay-rods and two are at the shell of the boiler. By the theory of continuous girders a uniform load on the channel-bar would be distributed among the five points of supports as follows: At each point of support at the shell 11/112, at each outer stay-rod 32/112, at the middle stay-rod 26/112. This would bring on each of the outer stay-rods

$$\tfrac{32}{112} \times 64{,}800 = 18{,}514 \text{ pounds.}$$

Now the load is not uniformly distributed, but is carried in part by the rivets and in part by the nuts and thick washers on the stay-rods; but the actual distribution will bring a less load on the two outer stays, so that the assumption of the load just found is on the side of safety, and it is conveniently calculated.

If we assume 9000 pounds for the working-stress in the stay-rods, we may calculate the diameter by the equation

$$\frac{\pi d^2}{4} = \frac{18,510}{9000},$$

which gives for the diameter something less than $1\frac{3}{8}$ of an inch. For simplicity all five stay-rods will be the same size, namely, $1\frac{3}{8}$ of an inch—that required for the two upper stay-rods. This is the diameter of the body of the rod; the ends are enlarged to $2\frac{1}{4}$ inches where the thread is cut for the nut.

Lower Channel-bar.—The determination of the actual stresses in the channel-bar, allowing for the effect of the nuts and thick washers on the stay-rods, is very uncertain. On the other hand, the application of the theory of continuous girders with a uniform load may not give us a stress as large as the actual maximum stress. We will therefore use an approximate method, which will give a stress at least as great as the greatest stress in the bar.

For this purpose we will assume that a piece of the channel-bar cut by the lines *ab* and *cd* (Fig. 140) may be treated as a simple beam. These lines *ab* and *cd* are drawn at one fourth of the diameter of the thick washers from the centre of the rod, or at

$$\tfrac{1}{4} \times 5\tfrac{1}{2} = 1\tfrac{3}{8}$$

of an inch. We will further assume that the load on the pair of rivets *A* and *B* is due to the pressure of the steam on the area *efgh*, bounded by lines drawn half-way between them and the nearest point of support. Thus *cg* is half-way between the rivets and the line *ab*, *gh* is half-way between the

BOILER DESIGN. 349

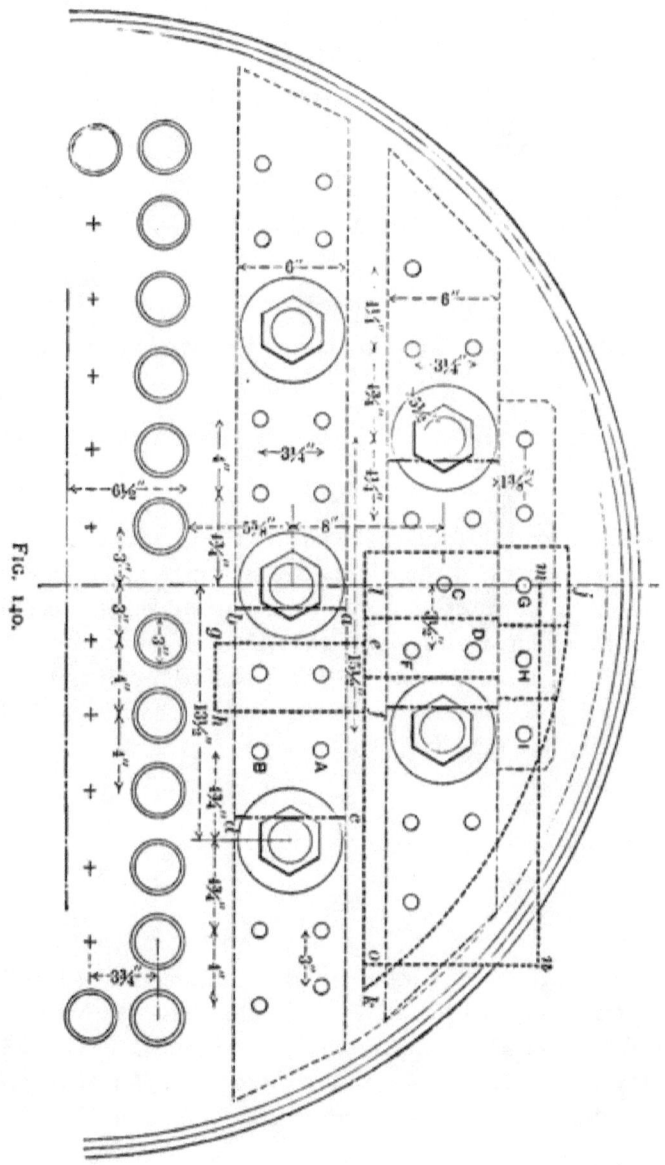

Fig. 140.

rivets and the line of support at the upper row of tubes, ef is half-way between the channel-bars, and fh is half-way to the next pair of rivets. The rivets are $4\frac{3}{4}$ inches from the nearest stay-rod, and are

$$4\tfrac{3}{4} - 1\tfrac{3}{8} = 3\tfrac{3}{8}$$

inches from the line ab; half of this is $1\frac{11}{16}$ of an inch. The two pairs of rivets are

$$(13\tfrac{1}{2} - 2 \times 4\tfrac{3}{4}) = 4$$

inches apart; half of this is 2 inches. The area of $efgh$ is

$$(1\tfrac{11}{16} + 2) \times 8 = 29\tfrac{1}{2}$$

square inches; and the steam-pressure on that area is

$$29\tfrac{1}{2} \times 150 = 4425 \text{ pounds.}$$

This is the load due to each pair of rivets between a pair of stay-rods; and since the rivets are symmetrically placed, this is also the supporting force at each end of the beam. Between the two pairs of rivets the beam is subjected to a uniform bending moment, equal to the load on a pair of rivets multiplied by their distance from the end of the beam; that is, the bending moment is

$$4425 \times 3\tfrac{3}{8} = 14934.$$

The theory of beams gives

$$M = \frac{fI}{y},$$

in which M is the bending moment, I is the moment of inertia of the section of the beam, y is the distance of the most strained fibre from the neutral axis, and f is the stress at that fibre. For rolled-steel channel-bars we may use, for f, 16,000 pounds, so that with the given value of M we have

$$14{,}934 = \frac{16{,}000\,I}{y}, \quad \text{or} \quad \frac{I}{y} = 0.933.$$

Now I and y depend on the form and size of the section of the beam, and, conversely, the size and form of beam required may be determined from them. But as the upper channel-bar is exposed to a greater bending moment and consequently must have a larger section than is required for the lower bar, we will defer the discussion of this matter, because it is convenient to make the bars of the same size.

Upper Stay-rods.—The flat surface of the boiler-head above the lower channel-bar is supported by the upper channel-bar aided by the angle-iron which is firmly riveted to it, and which will be assumed to act with and form a part of the channel-bar.

Following our general convention that the pressure on a portion of the head between two lines of support is divided equally between them, we will assume that the load on the upper channel-bar is due to the steam-pressure on an area bounded at the bottom by a line half-way between the upper and lower channel-bars, and at the top by an arc $3\frac{1}{4}$ inches inside the boiler-shell. On Fig. 140 half of this area is represented by jkl; the arc jk being about half-way between the root of the flange, shown by the outer dotted boundary line, and the adjacent rivets. In place of the area jkl we will take the rectangular area $lmno$, bounded at the end by a line at the middle of the end of the channel-bar, and at the top by a line mn so chosen as to make the rectangular area larger than the area it replaces. The width of this area, lm, is $9\frac{1}{4}$ inches, so that the load per inch of length is

$$9\frac{1}{4} \times 150 = 1387.5 \text{ pounds.}$$

The upper channel-bar may be assimilated to a continuous girder with three unequal spans; the middle span between the stay-rods is $15\frac{1}{2}$ inches, and the end spans between the stay-rods and the roots of the flange of the head are each $11\frac{1}{2}$ inches. This makes the end spans nearly 3/4 of the middle span. Now, a continuous girder uniformly loaded with w pounds per inch of length, which has a middle span l inches long, and two end spans $\tfrac{3}{4}l$ inches long, will have for the end-supporting forces $\tfrac{233}{864}wl$, and for the middle supporting forces $\tfrac{847}{864}wl$. The end supporting forces are provided by the shell, which is abundantly able to carry them. The stay-rods, which furnish the middle-supporting forces, must each carry

$$\tfrac{847}{864} \times 15\tfrac{1}{2} \times 1387.5 = 21{,}083 \text{ pounds.}$$

Assuming a working-stress of 9000 pounds per square inch for the stay, the area of the section for a stay is

$$21{,}083 \div 9000 = 2.34$$

square inches. The corresponding diameter is not quite $1\tfrac{11}{16}$ of an inch. As rods of this size are not regularly carried in stock, we will take the next larger regular size, namely, $1\tfrac{3}{4}$ of an inch. This is the size mentioned in connection with the discussion of the lower stay-rods.

Upper Channel-bar.—The calculation of the stress in the upper channel-bar will be made by an extension of the same approximate method used with the lower channel-bar. Since the middle span is wider than the end spans, it will be sufficient to make a calculation for it only. The calculation is made as for a simple beam supported at the ends, the points of support being at one fourth of the diameter of the thick washer from the middle stay-rod, that is, at the distance of $1\tfrac{3}{8}$ of an inch from the stay-rod. The distance between the upper stay-rods is $15\tfrac{1}{2}$ inches, so that the span of the beam is

$$15\tfrac{1}{2} - 2 \times 1\tfrac{3}{8} = 12\tfrac{3}{4} \text{ inches.}$$

The beam is assumed to be loaded with concentrated loads applied at the rivets C, D, E, F, G, and H (Fig. 140); the load on the rivet I is assumed to be carried by the stay-rod directly, and is not included in this calculation. The pair of rivets D and E, and the several rivets C, G, and H, are assumed to carry the load due to the pressure on the areas marked off by the dotted lines on Fig. 140, each line being drawn half-way between adjacent supporting points, except that the arc at the top is drawn $3\frac{1}{4}$ inches from the shell, as already said. The calculation of the loads on these rivets, of the supporting forces, and of the bending moments is simple and direct, but is tedious when stated in detail. We will therefore be contented to say that the bending moment at the middle of the beam is 37,390. Taking, as with the lower channel-bar, a working-stress of 16,000 pounds, we have

$$37{,}390 = \frac{16{,}000 I}{y}, \quad \text{or} \quad \frac{I}{y} = 2.17.$$

The makers of steel beams, channel-bars, and angle-irons publish handbooks which give the sizes and properties of the standard forms, including the moment of inertia I and the ratio $\frac{I}{y}$, which is called the moment of resistance. From such a handbook it appears that the moment of resistance of the channel-bar $6'' \times 2\frac{1}{2}'' \times \frac{1}{2}''$ is 1.08, and that the moment of resistance of the $3\frac{1}{2}'' \times 3''$ angle-iron is 1.55; the sum 2.63 is larger than the required moment of resistance given above. These forms are consequently used as shown on Plate I.

Brackets.—The boiler shown on Plate I is supported on four cast-iron brackets, each of which is 10 inches wide in the direction of the length of the boiler, and $15\frac{1}{2}$ inches long measured circumferentially. Each bracket is riveted to the shell by nine rivets 15/16 of an inch in diameter. Boilers over 16 feet long commonly have six brackets. The brackets are made wide and long in order that the local strains due to carrying the weight of the boiler may not be excessive. The rivets are larger than are used about the boiler, as the pitch is not restricted as in a calked seam.

The brackets are set above the middle line of the boiler so that the flanges may be protected by brickwork. In the case in hand they are 3½ inches above the middle; as much as 4½ inches is commonly used.

The brackets are arranged so that the weight of the boiler and accessories is equally divided among them, and so that there is as little bending-moment as possible on the shell of the boiler. When four brackets are used they may be somewhat less than a fourth of the length of a tube, from the tube-plates.

The load on the brackets may be estimated by calculating the weight of the boiler when entirely full of water, and adding the weight of all parts that are supported by the boiler, such as pipes, valves, and brickwork or covering, that may rest on the boiler. One fourth of this load is assigned to each bracket. This load on a bracket should be uniformly distributed over the bearing-surface of the flange, which is commonly 8 or 9 inches wide. But to guard against the effect of unequal bearing, it is well to assume the bracket to bear near the outer edge—say two inches from the edge. Such an assumption will bring the bearing-force on a bracket on Plate I, 10 inches from the shell. This bearing-force tends to rotate the bracket about its upper edge, and this tendency is resisted by the rivets under the flange, which must be large enough to resist the resulting pull on them. The other rivets are added to give sufficient resistance to shearing all the rivets. There are seldom less than nine rivets in a bracket, all as large as those below the flange, even though fewer would suffice. The bracket is usually made of cast iron, and the dimensions are commonly controlled as much by the conditions required for a sound casting as by calculations for strength. The strength may be calculated, treating it as a cantilever, allowing for the web connecting the flange to the body of the casting.

Specifications and Contract.—The engineer intrusted

with the design of a boiler prepares a set of working drawings and a set of specifications which give all necessary instructions concerning the material to be used and the methods of construction to be followed. The drawings and specifications form a part of the contract with the boiler-maker.

Boiler-makers commonly design standard forms of boilers, and in answer to inquiry will furnish a statement or set of specifications for a desired boiler in form of a letter, which letter forms the contract for the boiler. On the next page is given the contract and specifications for the boiler shown on Plate I.

.........IRON WORKS CO.,

....

Boston, MASS., *Feb. 1*, 1897.

....

Gentlemen :

....

Your letter of .... received. We will build One (1) Horizontal Tubular Boiler. *One* Boiler, viz., *Sixty* (60) inches diameter by *seventeen 2/12* (17 2/12) feet long. Containing *84* Tubes *3* inches diameter, by *sixteen* (16) feet long. Shell of Boiler of *O. H. Fire-box Steel*, 7/16" thick, *not less than 55,000 nor over 60,000 lbs. Tensile Strength. Not less than 50% reduction of area, and 25% elongation in 8"*.

Heads of Boiler of *O.H. Flange Steel* 9/16" thick. Longitudinal Seams BUTT JOINTED, with double covering-plates, TRIPLE RIVETED. *Rivet-holes drilled in place, i.e., Rivet-holes punched 1/4" small, courses rolled up, covering-plates bolted on courses. Heads in courses with all holes together perfectly fair. Then rivet-holes drilled to full size.*

Longitudinal braces without welds, *with upset screw ends.*

Two (2) or *three* (3) Lugs on each side, and to be provided with wall-plates and expansion-rolls. Manhole (internal frame) on top. *This frame a steel casting.*

....

Two (2) *5"* Nozzles on top, ...

A Hand-hole in each head, Fusible Safety Plug in back head. Bottom at back end reinforced and tapped for *2"* blowout .... Internal Feed Pipe placed in Boiler................Co.'s style,....................

With Boiler, Castings for setting, viz.: C. I., *Overhung* Front, Mouth-pieces, Division Plates, Grate Bars, *shaking pattern* 60" × 60". Grate Bearers, Ash-pit Door for the brickwork, Back Return Arched T Bars, the Anchor Bolts for Front. One (1) set of *six* (6) Buckstaves and Tie Rods *with the boiler*. With *the* Boiler One (1) *4"*............*Pop* Safety Valve, (3) *3/4"* Gauge Cocks, One (1) *6"* Steam Gauge, One (1) *3/4"* Water Gauge and One (1) Combination Column. ....Boiler tested 225 lbs. per square inch. Inspected and Insured in the sum of $400.00 for one year, by......... ...STEAM BOILER INSPECTION & INSURANCE CO..................................

The BOILER CASTINGS and FIXTURES as herein specified by name, delivered F. O. B. cars, or at vessel's wharf, or on sidewalk of building, Boston, Mass., for the sum of *six hundred and seventy* (670.00) dollars net.

Very respectfully yours,

............IRON WORKS CO.

P. S.—*Specimens will be furnished, one lengthwise and one crosswise, from each plate. To be at least 18" long and planed on edge 1" or 1½" wide. These specimens shall show no blowhole defects and shall bend double cold, at a red heat, and at a flanging heat.*

APPENDIX.

APPENDIX.

LOGARITHMS.

Nat. Nos.	0	1	2	3	4	5	6	7	8	9	Proportional Parts.								
											1	2	3	4	5	6	7	8	9
55	7404	7412	7419	7427	7435	7443	7451	7459	7466	7474	1	2	2	3	4	5	5	6	7
56	7482	7490	7497	7505	7513	7520	7528	7536	7543	7551	1	2	2	3	4	5	5	6	7
57	7559	7566	7574	7582	7589	7597	7604	7612	7619	7627	1	2	2	3	4	5	5	6	7
58	7634	7642	7649	7657	7664	7672	7679	7686	7694	7701	1	1	2	3	4	4	5	6	7
59	7709	7716	7723	7731	7738	7745	7752	7760	7767	7774	1	1	2	3	4	5	5	6	7
60	7782	7789	7796	7803	7810	7818	7825	7832	7839	7846	1	1	2	3	4	4	5	6	6
61	7853	7860	7868	7875	7882	7889	7896	7903	7910	7917	1	1	2	3	4	4	5	6	6
62	7924	7931	7938	7945	7952	7959	7966	7973	7980	7987	1	1	2	3	3	4	5	6	6
63	7993	8000	8007	8014	8021	8028	8035	8041	8048	8055	1	1	2	3	3	4	5	5	6
64	8062	8069	8075	8082	8089	8096	8102	8109	8116	8122	1	1	2	3	3	4	5	5	6
65	8129	8136	8142	8149	8156	8162	8169	8176	8182	8189	1	1	2	3	3	4	5	5	6
66	8195	8202	8209	8215	8222	8228	8235	8241	8248	8254	1	1	2	3	3	4	5	5	6
67	8261	8267	8274	8280	8287	8293	8299	8306	8312	8319	1	1	2	3	3	4	5	5	6
68	8325	8331	8338	8344	8351	8357	8363	8370	8376	8382	1	1	2	3	3	4	4	5	6
69	8388	8395	8401	8407	8414	8420	8426	8432	8439	8445	1	1	2	2	3	4	4	5	6
70	8451	8457	8463	8470	8476	8482	8488	8494	8500	8506	1	1	2	2	3	4	4	5	6
71	8513	8519	8525	8531	8537	8543	8549	8555	8561	8567	1	1	2	2	3	4	4	5	5
72	8573	8579	8585	8591	8597	8603	8609	8615	8621	8627	1	1	2	2	3	4	4	5	5
73	8633	8639	8645	8651	8657	8663	8669	8675	8681	8686	1	1	2	2	3	4	4	5	5
74	8692	8698	8704	8710	8716	8722	8727	8733	8739	8745	1	1	2	2	3	4	4	5	5
75	8751	8756	8762	8768	8774	8779	8785	8791	8797	8802	1	1	2	2	3	3	4	5	5
76	8808	8814	8820	8825	8831	8837	8842	8848	8854	8859	1	1	2	2	3	3	4	5	5
77	8865	8871	8876	8882	8887	8893	8899	8904	8910	8915	1	1	2	2	3	3	4	4	5
78	8921	8927	8932	8938	8943	8949	8954	8960	8965	8971	1	1	2	2	3	3	4	4	5
79	8976	8982	8987	8993	8998	9004	9009	9015	9020	9025	1	1	2	2	3	3	4	4	5
80	9031	9036	9042	9047	9053	9058	9063	9069	9074	9079	1	1	2	2	3	3	4	4	5
81	9085	9090	9096	9101	9106	9112	9117	9122	9128	9133	1	1	2	2	3	3	4	4	5
82	9138	9143	9149	9154	9159	9165	9170	9175	9180	9186	1	1	2	2	3	3	4	4	5
83	9191	9196	9201	9206	9212	9217	9222	9227	9232	9238	1	1	2	2	3	3	4	4	5
84	9243	9248	9253	9258	9263	9269	9274	9279	9284	9289	1	1	2	2	3	3	4	4	5
85	9294	9299	9304	9309	9315	9320	9325	9330	9335	9340	1	1	2	2	3	3	4	4	5
86	9345	9350	9355	9360	9365	9370	9375	9380	9385	9390	1	1	2	2	3	3	4	4	5
87	9395	9400	9405	9410	9415	9420	9425	9430	9435	9440	0	1	1	2	2	3	3	4	4
88	9445	9450	9455	9460	9465	9469	9474	9479	9484	9489	0	1	1	2	2	3	3	4	4
89	9494	9499	9504	9509	9513	9518	9523	9528	9533	9538	0	1	1	2	2	3	3	4	4
90	9542	9547	9552	9557	9562	9566	9571	9576	9581	9586	0	1	1	2	2	3	3	4	4
91	9590	9595	9600	9605	9609	9614	9619	9624	9628	9633	0	1	1	2	2	3	3	4	4
92	9638	9643	9647	9652	9657	9661	9666	9671	9675	9680	0	1	1	2	2	3	3	4	4
93	9685	9689	9694	9699	9703	9708	9713	9717	9722	9727	0	1	1	2	2	3	3	4	4
94	9731	9736	9741	9745	9750	9754	9759	9763	9768	9773	0	1	1	2	2	3	3	4	4
95	9777	9782	9786	9791	9795	9800	9805	9809	9814	9818	0	1	1	2	2	3	3	4	4
96	9823	9827	9832	9836	9841	9845	9850	9854	9859	9863	0	1	1	2	2	3	3	4	4
97	9868	9872	9877	9881	9886	9890	9894	9899	9903	9908	0	1	1	2	2	3	3	4	4
98	9912	9917	9921	9926	9930	9934	9939	9943	9948	9952	0	1	1	2	2	3	3	4	4
99	9956	9961	9965	9969	9974	9978	9983	9987	9991	9996	0	1	1	2	2	3	3	3	4

APPENDIX.

LOGARITHMS.

Nat. Nos.	0	1	2	3	4	5	6	7	8	9	\multicolumn{9}{c	}{Proportional Parts.}
											1	2	3	4	5	6	7	8	9
10	0000	0043	0086	0128	0170	0212	0253	0294	0334	0374	4	8	12	17	21	25	29	33	37
11	0414	0453	0492	0531	0569	0607	0645	0682	0719	0755	4	8	11	15	19	23	26	30	34
12	0792	0828	0864	0899	0934	0969	1004	1038	1072	1106	3	7	10	14	17	21	24	28	31
13	1139	1173	1206	1239	1271	1303	1335	1367	1399	1430	3	6	10	13	16	19	23	26	29
14	1461	1492	1523	1553	1584	1614	1644	1673	1703	1732	3	6	9	12	15	18	21	24	27
15	1761	1790	1818	1847	1875	1903	1931	1959	1987	2014	3	6	8	11	14	17	20	22	25
16	2041	2068	2095	2122	2148	2175	2201	2227	2253	2279	3	5	8	11	13	16	18	21	24
17	2304	2330	2355	2380	2405	2430	2455	2480	2504	2529	2	5	7	10	12	15	17	20	22
18	2553	2577	2601	2625	2648	2672	2695	2718	2742	2765	2	5	7	9	12	14	16	19	21
19	2788	2810	2833	2856	2878	2900	2923	2945	2967	2989	2	4	7	9	11	13	16	18	20
20	3010	3032	3054	3075	3096	3118	3139	3160	3181	3201	2	4	6	8	11	13	15	17	19
21	3222	3243	3263	3284	3304	3324	3345	3365	3385	3404	2	4	6	8	10	12	14	16	18
22	3424	3444	3464	3483	3502	3522	3541	3560	3579	3598	2	4	6	8	10	12	14	15	17
23	3617	3636	3655	3674	3692	3711	3729	3747	3766	3784	2	4	6	7	9	11	13	15	17
24	3802	3820	3838	3856	3874	3892	3909	3927	3945	3962	2	4	5	7	9	11	12	14	16
25	3979	3997	4014	4031	4048	4065	4082	4099	4116	4133	2	3	5	7	9	10	12	14	15
26	4150	4166	4183	4200	4216	4232	4249	4265	4281	4298	2	3	5	7	8	10	11	13	15
27	4314	4330	4346	4362	4378	4393	4409	4425	4440	4456	2	3	5	6	8	9	11	13	14
28	4472	4487	4502	4518	4533	4548	4564	4579	4594	4609	2	3	5	6	8	9	11	12	14
29	4624	4639	4654	4669	4683	4698	4713	4728	4742	4757	1	3	4	6	7	9	10	12	13
30	4771	4786	4800	4814	4829	4843	4857	4871	4886	4900	1	3	4	6	7	9	10	11	13
31	4914	4928	4942	4955	4969	4983	4997	5011	5024	5038	1	3	4	6	7	8	10	11	12
32	5051	5065	5079	5092	5105	5119	5132	5145	5159	5172	1	3	4	5	7	8	9	11	12
33	5185	5198	5211	5224	5237	5250	5263	5276	5289	5302	1	3	4	5	6	8	9	10	12
34	5315	5328	5340	5353	5366	5378	5391	5403	5416	5428	1	3	4	5	6	8	9	10	11
35	5441	5453	5465	5478	5490	5502	5514	5527	5539	5551	1	2	4	5	6	7	9	10	11
36	5563	5575	5587	5599	5611	5623	5635	5647	5658	5670	1	2	4	5	6	7	8	10	11
37	5682	5694	5705	5717	5729	5740	5752	5763	5775	5786	1	2	3	5	6	7	8	9	10
38	5798	5809	5821	5832	5843	5855	5866	5877	5888	5899	1	2	3	5	6	7	8	9	10
39	5911	5922	5933	5944	5955	5966	5977	5988	5999	6010	1	2	3	4	5	7	8	9	10
40	6021	6031	6042	6053	6064	6075	6085	6096	6107	6117	1	2	3	4	5	6	8	9	10
41	6128	6138	6149	6160	6170	6180	6191	6201	6212	6222	1	2	3	4	5	6	7	8	9
42	6232	6243	6253	6263	6274	6284	6294	6304	6314	6325	1	2	3	4	5	6	7	8	9
43	6335	6345	6355	6365	6375	6385	6395	6405	6415	6425	1	2	3	4	5	6	7	8	9
44	6435	6444	6454	6464	6474	6484	6493	6503	6513	6522	1	2	3	4	5	6	7	8	9
45	6532	6542	6551	6561	6580	6590	6599	6609	6618	1	2	3	4	5	6	7	8	9	
46	6628	6637	6646	6656	6665	6675	6684	6693	6702	6712	1	2	3	4	5	6	7	7	8
47	6721	6730	6739	6749	6758	6767	6776	6785	6794	6803	1	2	3	4	5	5	6	7	8
48	6812	6821	6830	6839	6848	6857	6866	6875	6884	6893	1	2	3	4	4	5	6	7	8
49	6902	6911	6920	6928	6937	6946	6955	6964	6972	6981	1	2	3	4	4	5	6	7	8
50	6990	6998	7007	7016	7024	7033	7042	7050	7059	7067	1	2	3	3	4	5	6	7	8
51	7076	7084	7093	7101	7110	7118	7126	7135	7143	7152	1	2	3	3	4	5	6	7	8
52	7160	7168	7177	7185	7193	7202	7210	7218	7226	7235	1	2	2	3	4	5	6	7	7
53	7243	7251	7259	7267	7275	7284	7292	7300	7308	7316	1	2	2	3	4	5	6	6	7
54	7324	7332	7340	7348	7356	7364	7372	7380	7388	7396	1	2	2	3	4	5	6	6	7

Explanation of the Table for Finding the Area of Segment of a Circle.—The areas given in the table are for a circle 1 inch in diameter. The diameter is divided into 1000 parts, and the area for segments of different heights can be taken directly from the table, since the ratio of the height of the segment to the diameter of the circle is the same as the height of the segment.

For a circle whose diameter is other than unity. Given the diameter of the circle and the height of segment. Required area of segment. Divide height of segment by diameter; find area given in the table opposite this ratio; multiply this area by the square of the diameter and the result is the required area.

Example.—Dia. of circle = 60″, height of segment = 18″.

$18 \div 60 = .30$; area in table opposite .30 is .19817.

$.19817 \times 60 \times 60 = $ area of segment $= 713.4$ sq. in.

Given the diameter of the circle and the area of a segment, to find the height.

Divide the area of the segment by the square of the diameter. Find in the table the area nearest to this, multiply the ratio corresponding to this by the diameter of the circle, and the result is the required height of the segment.

Example.—Area of segment = 713.4 sq. in.

Diameter of circle = 60″. Required the height of the segment.

$$\frac{713.4}{60 \times 60} = .19817.$$ Ratio opposite this is .300.

$.300 \times 60'' = 18''$, the required height.

Example —Area of segment = 640 sq. in.

Diameter of circle = 50″.

$$\frac{640}{50 \times 50} = .2560;$$ nearest ratio, .362.

$.362 \times 50 = 18.10''$, the required height.

APPENDIX.

TABLE FOR FINDING AREAS OF SEGMENTS OF A CIRCLE.

Ratio of Height of Segment to Diam. of Circle.	Area of Segment.	Ratio of Height of Segment to Diam. of Circle.	Area of Segment.	Ratio of Height of Segment to Diam. of Circle.	Area of Segment.	Ratio of Height of Segment to Diam. of Circle.	Area of Segment.	Ratio of Height of Segment to Diam. of Circle.	Area of Segment.
.210	.11990	.260	.16226	.310	.20738	.360	.25455	.410	.30319
1	.12071	1	.16314	1	.20830	1	.25551	1	.30417
2	.12153	2	.16402	2	.20923	2	.25647	2	.30516
3	.12235	3	.16490	3	.21015	3	.25743	3	.30614
4	.12317	4	.16578	4	.21108	4	.25839	4	.30712
.215	.12399	.265	.16666	.315	.21201	.365	.25936	.415	.30811
6	.12481	6	.16755	6	.21294	6	.26032	6	.30910
7	.12563	7	.16843	7	.21387	7	.26128	7	.31008
8	.12646	8	.16932	8	.21480	8	.26225	8	.31107
9	.12729	9	.17020	9	.21573	9	.26321	9	.31205
.220	.12811	.270	.17109	.320	.21667	.370	.26418	.420	.31304
1	.12894	1	.17198	1	.21760	1	.26514	1	.31403
2	.12977	2	.17287	2	.21853	2	.26611	2	.31502
3	.13060	3	.17376	3	.21947	3	.26708	3	.31600
4	.13144	4	.17465	4	.22040	4	.26805	4	.31699
.225	.13227	.275	.17554	.325	.22134	.375	.26901	.425	.31798
6	.13311	6	.17644	6	.22228	6	.26998	6	.31897
7	.13395	7	.17733	7	.22322	7	.27095	7	.31996
8	.13478	8	.17823	8	.22415	8	.27192	8	.32095
9	.13562	9	.17912	9	.22509	9	.27289	9	.32194
.230	.13646	.280	.18002	.330	.22603	.380	.27386	.430	.32293
1	.13731	1	.18092	1	.22697	1	.27483	1	.32392
2	.13815	2	.18182	2	.22792	2	.27580	2	.32491
3	.13900	3	.18272	3	.22886	3	.27678	3	.32590
4	.13984	4	.18362	4	.22980	4	.27775	4	.32689
.235	.14069	.285	.18452	.335	.23074	.385	.27872	.435	.32788
6	.14154	6	.18542	6	.23169	6	.27969	6	.32887
7	.14239	7	.18633	7	.23263	7	.28067	7	.32987
8	.14324	8	.18723	8	.23358	8	.28164	8	.33086
9	.14409	9	.18814	9	.23453	9	.28262	9	.33185
.240	.14494	.290	.18905	.340	.23547	.390	.28359	.440	.33284
1	.14580	1	.18996	1	.23642	1	.28457	1	.33384
2	.14666	2	.19086	2	.23737	2	.28554	2	.33483
3	.14751	3	.19177	3	.23832	3	.28652	3	.33582
4	.14837	4	.19268	4	.23927	4	.28750	4	.33682
.245	.14923	.295	.19360	.345	.24022	.395	.28848	.445	.33781
6	.15009	6	.19451	6	.24117	6	.28945	6	.33880
7	.15095	7	.19542	7	.24212	7	.29043	7	.33980
8	.15182	8	.19634	8	.24307	8	.29141	8	.34079
9	.15268	9	.19725	9	.24403	9	.29239	9	.34179
.250	.15355	.300	.19817	.350	.24498	.400	.29337	.450	.34278
1	.15441	1	.19908	1	.24593	1	.29435	1	.34378
2	.15528	2	.20000	2	.24689	2	.29533	2	.34477
3	.15615	3	.20092	3	.24784	3	.29631	3	.34577
4	.15702	4	.20184	4	.24880	4	.29729	4	.34676
.255	.15789	.305	.20276	.355	.24976	.405	.29827	.455	.34776
6	.15876	6	.20368	6	.25071	6	.29926	6	.34876
7	.15964	7	.20460	7	.25167	7	.30024	7	.34975
8	.16051	8	.20553	8	.25263	8	.30122	8	.35075
9	.16139	9	.20645	9	.25359	9	.30220	9	.35175

NATURAL TRIGONOMETRIC FUNCTIONS.

Deg.	Sine.	Tangent.	Cot.	Cos.	Deg.
0	.0000	.0000	Infinite	1.0000	90
1	.0175	.0175	57.290	.9998	89
2	.0349	.0349	28.636	.9994	88
3	.0523	.0524	19.081	.9986	87
4	.0698	.0699	14.301	.9976	86
5	.0872	.0875	11.430	.9962	85
6	.1045	.1051	9.5144	.9945	84
7	.1219	.1228	8.1443	.9925	83
8	.1392	.1405	7.1154	.9903	82
9	.1564	.1584	6.3138	.9877	81
10	.1736	.1763	5.6713	.9848	80
11	.1908	.1944	5.1446	.9816	79
12	.2079	.2126	4.7046	.9781	78
13	.2250	.2309	4.3315	.9744	77
14	.2419	.2493	4.0108	.9703	76
15	.2588	.2679	3.7321	.9659	75
16	.2756	.2867	3.4874	.9613	74
17	.2924	.3057	3.2709	.9563	73
18	.3090	.3249	3.0777	.9511	72
19	.3256	.3443	2.9042	.9455	71
20	.3420	.3640	2.7475	.9397	70
21	.3584	.3839	2.6051	.9336	69
22	.3746	.4040	2.4751	.9272	68
23	.3907	.4245	2.3559	.9205	67
24	.4067	.4452	2.2460	.9135	66
25	.4226	.4663	2.1445	.9063	65
26	.4384	.4877	2.0503	.8988	64
27	.4540	.5095	1.9626	.8910	63
28	.4695	.5317	1.8807	.8829	62
29	.4848	.5543	1.8040	.8746	61
30	.5000	.5774	1.7321	.8660	60
31	.5150	.6009	1.6643	.8572	59
32	.5299	.6249	1.6003	.8480	58
33	.5446	.6494	1.5399	.8387	57
34	.5592	.6745	1.4826	.8290	56
35	.5736	.7002	1.4281	.8192	55
36	.5878	.7265	1.3764	.8090	54
37	.6018	.7536	1.3270	.7986	53
38	.6157	.7813	1.2799	.7880	52
39	.6293	.8098	1.2349	.7771	51
40	.6428	.8391	1.1918	.7660	50
41	.6561	.8693	1.1504	.7547	49
42	.6691	.9004	1.1106	.7431	48
43	.6820	.9325	1.0724	.7314	47
44	.6947	.9657	1.0355	.7193	46
45	.7071	1.0000	1.0000	.7071	45
Deg.	Cos.	Cot.	Tangent.	Sine.	Deg.

CIRCLES

Diam. Inches.	Circumf. Inches.	Area, Sq. In.
12	37 5/8	113 1/8
14	44	154
16	50 1/4	201
18	56 1/2	254 1/2
20	62 7/8	314 1/8
22	69 1/8	380 1/8
24	75 3/8	452 3/8
26	81 5/8	531
28	88	615 3/4
30	94 1/4	706 7/8
32	100 1/2	804 1/4
34	106 3/4	907 7/8
36	113 1/8	1017 7/8
38	119 3/8	1134 1/8
40	125 5/8	1256 5/8
42	132	1385 1/2
44	138 1/4	1520 1/2
46	144 1/2	1661 7/8
48	150 3/4	1809 1/2
50	157 1/8	1963 1/2
52	163 3/8	2123 3/4
54	169 5/8	2290 1/4
56	175 7/8	2463
58	182 1/4	2642 1/8
60	188 1/2	2827 3/8
62	194 3/4	3019 1/8
64	201	3217
66	207 3/8	3421 1/4
68	213 5/8	3631 5/8
70	219 7/8	3848 1/2
72	226 1/8	4071 1/2
74	232 1/2	4300 7/8
76	238 3/4	4536 1/2
78	245	4778 3/8
80	251 3/8	5026 1/2
82	257 5/8	5281
84	263 7/8	5541 3/4
86	270 1/8	5808 7/8
88	276 1/2	6082 1/8
90	282 3/4	6361 3/4
92	289	6647 5/8
94	295 3/8	6939 3/4
96	301 5/8	7238 1/4
98	307 7/8	7543
100	314 1/8	7854
102	320 3/8	8171 1/4

ROUND RODS OF WROUGHT IRON.

Diameter in Inches.	Circumference in Inches.	Area in Sq. Inches.	Weight of Rod One Foot Long.	Diameter of Upset Screw End, Inches.	Diameter of Screw at Root of Thread, Inches.	Threads per Inch Number.	Excess of Effective Area of Screw End over Bar, Per Cent.
0							
1/16	.1963	.0031	.010				
1/8	.3927	.0123	.041				
3/16	.5890	.0276	.092				
1/4	.7854	.0491	.164				
5/16	.9817	.0767	.256				
3/8	1.1781	.1104	.368				
7/16	1.3744	.1503	.501				
1/2	1.5708	.1963	.654	3/4	.620	10	[illegible]
9/16	1.7671	.2485	.828	3/4	.620	10	21
5/8	1.9635	.3068	1.023	7/8	.731	9	37
11/16	2.1598	.3712	1.237	1	.837	8	48
3/4	2.3562	.4418	1.473	1	.837	8	25
13/16	2.5525	.5185	1.728	1 1/8	.940	7	34
7/8	2.7489	.6013	2.004	1 1/4	1.065	7	48
15/16	2.9452	.6903	2.301	1 1/4	1.065	7	29
1	3.1416	.7854	2.618	1 3/8	1.160	6	35
1/16	3.3379	.8866	2.955	1 3/8	1.160	6	19
1/8	3.5343	.9940	3.313	1 1/2	1.284	6	30
3/16	3.7306	1.1075	3.692	1 1/2	1.284	6	17
1/4	3.9270	1.2272	4.091	1 5/8	1.389	5 1/2	23
5/16	4.1233	1.3530	4.510	1 3/4	1.490	5	29
3/8	4.3197	1.4849	4.950	1 3/4	1.490	5	18
7/16	4.5160	1.6230	5.410	1 7/8	1.615	5	26
1/2	4.7124	1.7671	5.890	2	1.712	4 1/2	30
5/8	5.1051	2.0739	6.913	2 1/8	1.837	4 1/2	28
3/4	5.4978	2.4053	8.018	2 1/4	1.962	4 1/2	26
7/8	5.8905	2.7612	9.204	2 3/8	2.087	4 1/2	24
2	6.2832	3.1416	10.47	2 1/2	2.175	4	18
1/8	6.6759	3.5466	11.82	2 5/8	2.300	4	17
1/4	7.0686	3.9761	13.25	2 7/8	2.550	4	28
3/8	7.4613	4.4301	14.77	3	2.629	3 1/2	23
1/2	7.8540	4.9087	16.36	3 1/8	2.754	3 1/2	21
5/8	8.2467	5.4119	18.04	3 1/4	2.879	3 1/2	20
3/4	8.6394	5.9396	19.80	3 3/8	3.004	3 1/2	19
7/8	9.0321	6.4918	21.64	3 5/8	3.225	3 1/4	26
3	9.4248	7.0686	23.56	3 3/4	3.317	3	22

LAP-WELDED BOILER-TUBES.

Size.	External Diameter, Inches.	Internal Diameter, Inches.	Thickness, Inches.	Circumference, External (Inches).	Circumference, Internal (Inches).	Transverse Area, Sq. In., External.	Transverse Area, Sq. In., Internal.	Length, per Sq. Ft. of Surface, External, Ft.	Length, per Sq. Ft. of Surface, Internal, Ft.	Surface, per Ft. of Length, External.	Surface, per Ft. of Length, Internal.	Weight per Foot.
1	1	.86	.072	3.14	2.69	.78	.57	3.82	4.46	.26	.22	.71
1¼	1¼	1.11	.072	3.93	3.47	1.23	.96	3.06	3.45	.33	.29	.89
1½	1½	1.33	.083	4.71	4.19	1.77	1.40	2.55	2.86	.39	.35	1.24
1¾	1¾	1.56	.095	5.50	4.90	2.40	1.91	2.18	2.45	.46	.41	1.66
2	2	1.81	.095	6.28	5.69	3.14	2.57	1.91	2.11	.52	.47	1.91
2¼	2¼	2.06	.095	7.07	6.47	3.98	3.33	1.70	1.85	.59	.54	2.16
2½	2½	2.28	.109	7.85	7.17	4.91	4.09	1.53	1.67	.65	.60	2.75
2¾	2¾	2.53	.109	8.64	7.95	5.94	5.03	1.39	1.51	.72	.66	3.04
3	3	2.78	.109	9.42	8.74	7.07	6.08	1.27	1.37	.79	.73	3.33
3¼	3¼	3.01	.120	10.21	9.46	8.30	7.12	1.17	1.26	.85	.79	3.96
3½	3½	3.25	.120	11.00	10.24	9.62	8.35	1.09	1.17	.92	.85	4.28
3¾	3¾	3.51	.120	11.78	11.03	11.04	9.68	1.02	1.09	.98	.92	4.60
4	4	3.73	.134	12.57	11.72	12.57	10.04	.95	1.02	1.05	.98	5.47
4½	4½	4.23	.134	14.14	13.29	15.90	14.07	.85	.90	1.18	1.11	6.17
5	5	4.70	.148	15.71	14.78	19.63	17.38	.76	.81	1.31	1.23	7.58
6	6	5.67	.165	18.85	17.81	28.27	25.25	.64	.67	1.57	1.48	10.16
7	7	6.67	.165	21.99	20.95	38.48	34.94	.55	.57	1.83	1.75	11.90
8	8	7.67	.165	25.13	24.10	50.27	46.20	.48	.50	2.09	2.01	13.05
9	9	8.64	.180	28.27	27.14	63.62	58.63	.42	.44	2.35	2.26	16.76
10	10	9.59	.203	31.42	30.14	78.54	72.20	.38	.40	2.62	2.51	20.99
11	11	10.56	.220	34.56	33.17	95.03	87.58	.35	.36	2.88	2.76	25.03
12	12	11.54	.229	37.70	36.26	113.10	104.63	.32	.33	3.14	3.02	28.46

SCREW-THREADS.

Angle of thread 60°. Flat at top and bottom = ⅛ of pitch.

Diameter of Screw, Inches.	Diameter at Root of Thread, Inches.	Threads per Inch, No.	Diameter of Screw, Inches.	Diameter at Root of Thread, Inches.	Threads per Inch, No.
¼	.185	20	2	1.712	4½
5/16	.240	18	2¼	1.962	4½
⅜	.294	16	2½	2.175	4
7/16	.344	14	2¾	2.425	4
½	.400	13	3	2.629	3½
9/16	.454	12	3¼	2.879	3½
⅝	.507	11	3½	3.100	3¼
¾	.620	10	3¾	3.317	3
⅞	.731	9	4	3.567	3
1	.837	8	4¼	3.798	2⅞
1⅛	.940	7	4½	4.028	2¾
1¼	1.065	7	4¾	4.255	2⅝
1⅜	1.160	6	5	4.480	2½
1½	1.284	6	5¼	4.730	2½
1⅝	1.389	5½	5½	5.053	2⅜
1¾	1.490	5	5¾	5.203	2⅜
1⅞	1.615	5	6	5.423	2¼

WROUGHT-IRON WELDED STEAM-, GAS-, AND WATER-PIPE.

Diameter.			Thickness.	Transverse Areas.		Nominal Weight per Foot.	Number of Threads per Inch of Screw.
Nominal Internal	Actual External.	Actual Internal.		External.	Internal.		
Inches.	Inches.	Inches.	Inches.	Sq. In.	Sq. In.	Pounds.	
⅛	.405	.27	.068	.129	.0573	.241	27
¼	.543	.364	.088	.229	.1041	.42	18
⅜	.675	.494	.091	.358	.1917	.559	18
½	.84	.623	.109	.554	.3048	.837	14
¾	1.05	.824	.113	.866	.5333	1.115	14
1	1.315	1.048	.134	1.358	.8626	1.668	11½
1¼	1.66	1.38	.14	2.164	1.496	2.244	11½
1½	1.9	1.611	.145	2.835	2.038	2.678	11½
2	2.375	2.067	.154	4.43	3.356	3.609	11½
2½	2.875	2.468	.204	6.492	4.784	5.739	8
3	3.5	3.067	.217	9.621	7.388	7.536	8
3½	4.	3.548	.226	12.566	9.887	9.001	8
4	4.5	4.026	.237	15.904	12.73	10.665	8
4½	5.	4.508	.246	19.635	15.961	12.34	8
5	5.563	5.045	.259	24.306	19.99	14.502	8
6	6.625	6.065	.28	34.472	28.888	18.762	8
7	7.625	7.023	.301	45.664	38.738	23.271	8
8	8.625	7.982	.322	58.426	50.04	28.177	8
9	9.625	8.937	.344	72.76	62.73	33.701	8
10	10.75	10.019	.356	90.763	78.839	40.065	8
11	12	11.25	.375	113.098	99.402	45.95	8
12	12.75	12	.375	127.677	113.098	48.985	8
13	14	13.25	.375	153.938	137.887	53.921	8
14	15	14.25	.375	176.715	159.485	57.893	8
15	16	15.25	.375	201.062	182.655	61.77	8
....	18	17.25	.375	254.47	233.706	69.66	....
....	20	19.25	.375	314.16	291.04	77.57	....
....	22	21.25	.375	380.134	354.657	85.47	....
....	24	23.25	.375	452.39	424.558	93.37	....

WROUGHT-IRON WELDED EXTRA STRONG PIPE.

⅛	.405	.205	.1	.129	.033	.29	27
¼	.54	.294	.123	.229	.068	.54	18
⅜	.675	.421	.127	.358	.139	.74	18
½	.84	.542	.149	.554	.231	1.09	14
¾	1.05	.736	.157	.866	.452	1.39	14
1	1.315	.951	.182	1.358	.71	2.17	11½
1¼	1.66	1.272	.194	2.164	1.271	3	11½
1½	1.9	1.494	.203	2.835	1.753	3.63	11½
2	2.375	1.933	.221	4.43	2.935	5.02	11½
2½	2.875	2.315	.28	6.492	4.209	7.67	8
3	3.5	2.892	.304	9.621	6.569	10.25	8
3½	4	3.358	.321	12.566	8.856	12.47	8
4	4.5	3.818	.341	15.904	11.449	14.97	8
5	5.563	4.813	.375	24.306	18.193	20.54	8
6	6.625	5.75	.437	34.472	25.967	28.58	8

APPENDIX.

HEAT OF THE LIQUID—WATER.

Temp. Deg. F. t	Heat of Liquid. q	Temp. Deg. F. t	Heat of Liquid. q	Temp. Deg. F. t	Heat of Liquid. q	Temp. Deg. F. t	Heat of Liquid. q	Temp. Deg. F. t	Heat of Liquid. q
32	0	69	37.12	106	74.0	143	111.2	180	148.5
33	1.01	70	38.11	107	75.0	144	112.2	181	149.5
34	2.01	71	39.11	108	76.0	145	113.3	182	150.6
35	3.02	72	40.11	109	77.0	146	114.3	183	151.6
36	4.03	73	41.11	110	78.0	147	115.3	184	152.6
37	5.04	74	42.11	111	79.0	148	116.3	185	153.6
38	6.04	75	43.11	112	80.0	149	117.3	186	154.6
39	7.05	76	44.11	113	81.0	150	118.3	187	155.6
40	8.06	77	45.10	114	82.0	151	119.3	188	156.6
41	9.06	78	46.10	115	83.0	152	120.3	189	157.6
42	10.07	79	47.09	116	84.0	153	121.3	190	158.6
43	11.07	80	48.09	117	85.0	154	122.3	191	159.6
44	12.08	81	49.08	118	86.0	155	123.3	192	160.6
45	13.08	82	50.08	119	87.0	156	124.3	193	161.6
46	14.09	83	51.07	120	88.1	157	125.4	194	162.6
47	15.09	84	52.07	121	89.1	158	126.4	195	163.7
48	16.10	85	53.06	122	90.1	159	127.4	196	164.7
49	17.10	86	54.06	123	91.1	160	128.4	197	165.7
50	18.10	87	55.05	124	92.1	161	129.4	198	166.7
51	19.11	88	56.05	125	93.1	162	130.4	199	167.7
52	20.11	89	57.04	126	94.1	163	131.4	200	168.7
53	21.11	90	58.04	127	95.1	164	132.4	201	169.7
54	22.11	91	59.03	128	96.1	165	133.4	202	170.7
55	23.11	92	60.03	129	97.1	166	134.4	203	171.7
56	24.11	93	61.03	130	98.1	167	135.4	204	172.7
57	25.12	94	62.02	131	99.1	168	136.4	205	173.7
58	26.12	95	63.02	132	100.2	169	137.4	206	174.7
59	27.12	96	64.01	133	101.2	170	138.5	207	175.8
60	28.12	97	65.01	134	102.2	171	139.5	208	176.8
61	29.12	98	66.01	135	103.2	172	140.5	209	177.8
62	30.12	99	67.01	136	104.2	173	141.5	210	178.8
63	31.12	100	68.01	137	105.2	174	142.5	211	179.8
64	32.12	101	69.01	138	106.2	175	143.5	212	180.8
65	33.12	102	70.00	139	107.2	176	144.5		
66	34.12	103	71.00	140	108.2	177	145.5		
67	35.12	104	72.0	141	109.2	178	146.5		
68	36.12	105	73.0	142	110.2	179	147.5		

VOLUME AND WEIGHT OF DISTILLED WATER.

Temp. Degrees Fahr.	Weight of a Cubic Foot in Pounds.	Temp. Degrees Fahr.	Weight of a Cubic Foot in Pounds.	Temp. Degrees Fahr.	Weight of a Cubic Foot in Pounds.
32	62.417	90	62.119	160	61.007
39.1	62.425	100	62.000	170	60.801
40	62.423	110	61.867	180	60.587
50	62.409	120	61.720	190	60.366
60	62.367	130	61.556	200	60.136
70	62.302	140	61.388	210	59.894
80	62.218	150	61.204	212	59.707

PROPERTIES OF SATURATED STEAM.
English Units.

Pressure, Pounds per Sq.In. p	Temperature, Degrees Fahr. t	Heat of the Liquid. q	Total Heat. λ	Heat of Vaporization. r	Volume in Cu. Ft. of 1 Pound. s	Weight in Pounds of 1 Cubic Foot d	Pressure, Pounds per Sq In. p
5	162.34	130.7	1131.5	1000.8	73.22	0.01366	5
10	193.25	161.9	1140.9	979.0	38.16	0.02621	10
15	213.03	181.8	1146.9	965.1	26.15	0.03826	15
20	227.95	196.9	1151.5	954.6	19.91	0.05023	20
25	240.04	209.1	1155.1	946.0	16.13	0.06199	25
30	250.27	219.4	1158.3	938.9	13.59	0.07360	30
35	259.19	228.4	1161.0	932.6	11.75	0.08508	35
40	267.13	236.4	1163.4	927.0	10.37	0.09644	40
45	274.29	243.6	1165.6	922.0	9.287	0.1077	45
50	280.85	250.2	1167.6	917.4	8.414	0.1188	50
55	286.89	256.3	1169.4	913.1	7.696	0.1299	55
60	292.51	261.9	1171.2	909.3	7.096	0.1409	60
65	297.77	267.2	1172.7	905.5	6.583	0.1519	65
70	302.71	272.2	1174.3	902.1	6.144	0.1628	70
75	307.38	276.9	1175.7	898.8	5.762	0.1736	75
80	311.80	281.4	1177.0	895.6	5.425	0.1843	80
85	316.02	285.8	1178.3	892.5	5.125	0.1951	85
90	320.04	290.0	1179.6	889.6	4.858	0.2058	90
95	323.89	294.0	1180.7	886.7	4.619	0.2165	95
100	327.58	297.9	1181.9	884.0	4.403	0.2271	100
105	331.13	301.6	1182.9	881.3	4.206	0.2378	105
110	334.56	305.2	1184.0	878.8	4.026	0.2484	110
115	337.86	308.7	1185.0	876.3	3.862	0.2589	115
120	341.05	312.0	1186.0	874.0	3.711	0.2695	120
125	344.13	315.2	1186.9	871.7	3.572	0.2800	125
130	347.12	318.4	1187.8	869.4	3.444	0.2904	130
135	350.03	321.4	1188.7	867.3	3.323	0.3009	135
140	352.85	324.4	1189.5	865.1	3.212	0.3113	140
145	355.59	327.2	1190.4	863.2	3.107	0.3218	145
150	358.26	330.0	1191.2	861.2	3.011	0.3321	150
155	360.86	332.7	1192.0	859.3	2.919	0.3426	155
160	363.40	335.4	1192.8	857.4	2.833	0.3530	160
165	365.88	338.0	1193.6	855.6	2.751	0.3635	165
170	368.29	340.5	1194.3	853.8	2.676	0.3737	170
175	370.65	343.0	1195.0	852.0	2.603	0.3841	175
180	372.97	345.4	1195.7	850.3	2.535	0.3945	180
185	375.23	347.8	1196.4	848.6	2.470	0.4049	185
190	377.44	350.1	1197.1	847.0	2.408	0.4153	190
195	379.61	352.4	1197.7	845.3	2.349	0.4257	195
200	381.73	354.6	1198.4	843.8	2.294	0.4359	200
205	383.82	356.8	1199.0	842.2	2.241	0.4461	205
210	385.87	358.9	1199.6	840.7	2.190	0.4565	210
215	387.88	361.0	1200.2	839.2	2.142	0.4669	215
220	389.84	363.0	1200.8	837.8	2.096	0.4772	220
225	391.79	365.1	1201.4	836.3	2.051	0.4876	225
230	393.69	367.1	1202.0	834.9	2.009	0.4979	230
235	395.56	369.0	1202.6	833.6	1.968	0.5082	235
240	397.41	371.0	1203.2	832.2	1.928	0.5186	240
245	399.21	372.8	1203.7	830.9	1.891	0.5289	245
250	400.99	374.7	1204.2	829.5	1.854	0.5393	250

INDEX.

	PAGE
Accumulators	292, 293
Acetic acid	71
Adamson joints	210
Air for combustion	48, 51
dilution	51
loss from excess	61
per pound of coal	53
supply for boiler, measurement of	318
Almy boiler	34
Angle-valves	237
Anthracite coal	37
Area of circles	362
steam-pipe	271
uptake	341
Areas of segments of circles	360, 361
Ash-pit	5
doors	5
Assembling and riveting boilers	285
Atmosphere, composition of	49
Atomic weight	45
Babcock & Wilcox	22
Baird's steam-trap	258
Belleville boiler	30
Belpaire fire-box	20, 158
Berryman feed-water heater	264
Blow-off pipes	4, 266
Blowing out brine	88
Blue heat	177
Board of Trade, rules for flues	218
Boilers, Almy	34
Babcock & Wilcox	22
Belleville	30

INDEX.

	PAGE
Boilers, Cahall	27
Cornish	7
cylindrical tubular	2
double-ended	16
fire-engine	13
Galloway	7
gunboat	18
Heine	25
Lancashire	6
Leavitt	20
locomotive	18, 20
Manning	9, 10
marine	14
(water-tube)	30
plain cylindrical	6
Root	25
Stirling	27
Thornycroft	32
two-flue	5
vertical	9, 11, 12, 13
water-tube	21
(marine)	30
Yarrow	34
Boiler accessories	235
design	323
explosions	227
front	5
horse-power	135
Boilers, life of	230
methods of supporting	166
proportions of	138
Boiler-setting	91
shop, plan and description of	273
testing (evaporative)	300
tubes, size and surface	364
Boilers, U. S. S. Brooklyn	146
Boring-mill	279
Brackets	166, 353
Brass	180
Bridge-wall	2
Bronze	180
Buck-staves	94
Bumped-up head	186
Bundy steam-trap	260
Butt-joint	199, 201

INDEX.

	PAGE
Cahall boiler	27
Calculation of riveted joints	192
stay-rods	347, 351
Calking	298
Calorimeters for steam, Carpenter	312
Peabody	309
Cam and toggle riveting-machine	288
Carbon, heat of combustion of	43, 44
Carbonate of lime	68
Cast iron	179
Chapman valves	238
Channel-bars, calculation of	348, 352
Charcoal	39
Check-valves	239
Chemistry of combustion	44
Chimney draught	112
forms of	125
height of	125
stability of	127
Chimneys	112
Kent's table	122, 124
Circles, area of	362
circumference of	362
Circumference of circles	362
Cleaning fires	110
Coals, anthracite	37
bituminous	38
caking, bituminous	38
dry bituminous	38
long-flaming bituminous	38
semi-bituminous	38
Coal, air per pound of	53
value of	131
Cold-water test	299
Collapsing pressure	207, 210–216
Coke	38
Combination	249
Combustion, air required for	48, 51
chemistry of	44
heat of	42–44
incomplete	60
rate of	136
temperature of	63
Composition	180
of atmosphere	49

372 INDEX.

	PAGE
Composition of fuels	40
Compression	175
Conical through-tubes	7
Copper	179
Cornish boiler	7
Corrosion	84
Corrugated furnace	211
Crane-lifts	277
Crown-bars	20, 157
Crowfeet	152
Crushing of rivets	191
Curtis steam-trap	260
Cylinder, strength of	183, 184
Cylindrical tubular-boiler setting	91
staying of	148
Damper regulator	255
Density of gases	45
Detachable brackets	167
Diameter of boiler	327
of rivet	192
Diagonal stays	182
Du Long's formula	47
Double-ended boiler	16
Down-draught furnaces	105
Draught of chimneys	112
gauge	316
Howdens system	110
split	9
wheel	8
Drill for tube-holes	278
Drilled or punched plates	190
Dry pipe	164
Dudgeon tube-expander	297
Economizer, Green's	111
Efficiency of riveted joints	188
Elastic limit	174
Elasticity, modulus of	174
Equivalent evaporation	133
Evaporative test of boiler	300
Excess of air, loss from	61
Expanders for tubes	296, 297
Expansion pads	20
Explosions of boilers	227

INDEX.

	PAGE
Factor of safety	224, 329
Farnley furnace	212
Feed-pipes	4, 264
pumps	265
water filter	77
Feed-water heaters (lime-extracting)	72
Berryman	264
Hoppes	72
Wainwright	263
impurities in (table)	66
organic impurities	81
Filter, feed-water	77
Finishing flanges	279
Fire-doors	5
engine boiler	13
tubes	2, 220
Firing	100
Flange-punch	277
Flanging heads	275
Flat plates	222
Flexible tube	252
Flow of steam	314
Flue-gases	316
Flues	206
collapsing, pressure of	207, 210–216
discussion of tests	217
rules for	218
strengthened	208
Flynn steam-trap	259
Forms of test-pieces	171
Foundation-ring	9
Fox's corrugated furnace	211
Friction of riveted joints	191
Fuel, artificial	39
standard	130
Fuels, composition of	40, 41, 42
Furnace, corrugated	211–216
Farnley's	212
Holmes	213
Morison's	216
Purve's	214, 215
Furnaces	95
down-draught	105
oil-burning	108
Fusible plugs	253

	PAGE
Gas apparatus, Orsat's	54
natural	39
Galloway boiler	7
General proportions of horizontal multitubular boiler	324
Girders	220
Globe valves	235
Grate-area	325
bars	98
water	107
Grates, rocking	99
Green's economizer	111
Grooving	87
Gun boat boiler	18
Gun iron	179
Gusset stays	162, 183
Hand-holes	4, 166, 342
Hand-riveting	295
Hangers for pipe	270
Heat of combustion	44
(carbon)	43, 44
calculation of	46
(fuels)	42
the liquid (water)	366
Heating-surface	5, 136, 328
of boiler-tubes (table)	364
value of	137
Heine boiler	25
Holmes' furnace	213
Hollow cylinder	183
Hoppe's purifier	72
Horse-power of boilers	135
Howden's system	110
Hydraulic accumulators	292, 293
riveting-machine	288
with cam and toggle	293
test	224, 227
of boiler	299
Incomplete combustion, loss from	60
Iron rods, weight of	363
Kent's table of chimneys	122
Kerosene oil	84

	PAGE
Lancashire	6
Lap	191, 337
joints	192, 194
Lap-joint with welt	196, 198
Laying out plates	279
stays	345
Leavitt boiler	20
Length of sections	341
Lever safety-valve	242
Lewes (marine-boiler scale)	74
Life of boilers	230
Lifting-dogs	276
Lignite	38
Lime-extracting feed-water heater	72
Limit of elasticity	174
Lloyd's rules for flues	218
Locke damper regulator	256
Locomotive-boiler	18
staying of	155
type	20
Logarithms	358
Longitudinal joint	331
Mahler's composition of fuels	41
formula	48
Malleable iron	179
Manholes	4, 165, 342
Manning boilers	9, 10
Marine boilers	14
water-tube	30
proportions of	140
scale	74, 79
staying of	160
Materials	170
McDaniels trap	257
Mechanical stokers	102
Methods of failure of riveted joints	189
of making boiler-tests	300
of testing plate	172
Mineral impurities	67
matter in water (table)	66
oil	39
Modulus of elasticity	174
Morison's furnace	216

	PAGE
Natural sines, cos, and tan	362
trigonometric functions	362
Naval boilers, proportions of	141, 142
Nozzles	342
Oil-burning furnaces	108
Organic impurities in feed-water	81
Orsat's gas apparatus	54
Peat	38
Peclet's chimney theory	116
Peet valve	238
Petroleums, composition of	40
Pipe, arrangement of steam	267
area and size of	(Appendix)
blow-off	266
feed	265
hangers for	270
size for given horse-power	271
sketches for ordering	269
Piping	267
Pitch of rivets	192
Pitting	85
Plain cylindrical boiler	6
Plan of boiler-shop	274
Planing-machine	282
plates	282
Plate planer	282
rolls	282
Pop safety-valve	246
Portable riveting-machine	289
Power-pump for riveter	290
Priming	131, 308
Proportions of boilers	138
Properties of saturated steam	367
of steel	176
Prosser tube-expander	296
Pumps	265
Pump for hydraulic riveter	290
Punch	282
and holder	278
for tube-holes	278
Punched or drilled plates	190
Purve's furnace	214, 215
Pyrometers	317

INDEX. 377

	PAGE
Rate of combustion	136
Reach of a riveting-machine	289
Reducing-valve	254
Reduction of area	175
Return steam-traps	260
Ring-seam	336
Rivet, diameter of	192
Riveted joints, calculation of	192
designing	202
efficiency of	188
friction of	191
limitations	205
methods of failure	189
Riveting-machines, cam and toggle	288
hydraulic	288
with cam and toggle	293
portable	289
steam	294
Rivets	178
pitch of	192
shearing and crushing	191
Rocking-grates	99
Rolls for plate	282
Roney stoker	103
Root boiler	25
Safety plugs	253
valve	240
Sample boiler-test blank	319
Scale, marine-boiler	74, 79
Scarfing	282
Screw-threads (table)	364
Sea-water, composition of	74
Segments of circles	360, 361
Selection of type of boiler	323
Semi-bituminous coal	38
Separator	262
Shearing	175
of rivets	191
plates	281
Shears	281
Shop-practice	272
Size and surface of boiler-tubes	364
of steam-pipe	271
Sizes of steam, gas, and water pipe (table)	365

INDEX.

	PAGE
Smoke-box	5
prevention	104
Snap-riveting	295
Soda	70
Specific heat	45
of superheated steam	307
volumes	45
Specifications and contract for boiler	354, 356
Sphere, strength of	185
Spherical ends	163
Split-draught	9
Stability of chimneys	127
Stay-bolts	156, 181
Stay-rods	182
calculation of	347, 351
Stayed flat plates	222
Staying	148
(calculation of)	343
cylindrical tubular boiler	148
laying out	345
locomotive-boiler	155
of marine boiler	160
Stays, diagonal	182
Steam-dome	163
flow of	314
gas, and water pipe (table)	365
gauges	251
nozzle	165
piping	267
quality of	131
riveting-machine	294
space	326
tables	367
traps, Baird's	258
Bundy	260
Curtis	260
Flynn	259
McDaniel's	257
return	260
Walworth	258
Steel	176
Stirling boiler	27
Stokers, mechanical	102
Strain	174
Stratton separator	262

	PAGE
Stress	174
Stretch limit	174
Submerged tube-sheet	12
Sunken tube-sheet	12
Superheating surface	11
Surface blow	82
Table of logarithms	358
Tables of properties of saturated steam	367
Tannic acid	71
Temperature of combustion	63
Test on furnace flues	207, 210–216
Testing boilers for evaporation	300
Testing-machines	170
Test-pieces	171
Testing plate, methods of	172
Thickness of shell	331
Thornycroft boiler	32
Throttling calorimeter	309
Through-stays	2
Trowbridge's table of chimneys	125
Tube-expanders	296, 297
Tube-holes, drills for	278
punch for	278
plates	2
sheet	339
Tubes	325
after expanding	297
Two flue boiler	5
Type of boiler, selection of	323
Ultimate elongation	175
strength	174
Uptake	2
area of	341
U. S. Inspectors' rules for flues	218
Valves	235
angle	236
Chapman	238
check	239
gate	238
globe	235
Peet	238
reducing	254

Valves safety lever	242
pop	246
Vertical boilers	9, 11, 12, 13
rolls for plate	284
Volumes, specific	45
Wainwright feed-water heater	263
Walworth steam-trap	258
Wash-out plugs	166
Water column	249
grate	107
heat of the liquid	366
level	328
tube boilers	21
boiler-setting	94
marine boilers	30
weight and volume of (table)	366
Wheel-draught	8
William's composition of fuels	41
Wind pressure	127
Wood	39
Wrought iron	177
steam, gas, and water pipe	365
Yarrow boiler	34
Zinc in boilers	77

MAN

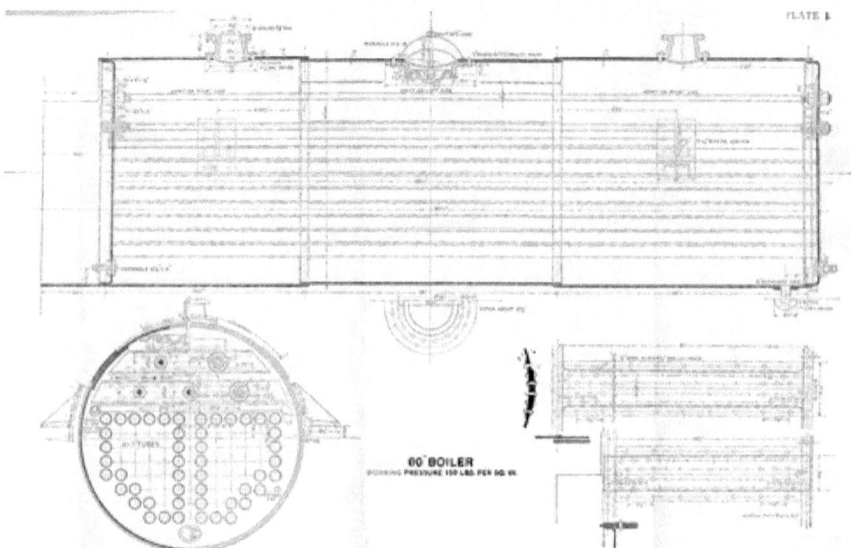

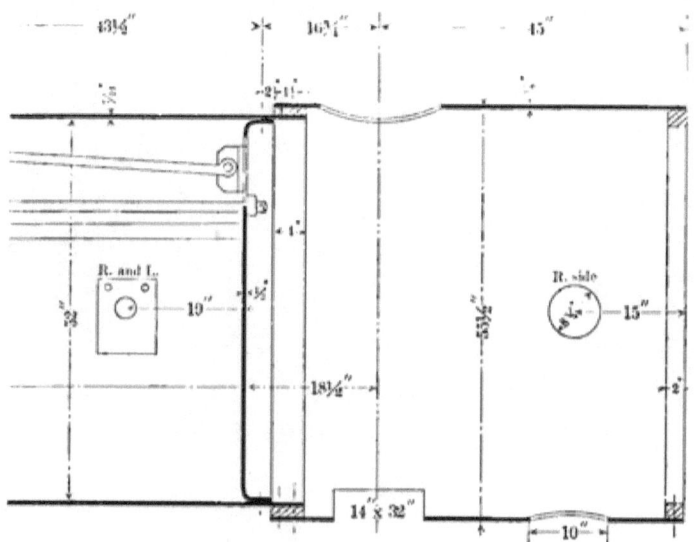

Fig. 3

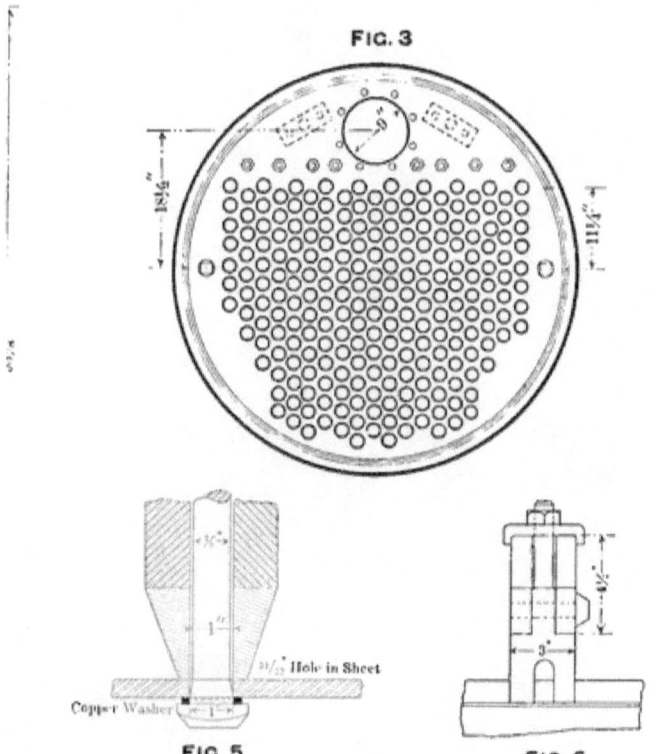

Fig. 5   Fig. 6

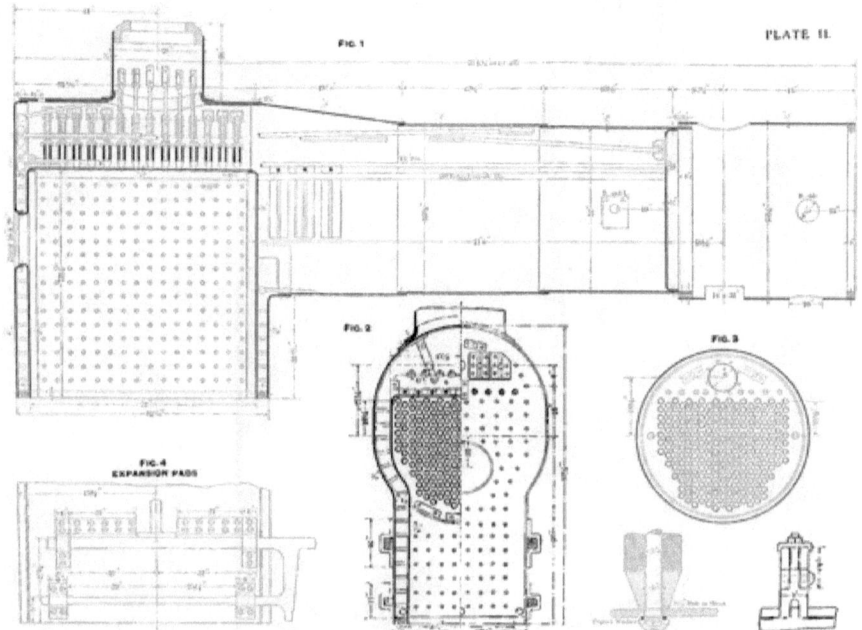

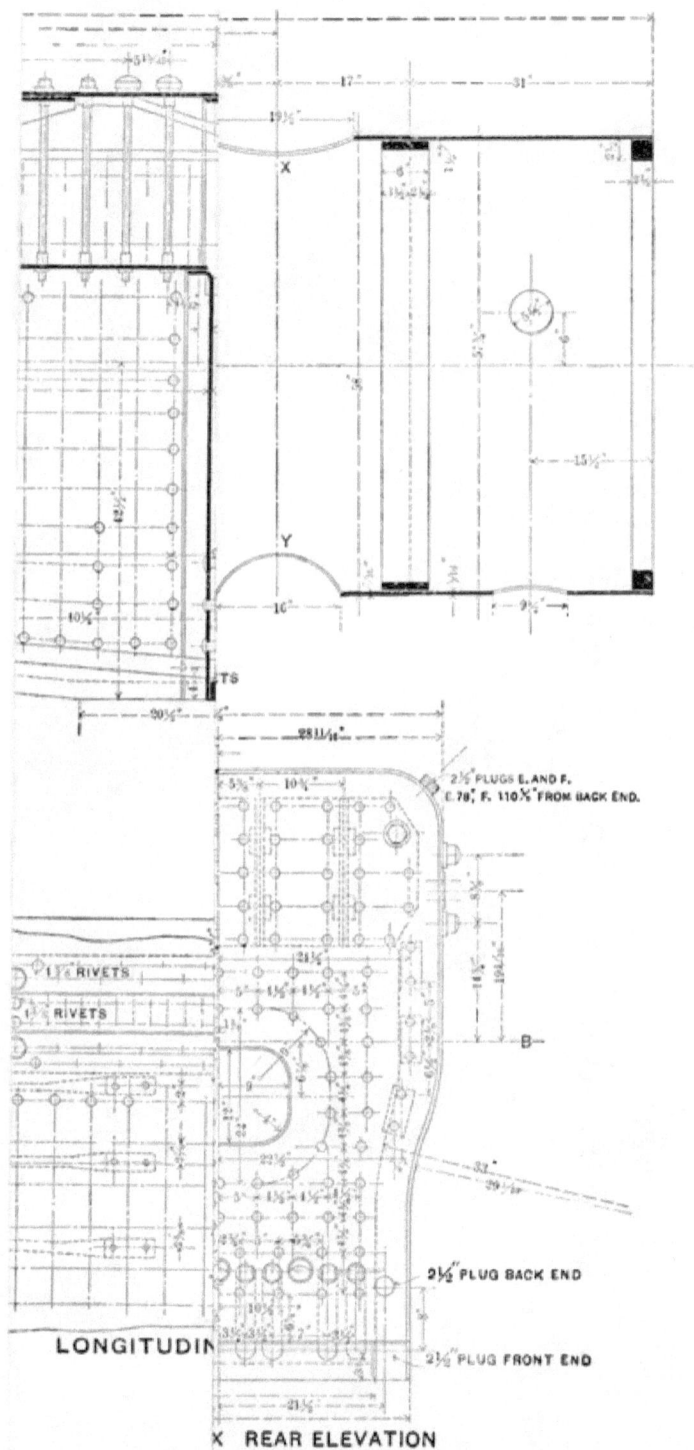

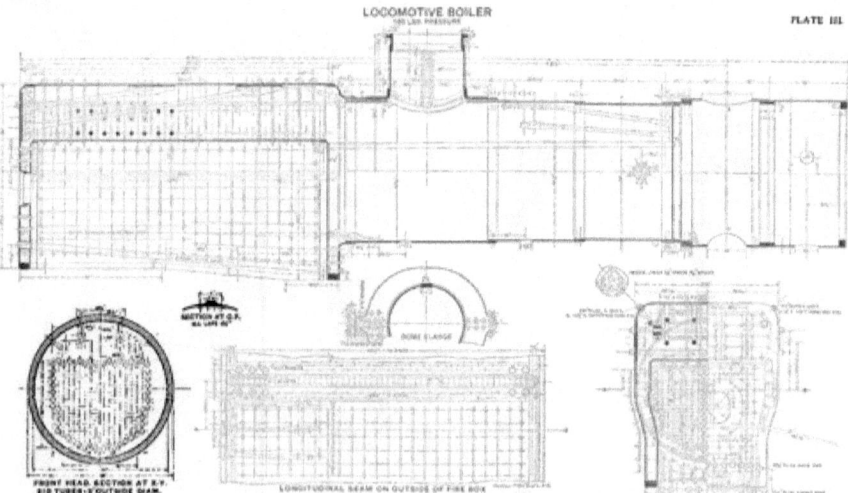

PLATE IV.

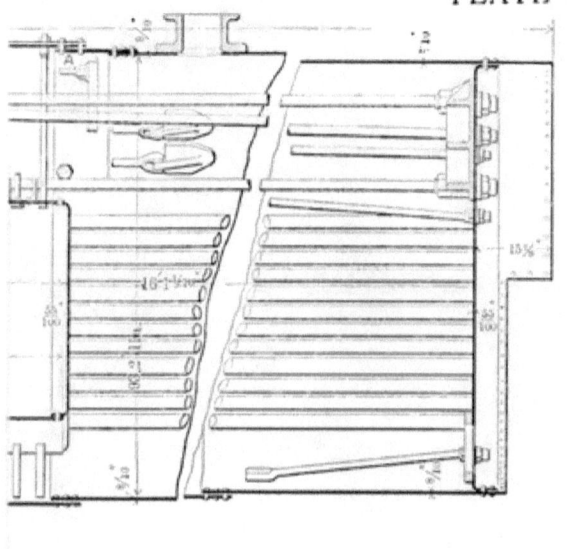

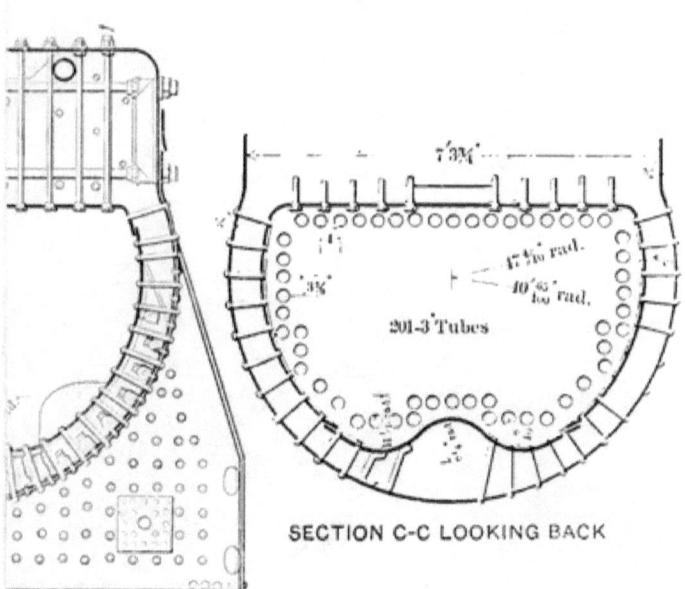

SECTION C-C LOOKING BACK

NG FRONT.

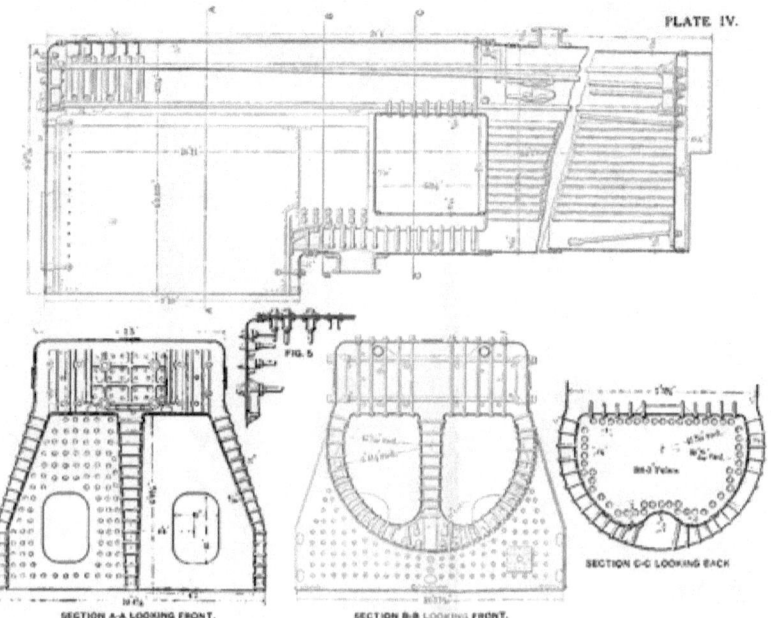

www.ingramcontent.com/pod-product-compliance
Lightning Source LLC
Chambersburg PA
CBHW030422300426
44112CB00009B/816